Heat Transfer Engineering

Heat Transfer Engineering
Fundamentals and Techniques

C. Balaji
Indian Institute of Technology Madras,
Chennai, Tamil Nadu, India

Balaji Srinivasan
Indian Institute of Technology Madras,
Chennai, Tamil Nadu, India

Sateesh Gedupudi
Indian Institute of Technology Madras,
Chennai, Tamil Nadu, India

ELSEVIER

ACADEMIC PRESS
An imprint of Elsevier

Academic Press is an imprint of Elsevier
125 London Wall, London EC2Y 5AS, United Kingdom
525 B Street, Suite 1650, San Diego, CA 92101, United States
50 Hampshire Street, 5th Floor, Cambridge, MA 02139, United States
The Boulevard, Langford Lane, Kidlington, Oxford OX5 1GB, United Kingdom

Library of Congress Cataloging-in-Publication Data
A catalog record for this book is available from the Library of Congress

British Library Cataloguing-in-Publication Data
A catalogue record for this book is available from the British Library

ISBN: 978-0-12-818503-2

For information on all Academic Press publications
visit our website at https://www.elsevier.com/books-and-journals

Publisher: Joe Hayton
Acquisitions Editor: Graham Nisbet
Editorial Project Manager: Mona Zahir
Production Project Manager: Poulouse Joseph
Designer: Miles Hitchen

Typeset by Thomson Digital

Working together to grow libraries in developing countries

www.elsevier.com • www.bookaid.org

To the Almighty, who alone is.

C. Balaji

*To my teachers, parents, Priya, Rukmini, and Bharat, who gave me
everything but asked for nothing in return.*

Balaji Srinivasan

To my little ones, Kirti Priya and Yeshaswin.

Sateesh Gedupudi

Contents

For additional information on the topics covered in the book, visit the companion site: https://www.elsevier.com/books-and-journals/book-companion/9780128185032

Preface

This book is an outgrowth of teaching the heat transfer course over 25 years (CB) and over a decade of teaching (BS|SG). Additionally, during the past 4 years, the three of us have co-taught the undergraduate heat transfer course at the Indian Institute of Technology Madras, India in close coordination, with shared tutorials, assignments, and exams. It is our hope that, in view of the above, the readers do not find any jerk in the narration and presentation of core ideas of heat transfer, and in the mathematical treatment peppered with insightful physics. The nuances of the underlying physics are usually easy to miss and we have earnestly strived to get these across to the students, all through the book.

Heat transfer is not only a major engineering tool in the hands of a mechanical, chemical, aerospace, automotive, materials and electrical engineer, but is a discipline by itself, with its principles and techniques finding direct use in power generation, heating, ventilation, air conditioning, insulation systems, apart from applications in critical technologies, such as electronic cooling, combustion and propulsion, satellite cooling, data center cooling, heat removal from nuclear fuel rods in nuclear power plants, and so on. The list is endless. Heat transfer has witnessed spectacular growth in the last 100 years with accelerated development since World War II and has come out of the shadows of being an "inferior" interaction at a system boundary compared to work, as traditionally studied, taught and prescribed in an equally game-changing and fascinating subject of thermodynamics.

Even so, heat transfer has grown to be an independent discipline with numerous applications, where the primary objective (unlike in thermodynamics) is not the conversion of heat to work. Many applications, as already mentioned, demand a knowledge of the heat transfer physics, the rate and the means to increase or decrease the heat transfer rate as the case may be, along with an equally important engineering objective of keeping temperatures under check, across equipment in multifarious engineering disciplines.

In this book, we have endeavored to give a fresh flavor to heat transfer, always trying to keep the above objectives right through the text. The exposition in every chapter is aimed at an end-to-end experience, starting from the physics of the process to the engineering of the equipment that accomplishes the heat transfer. The two cardinal pillars we have strictly adhered to are (1) Fundamentals and (2) Techniques, as reflected in the title of the book.

The book is divided into 12 chapters. Chapter 1 deals with an introduction to the field of heat transfer and lists the possible applications of the subject. In Chapters 2 and 3, we elaborately present conduction heat transfer in several engineering systems, covering both steady and unsteady heat transfer. In Chapter 4, we present the basic ideas of convection and derive the equations governing the flow and the heat transfer. Chapter 5 provides an exhaustive treatment of forced convection involving both external and internal flow and also presents a quick introduction to turbulence.

Prandtl's legendary boundary-layer simplifications and the integral solution to the problem of convection heat transfer are elucidated. The powerful Reynolds analogy is presented together with a brief presentation of forced convection in other geometries of interest. In so far as internal flow is concerned, the concept of fully developed flow is elaborated, followed by a treatment of the analytical approach to getting the Nusselt number for a constant heat flux case. Correlations for turbulent flow and heat transfer are given for a few cases of engineering interest.

Chapter 6 discusses free convection in detail. Boundary layer simplifications, integral solutions, and correlations are presented for a few representative geometries.

Chapter 7 deals with the analysis and design of heat exchangers. Two powerful techniques, namely the logarithmic mean temperature difference and the effectiveness-number of transfer units (NTU), are detailed and specific instances of using one method over the other in an engineering problem are driven home through examples. Ideas on the overall heat transfer coefficient are reinforced in this chapter.

Chapter 8 presents an overview of basic radiation laws, black body behavior, and characteristics of real surfaces from the viewpoint of radiation. The enclosure theory is elaborately presented together with a detailed treatment of view factors to enable the calculation of radiation heat transfer between surfaces that are separated by a transparent medium. The chapter also gives a breezy introduction to gas radiation and its engineering treatment.

Chapter 9 provides an introduction to numerical heat transfer. There is a detailed discussion of the finite difference method for heat transfer, along with applications to problems handled analytically in earlier chapters. The chapter also discusses the need and indispensability of numerical methods for engineering heat transfer and the practical issues that a heat transfer engineer must consider in industrial problems.

Chapter 10 presents an introduction to applications of machine learning in heat transfer. We discuss the relevance and importance of this topic in modern day heat transfer and provide an overview of various machine learning algorithms in the context of possible applications in heat transfer. An elaboration of the learning process and neural networks is provided. Possible future applications of this topic are also given at the end of this chapter.

Chapter 11 presents the commonly observed regimes of pool boiling and flow patterns in flow boiling and the correlations used to predict the heat transfer coefficients and the critical heat flux. The wall superheat required for nucleation from a heating surface is briefly discussed. The theory of film condensation on flat plates and horizontal tubes is presented, along with the respective heat transfer coefficient correlations. An introduction to the prediction of two-phase pressure drop is also given.

Chapter 12 discusses the laws that govern diffusion and convective mass transfer, and the analogy between mass transfer and heat transfer. Equations for the determination of mass transfer coefficients for a gas flow over a volatile liquid or solid surface are presented.

It is our trust and hope that the material presented in the book, together with the worked examples and the end of chapter problems, will be more than adequate for coverage in a one-semester course either at the third-year undergraduate level or at the first-year Masters level. For the undergraduate course, instructors may at their discretion, not dwell too much on advanced topics like gas radiation or machine learning. Even so, these are important, and the decision is best left to the instructor/reader.

We would like to thank IIT Madras for providing us with an academically stimulating environment to teach, practice, and research the ever-glorious subject of heat transfer. Thanks are due to all our teachers who have been a key source of motivation and to thousands of students over the years; who have often surprised and humbled us with matter-of-fact yet deep questions that have quite a few times left us startled and speechless.

To all at Elsevier starting from Gaelle Hull, Publisher, Thermal and Fluids Engineering to Nisbet Graham, Senior Acquisitions Editor, Energy, Mona Zahir, Editorial Project Manager and Poulouse Joseph, Senior Project Manager, for making this happen and for the relentless follow-up and back by Poulouse and Mona.

We are extremely grateful to our scholars, Girish, Rajesh, Sandeep, Sangamesh, Suraj, Harish, Kiran, Vikas Dwivedi, Gaurav Yadav, Rishi Mishra, Akhil Dass and Prasanna Jayaramu, for their help in typing, preparation of figures, and proof corrections. They have been pillars of support to us in the entire project.

Finally, we are ever grateful to our families for their support, encouragement, and forbearance.

Chennai, India   C. Balaji
August 2020   Balaji Srinivasan
                                                Sateesh Gedupudi

Introduction

1.1 Thermodynamics and heat transfer

Heat transfer may be defined as one of the two interactions that takes place at the boundary of a thermodynamic system or a control volume by a virtue of a temperature difference. Esoteric as it sounds, the above definition can be more easily understood if we write down the first law of thermodynamics for a system as given below.

$$Q - W = \Delta E \tag{1.1}$$

The second term on the left is work, and the term on the right-hand side represents the change in the energy of the system. The units of all the terms in Eq. (1.1) are J or kJ, and on a rate basis, they are all in W or kW. Now let us turn our attention to the first term on the left-hand side, often denoted by Q. This represents the heat interaction or the heat transfer, which takes place at the boundary of a system (or a control volume).

Invariably, a temperature difference is required to cause this heat interaction, though later we will look at particular situations where an isothermal heat transfer is possible. The work that appears in Eq. (1.1) is thermodynamic work, which is usually defined in terms of positive work. Positive work is said to be done by a system during an operation when the sole effect external to the system can be reduced to the rise of a weight. This is the classic definition of work and resonates with the early efforts of man to convert heat to mechanical work. Work was and is considered to be superior to heat, as work can help us operate machines, can be used to generate electricity, and in general be used to reduce human toil. An equally perplexing definition of work is that "work is any interaction that is not heat." This does not help a student of heat transfer, as it does not tell us anything about how to calculate or compute or estimate heat transfer independently in an engineering situation.

From the preceding discussion, it is clear that work or thermodynamic work and its calculation or estimation is central to the study of thermodynamics, as is energy change (please look at the term on the right-hand side of Eq. 1.1). In a typical study of thermodynamics, neither the mechanisms of heat transfer nor the methods required to calculate it are discussed. Heat transfer is often an "assumed" input in the thermodynamic analysis. On the other hand, the question "how do we calculate heat transfer in a situation?" is fundamental, and through a study of heat transfer, we try to answer this question. There are many situations where the original idea of knowing heat transfer in order to calculate how much heat is converted to work is not applicable.

Consider, for example, an electric transformer or a desktop computer. Heat is generated in both these applications, and the challenge is to dissipate the heat efficiently and keep temperatures under check. Hence, the fundamental objective in a heat transfer study is to be able to calculate it, with a view to either increasing or decreasing the heat transfer in a situation or controlling the temperature in an application.

Let us now consider two somewhat familiar examples—(1) An oil cooler and (2) heat treatment of a ball bearing—to obtain better insight into the scope of a heat transfer study vis-a-vis a thermodynamic analysis.

Example 1: Oil cooler

Consider an oil cooler. Let $\dot{m}_h, c_{p_h}, T_{h_i}, T_{h_o}$ be the mass flow rate, specific heat, inlet temperature, and outlet temperatures of the hot fluid (oil), respectively. Similarly let $\dot{m}_c, c_{p_c}, T_{c_i}, T_{c_o}$ be the mass flow rate, specific heat, inlet temperature, and outlet temperature of the cold fluid (coolant) respectively. The inlet and outlet conditions are represented through subscripts 'i' and 'o' respectively. A schematic representation of the oil cooler is shown in Fig. 1.1.

Due to a temperature difference, heat transfer occurs between the two fluids. From the first law of thermodynamics, we have the following expression.

$$\text{Enthalpy lost by the hot fluid} = \text{Enthalpy gained by the cold fluid} \tag{1.2}$$

$$Q = \dot{m}_h c_{p_h}(T_{h_i} - T_{h_o}) = \dot{m}_c c_{p_c}(T_{co} - T_{c_i}) \tag{1.3}$$

If T_{hi}, T_{ci} and T_{co} are known then, in Eq. (1.3), the only unknown is T_{h_o}. Using Eq. (1.3), the outlet temperature of the hot fluid can be calculated. This sounds very good!

However, the key questions to ask are: To achieve the heat duty given in Eq. (1.3), what is the size of the heat exchanger or the oil cooler required? Stated explicitly, how many tubes are required, and what will be their diameters and lengths if we employ an equipment with tubes and a shell enclosing the tubes? These questions, unfortunately, cannot be answered by thermodynamics.

From heat transfer, we add one more equation to Eq. (1.3), as given below.

$$\dot{m}_h c_{p_h}(T_{h_i} - T_{h_o}) = \dot{m}_c c_{p_c}(T_{co} - T_{c_i}) = UA\Delta T_{mean} \tag{1.4}$$

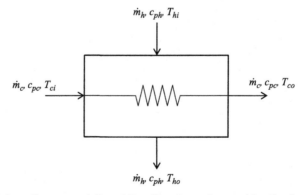

$\dot{m}_h, c_{ph}, T_{hi}$

$\dot{m}_c, c_{pc}, T_{ci}$

$\dot{m}_c, c_{pc}, T_{co}$

$\dot{m}_h, c_{ph}, T_{ho}$

FIGURE 1.1 Schematic representation of the oil cooler under consideration in example 1.

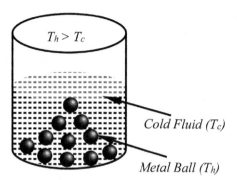

$T_h > T_c$

Cold Fluid (T$_c$)

Metal Ball (T$_h$)

FIGURE 1.2 A schematic representation of cooling of hot ball bearings in a cold fluid.

In Eq. (1.4), U is the overall heat transfer coefficient (W/m^{2}K), ΔT_{mean} is a "mean" temperature difference between the two fluids (this needs to and will be quantified in a later chapter), and A is the surface area of the heat exchanger or an engineering equipment that accomplishes the transfer of heat (m^2). Once the area is calculated from Eq. (1.4), one can calculate the number of tubes and so on for the heat exchanger. The catch here is that the calculation of 'U' is formidable and is often the subject of study for a lifetime!

Example 2: Cooling of ball bearings (heat treatment)

Let us consider stainless ball bearings, initially at 300 °C, that are to be heat treated. They are immersed in a tank containing a fluid that is at 30 °C. A very simplified schematic is shown in Fig. 1.2. The ball bearings are now dipped inside the pool for heat treatment. With thermodynamics, using Eq. (1.3), one can calculate the final temperature of the bearings, which is the steady-state temperature value. However, the question to be answered is, "What is the time required for the ball to reach 200 °C, so that we can go for some other heat-treatment process?"

The above is a question that again cannot be answered by thermodynamics.

From the above examples, one can see that thermodynamics deals with energy transfer and work transfer, and though heat transfer is very much an entity in thermodynamics, the subject per se cannot help us answer the above questions (i.e., What is the area, or how much time does it take to reach a particular temperature?).

1.2 Heat transfer and its applications

Heat transfer plays a critical role in many engineering, technological, environmental, and industrial problems (Incropera et al., 2013). Some of these are:

1. Fossil fuel-based power plants—boiler, economizer, superheater, condenser, and cooling tower.

2. Nuclear fission reactors—all of the above except that instead of using the chemical energy of fossil fuel, the heat generated by nuclear fission is used to produce motive steam.
3. Internal combustion engines.
4. Gas turbine engines.
5. Rocket motors.
6. Cryogenic storage systems.
7. Solar thermal systems.
8. Land and sea breezes.
9. Prediction of cyclones and typhoons.
10. Prediction of monsoons.
11. Global warming.
12. Cooling of electronics.
13. Cooling of data centers.
14. Design of insulation systems.
15. Refrigeration and air conditioning equipment.
16. Climate control of buildings.
17. Biological systems.

1.3 Modes of heat transfer

There are two fundamental modes of heat transfer—namely, conduction and radiation. Both of these are independent and are based on completely different mechanisms. Convection is an enhanced or modified form of conduction, in which a bulk motion of the medium is additionally present. Convection has its underpinnings in conduction and thermodynamic laws applicable to bulk transport. Because of its importance in heat transfer engineering, it is common to declare convection as the third mode of heat transfer. Let us look at the rate laws governing these heat transfer modes and their underlying physics.

1.4 Conduction

The conduction mode of heat transfer is related to the atomic and molecular activity, where energy transfer takes place from more energetic particles to less energetic particles.

Consider a gas between two plates at T_1 and T_2, as shown in Fig. 1.3. Assume that there is no bulk motion. The temperature at any point is associated with the energy of the gas molecules in the vicinity of the point. Energy is made up of translational, vibrational, and rotational energy of the molecules. Due to molecular collision, there is a constant transfer of energy as higher temperature molecules have high energy. In the presence of a temperature gradient, energy transfer by conduction must then occur in the direction of decreasing temperature. The point here is that at any plane A-A, the number of molecules moving from the left to the right is the same as the

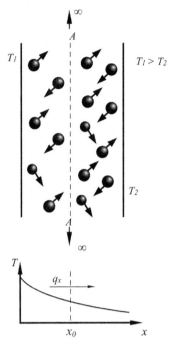

FIGURE 1.3 Schematic depiction of conduction across a gas layer between two infinitely long plates.

number moving from the right to the left, when averaged over a reasonable length of time. However, molecules to the left of the plane A-A have higher energy consequent upon their being at a higher temperature. Hence, even though there is no net movement of molecules across A-A, there is a net transfer of energy in the direction of decreasing temperature. Hence, we can say that there is a diffusion of energy. The situation is the same in the case of liquids, though the molecules are more closely spaced and the molecular interaction is stronger and more frequent.

In a solid material, the energy transfer is induced more or less entirely through lattice waves caused due to atomic motion. In insulators, the transfer is exclusively through lattice waves. In a material, the transfer is also due to the translational motion of the free electrons.

When there is a temperature difference across the wall, heat transfer will take place through the wall, as shown in Fig. 1.4.

Joseph Fourier proposed the following, based on experimental measurements.

$$Q \propto (T_1 - T_2) \tag{1.5}$$

$$Q \propto A_n \tag{1.6}$$

$$Q \propto (1/L) \tag{1.7}$$

In Eq. (1.6), A_n is the area normal to the flow of heat.

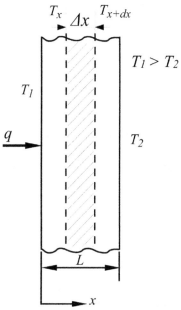

FIGURE 1.4 Schematic of conduction in the horizontal direction, x across a simple plane wall of thickness L.

Combining the above three equations, we have

$$Q \propto (T_1 - T_2) A_n (1 / L) \tag{1.8}$$

To remove the proportionality, a constant "k" is introduced.

$$Q = -kA_n \frac{T_2 - T_1}{L} \tag{1.9}$$

The above equation is known as Fourier's law of heat conduction. In Eq. (1.9), the negative sign is consistent with the second law of thermodynamics so that when $T_2 < T_1$, "positive" heat transfer takes place along x. As $(T_2 - T_1)$ is −ve, the −ve sign ensures that Q is positive. From figure 1.4 and eqn. 1.9 it can be seen that the last term on the right hand side of eqn 1.9 is the temperature gradient.

To determine the proportionality constant, Fourier conducted experiments and found that this value changes from material to material. Hence "k" is named as thermal conductivity, which is a property of a material. The units of "k" are W/mK. Table 1.1 shows the thermal conductivity values for select materials at room temperature.

In the case of solids, with an increase of temperature the thermal conductivity ("k") reduces, except for some materials like aluminum and uranium. In the case of gases, with an increase of "T," "k" too increases. For liquids, "k" decreases with temperature. Readers are advised to refer to advanced texts on conduction for a fuller discussion on the topic of variation of thermal conductivity of materials with temperature.

Table 1.1 Thermal conductivity of select materials at room temperature.

Material	Thermal conductivity k (W/m K)
Copper	400
Aluminum	205
Mild steel	45
Stainless steel	15
Water	0.6
Insulators	0.02–1
Air	0.03

1.5 Convection

When conduction is superimposed with a bulk motion of the medium (advection), we call it convection. Convection can be free or forced. In natural or free convection heat transfer, the heated molecules will move in an upward direction when gravity is acting downward, and the cooled molecules move in a downward direction due to density differences. The upward movement of heated molecules and downward movement of cooled molecules is called a natural convection current. Forced convection, on the other hand, requires mechanical equipment for accomplishing the flow. This may be done through a fan, blower, or compressor.

The movement of molecules in convection often can be seen with the human eye (e.g., water heating with an immersion heater). Hence, it is a macroscopic phenomenon. For convection heat transfer to take place, the existence of a temperature difference is mandatory.

It is intuitive that the convection heat transfer rate must be proportional to the surface area and the temperature difference. Mathematically, they can be represented as

$$Q \propto A_s \tag{1.10}$$

$$Q \propto T_w - T_\infty \tag{1.11}$$

By combining the above two equations and replacing the proportionality by equality, we obtain

$$Q = hA_s (T_w - T_\infty) \tag{1.12}$$

Eq. (1.12) is known as Newton's law of cooling and is the rate law for convection. In Eq. (1.12), h is the heat transfer coefficient and has the units W/m^{2}K. Typical values of convection heat transfer coefficient are given in Table 1.2. Eq. (1.12) suggests that when $h \rightarrow \infty$, $\Delta T \rightarrow 0$. Though this is a mathematical possibility, it is an engineering impossibility.

Table 1.2 Typical heat transfer coefficients for a few commonly encountered heat transfer processes (Incropera et al., 2013).

Process	Heat transfer coefficient h (W/m² K)
Free convection (no fan/pump)	2–25 (gases) 50–1000 (liquids)
Forced convection	25–250 (gases) 50–20000 (liquids)
Convection with phase change, boiling/condensation	2500–100000

1.5.1 Mechanism of convection

Consider a heated horizontal plate as shown in Fig. 1.5A and B. A fluid with a free stream velocity of u_∞ flows from the left to the right. Now consider Fig. 1.5A. On the surface of the plate, the fluid has to be stationary relative to the plate. This is referred to as the no-slip condition. Now consider any vertical section A-A along x. At $y = 0$ on this section, $u = 0$. As $y \to \infty$, $u \to u_\infty$. Suppose we ask the question, "What is the height at which $u \to 0.99u_\infty$?"

Let us say that this value is δ, and it stands to reason that $\delta = f(x)$. The locus of all the δs across x is known as the hydrodynamic boundary layer; δ_x is itself known as the boundary layer thickness. We will see later than $\delta_x = f(Re_x)$, where Re_x is the Reynolds number defined as $u_\infty x / \nu$. The boundary layer demarcates the flow into two regions. These are (i) the region between the plate ($y = 0$) and the edge of boundary layer ($y = \delta$): the region in which the effects of fluid viscosity are felt, and (ii) the

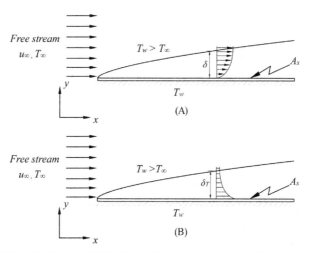

FIGURE 1.5 Pictorial depiction of fluid flow and heat transfer over a flat plate.

(A) velocity profile and (B) temperature profile.

region beyond $y = \delta$ at all x: the region in which the effects of fluid viscosity are not felt. The actual thickness of the boundary layer, namely δ, directly affects the wall shear stress and pumping power and is intuitively expected to affect the convective heat transfer.

We can now extend this argument to study how the temperature of the fluid varies across y. Consider the same section, A-A, again. At $y = 0$, $T = T_W$ (see Fig. 1.5B). We can now mark a point on A-A where $T = T_\infty + 0.01(T_w - T_\infty)$. Stated explicitly, we are looking at the height y at which the temperature difference imposed in the problem between the wall and fluid, that is, $(T_W - T_\infty)$, is reduced to 1% of its value. The locus of all the points along with x at which $T = T_\infty + 0.01(T_w - T_\infty)$ is called the thermal boundary layer. Again, one can say that all the action arising out of T_W being not equal to T_∞ is restricted to this layer, often only a few millimeters thick. It stands to reason that the value of δ_T at any section x gives the idea of the heat transfer at that value of x. The higher the δ_T, the lower the value of heat flux, as is intuitively apparent if we invoke the analogy of conduction.

In other words, taking a cue from Fourier's law of heat conduction, we can write an expression for local heat flux as follows.

$$q_x \cong k \frac{\Delta T}{\Delta y} \cong k \frac{\Delta T}{\delta_T} \tag{1.13}$$

$$\therefore q_x \propto \frac{1}{\delta_T} \tag{1.14}$$

The above is a key engineering result. If one can measure δ_T experimentally or obtain or even estimate it through theory, we are home. At the surface, the convection heat flux is equal to conduction heat flux, and if we have information on $\delta_T(x)$, we have it all!

The preceding discussion is also one of the principal reasons why the study of boundary layers or boundary layer theory has been the cornerstone of both fluid mechanics and convective heat transfer.

1.6 **Thermal radiation**

Any body above a temperature of 0 Kelvin emits radiation. This is a basic law of nature and was first proposed by Pierre Prevost in 1791. Two widely accepted theories to describe and characterize radiation are

- Electromagnetic theory

  According to this, radiation travels as an electromagnetic wave and is characterized by speed, c, wavelength, λ, and frequency, v, that are related as given below.

$$c = v\lambda \tag{1.15}$$

- Quantum theory

Insofar as quantum theory is concerned, the proposal is that radiation travels in the form of energy packets called quanta, and this energy is given by

$$E = h\nu \tag{1.16}$$

where h is Planck's constant, which is equal to 6.629×10^{-34} Js.

Some properties of radiation are best described and understood by electromagnetic theory, while black body behavior can be described only by the quantum theory. In a sense, both are right.

A black body is one that emits the maximum amount of radiation at a given temperature. The emissive power of a black body is given by

$$E_b = \sigma T^4 \tag{1.17}$$

In Eq. (1.17) σ is the Stefan-Boltzmann constant (5.67×10^{-8} W/m² K⁴) which has its foundations in thermodynamics (that part of result where E_b was proved to be proportional to T⁴. The constant σ was determined by matching measurements with general expression $E = cT^4$ where c is the multiplicative constant.)
The emissive power from a real surface is given by

$$E_R = \varepsilon \sigma T^4 \tag{1.18}$$

In Eq. (1.18), ε is the emissivity of the surface, and it varies between 0 and 1.

The net radiation between the two surfaces when the two surfaces face each other completely, or if a small surface is enclosed in large surroundings at T_∞, is given by

$$Q_{net} = \varepsilon \sigma A(T^4 - T_\infty^4) \tag{1.19}$$

$$\frac{Q_{net}}{A} = q = \varepsilon \sigma (T^4 - T_\infty^4) \tag{1.20}$$

$$\frac{Q_{net}}{A} = q = \varepsilon \sigma (T^2 - T_\infty^2)(T^2 + T_\infty^2) \tag{1.21}$$

$$q = \varepsilon \sigma (T + T_\infty)(T^2 + T_\infty^2)(T - T_\infty) \tag{1.22}$$

$$q = h_r(T - T_\infty) \tag{1.23}$$

In Eq. (1.23) h_r is the radiation heat transfer coefficient and is given by

$$h_r = \varepsilon \sigma (T + T_\infty)(T^2 + T_\infty^2) \tag{1.24}$$

If $T \cong T_\infty$, then Eq. (1.24) reduces to

$$h_r = 4\varepsilon \sigma T_\infty^3 \tag{1.25}$$

Radiation heat transfer is a surface phenomenon in opaque bodies and a volumetric one in transparent bodies. Eq. (1.25) needs to be used with caution and is not recommended, unless a quick back of the envelope estimate is required in an engineering problem.

Radiation heat transfer can take place without temperature difference too. The key is the existence of temperature, which for all practical purposes is more or less guaranteed.

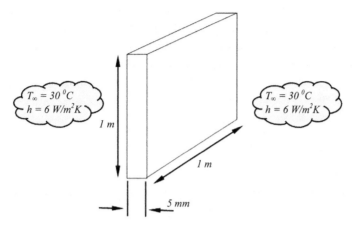

$T_\infty = 30\,^\circ C$
$h = 6\ W/m^2 K$

$1\ m$

$1\ m$

$5\ mm$

$T_\infty = 30\,^\circ C$
$h = 6\ W/m^2 K$

FIGURE 1.6 Pictorial depiction of the plane wall problem.

Example 1.1 *Consider a vertical plate 1 m in height, and 1 m in the direction perpendicular to the plane of the paper and 5 mm thick. The plate is at 100 °C and the surroundings at 30 °C. Free convection from both sides of the plate gives a h value of 6 W/m² K. The emissivity of the plate is 0.9. Calculate (1) convection heat transfer from both the sides, (2) radiation heat transfer from both the sides, and (3) total heat transfer and the fraction of heat transferred by radiation.*

Solution:

A pictorial depiction of the given problem is shown in Fig. 1.6

1. The convection heat transfer is given by

$$
\begin{aligned}
Q_{convection} &= hA_s(T_w - T_\infty) \\
Q &= 6 \times 2 \times 1 \times 1(100 - 30) \\
&= 840\ \text{W}
\end{aligned}
\tag{1.26}
$$

2. The radiation heat transfer is given by

$$
\begin{aligned}
Q_{radiation} &= \varepsilon\sigma A\left(T^4 - T_\infty^4\right) \\
Q &= 0.9 \times 5.67 \times 10^{-8} \times 2 \times 1 \times 1(373^4 - 303^4) \\
&= 1115.3\ \text{W}
\end{aligned}
\tag{1.27}
$$

3. The total heat transfer from the surface is given by

$$
\begin{aligned}
Q_{total} &= Q_{convection} + Q_{radiation} \\
&= 1955.3\ \text{W}
\end{aligned}
\tag{1.28}
$$

The fraction of heat transferred by radiation is

$$
= \frac{1115.3}{1955.3} = 57\%
$$

1.7 Combined modes of heat transfer

There are many situations where more than one mode of heat transfer is involved. The ubiquitous problem of condensation of the working medium in a refrigerator or an air conditioner involves combined surface radiation and convection. Exercise problem 1.1, which concerns heat transfer in a filament bulb, is a classic example of multimode heat transfer. The challenge in these problems is that the resultant heat transfer is not additive. The individual modes of heat transfer often interact, making the resulting problem quite formidable to solve.

1.8 Phase-change heat transfer

Heat transfer processes can also occur isothermally in principle when phase change takes place. This can be solid-liquid or liquid-solid or liquid-vapor or vapor-liquid change of phase. Boiling and condensation are two-phase-change processes widely seen in nature, as well as in several engineering applications. The hydrological processes in the Earth's atmosphere involve evaporation of water and condensation back into rainfall. Water is converted into steam in a boiler of a thermal power plant, and this is converted to useful mechanical work in a turbine. The expanded steam is then condensed in a condenser to complete the cycle.

Phase change heat transfer can also occur in conduction problems involving melting and solidification. A classic example is a heat sink with phase change material used for the cooling of electronics (Baby and Balaji, 2019). The phase change material in the heat sink absorbs the heat isothermally at its melting temperature and undergoes a phase change (solid to liquid). As a consequence, the electronic device is maintained at a particular temperature. Another widely used device involving a phase change heat transfer in convection, for application of electronic cooling, is a heat pipe. In general, a heat pipe is a hollow tube that contains an evaporator section, an adiabatic section, and a condenser section and is filled with the working fluid at saturation temperature and corresponding saturation pressure. The working fluid at the evaporator section takes the heat, is converted into vapor and travels to the condenser section. At the condenser section, the vapor releases heat and is condensed into a liquid and travels back to the evaporator through gravity or a wick (in the case of a wicked heat pipe).

1.9 Concept of continuum

Much of the heat transfer theory that we know and are going to study through this book critically hinges on a key assumption known as the "continuum hypothesis." According to this, we model the behavior of materials or substances or media based on the assumption that they are a continuous mass rather than discrete particles. This assumption is fine as long as the particle density is high enough so that even though particles enter or leave a system boundary, the number of particles inside the system may be deemed to be constant. In convection, the continuum hypothesis breaks down in highly rarefied flows. In radiation, for example, the continuum breaks down in the atmosphere beyond 50 km from the mean sea level.

Problems

1.1. Consider the heat transfer from an evacuated filament bulb. List all the pertinent heat transfer processes associated with the bulb.

1.2. A widely used method to determine the thermal conductivity of solid material is to make a thin sheet out of it, thereby reducing the heat transfer across the material to be one-dimensional. A smart way of measuring the thermal conductivity would be to use a foil heater and sandwich it between two identical samples of the material, as shown in Fig. 1.7. The four sides (left, right, front, and back) are insulated to reduce heat losses.

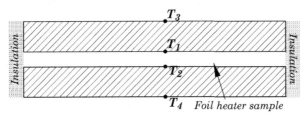

FIGURE 1.7 Foil heater and sample arrangement.

Consider an experiment where two 6 mm plates of the material whose thermal conductivity is to be estimated are used. The sides of the plates are 20 cm by 20 cm. The heater is energized, and temperatures $T_1, T_2, T_3,$ and T_4 are recorded. After some time, steady-state is reached and the following temperatures are recorded (see Table 1.3).

Assuming that the thermal conductivity of the material is invariant with respect to temperature, determine the thermal conductivity of the material when the electric power used is 40 W.

1.3. Consider a typical freezer compartment that is 1.65 m high, 0.75 m wide, and 0.75 m deep. The freezer is insulated with the help of polyurethane foam with a thermal conductivity of 0.028 W/m/K. Consider insulation with this material with a thickness of 200 mm all around the freezer except on the bottom side. If the outside and inside of the insulation are at 40 and $-12\,°C$, what is the heat leak into the freezer?

Table 1.3 Recorded temperature data for problem 1.2.

S. No.	Quantity	Temperature (°C)
1	T_1	90.4
2	T_2	90.5
3	T_3	81.6
4	T_4	81.7

1.4. Consider a plane wall whose one side is maintained at a temperature of 150 °C. The other side is exposed to free convection with a heat transfer coefficient of 6 W/m².K, and T_∞ is equal to 30 °C. The thermal conductivity of the wall is 1.7 W/m.K, and the thickness of the wall is 50 cm. Determine the heat transfer through the wall in W/m². Also determine the temperature at the side of the wall exposed to the convection environment. When will this temperature approach the free stream temperature? When will it approach 150 °C?

1.5. In a coal-fired steam power plant, hot steam at 540 °C, 160 bar is transported from a boiler to the turbine in a pipe. The internal diameter of the pipe is 400 mm, and the outer diameter of the pipe is 410 mm and is made of stainless steel (SS-304) with $k = 15$ W/m.K.

   a. Determine the heat loss per meter length of the pipe if natural convection exists on the outside with $h = 6$ W/m².K and $T_\infty = 40$ °C if the pipe is assumed to be at 540 °C.

   b. If the conduction inside the pipe is considered, what is the approach to determine the temperature at the surface of the pipe. What would be an engineering solution to say bring down the outer surface temperature to 55 °C or less?

1.6. The solar flux in W/m² falling on the Earth has a value of 1368 W/m². Assuming that the Earth absorbs 70% of this and is in radiative equilibrium. Determine the equivalent black body temperature of the earth if the radius of the earth is 6370 km.

1.7. A 1-m-long mild steel plate has a thickness of 5 mm and is 250 mm wide. It is suspended vertically in still air and is energized by a foil heater with a uniform power of Q Watts. The emissivity of the plate is 0.85, and the heat transfer coefficient associated with natural convection is 6 W/m².K. The ambient temperature is 30 °C, and this can also be considered to be the temperature of the surroundings for radiative heat transfer. The plate loses heat by convection and radiation from both the sides.

   a. Write down the equation governing for the variation of temperature with time of the plate, assuming the whole plate to be spatially isothermal.

   b. Determine the value of Q, for which the plate reaches a steady-state temperature of 80 °C.

   c. If the thermal conductivity of the plate is 45 W/m.K, and its density and specific heat capacity are 7850 kg/m³ and 500 J/kg.K respectively, what is the cooling rate of the plate when the power is switched off after the steady-state is reached?

   d. If you neglect radiation, what will be the error in your estimate of the cooling rate?

References

Baby, R., Balaji, C., 2019. Thermal Management of Electronics, Volume I: Phase Change Material-Based Composite Heat Sinks - An Experimental Approach, vol.1 Momentum Press, pp. 1–165.

Incropera, F.P., Lavine, A.S., Bergman, T.L., DeWitt, D.P., 2013. Principles of Heat and Mass Transfer, seventh ed. Wiley, pp. 1–1076.

One-dimensional, steady state heat conduction

2.1 Introduction

In this chapter, we first derive the general three-dimensional heat conduction equation in Cartesian coordinates with the different kinds of boundary conditions associated with it. The solution to the heat conduction equation gives us the temperature field, $T(x,y,x,t)$, in Cartesian coordinates. Once we have this, we can calculate the heat flux anywhere by using the pertinent rate law, which, in this case, happens to be the Fourier's law of heat conduction. Following this derivation, we look at engineering problems that one can solve with a one-dimensional heat conduction approach. Furthermore, we will be looking at situations involving steady state conditions, that is, when the temperature does not change with time and end the chapter with an elaborate treatment of fins.

2.2 Three-dimensional conduction equation

In what follows, the three-dimensional conduction equation is derived from first principles. The goal is to relate all fluxes and energy exchanges to temperature or its gradients, so that we finally get an equation in temperature. A schematic representation of a three-dimensional control volume is shown in Fig. 2.1.

The major assumptions are

1. Homogeneous material-i.e., material properties do not change in space
2. No bulk motion-this removes convective effects
3. The material obeys Fourier's law of heat conduction

In Fig. 2.1 Q_i's represents the rate of heat transfer in a given direction $Q_x = q_x.A$. $Q_y = q_y.A$ and $Q_z = q_z.A$.

At any instant of time, by invoking the first law of thermodynamics with $\dot{E}$ denoting energy rate, we have

$$\dot{E}_{in} + \dot{E}_{generated} = \dot{E}_{out} + \dot{E}_{stored} \tag{2.1}$$

In Eq. (2.1) the subscripts are self-explanatory and it is instructive to observe that the units of all the terms in Eq. (2.1) is W, since this is a heat rate equation.

The individual terms in Eq. (2.1) can be written

$$\dot{E}_{in} = (q_x + q_y + q_z)A \tag{2.2}$$

Heat Transfer Engineering. http://dx.doi.org/10.1016/B978-0-12-818503-2.00002-2

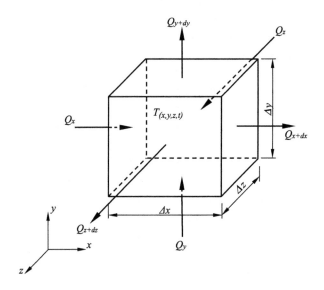

FIGURE 2.1

Schematic representation of a three-dimensional control volume.

$$\dot{E}_{generated} = q_v \Delta x \Delta y \Delta z \tag{2.3}$$

Here q_v is heat generated per unit volume (W/m^3), usually assumed to be uniform.

$$\dot{E}_{out} = (q_{x+dx} + q_{y+dy} + q_{z+dz})A \tag{2.4}$$

In Eq. (2.4), q_{x+dx} can be expanded using the first term of Taylor's series as follows (by neglecting higher order terms)

$$q_{x+dx} \approx q_x + \Delta x \frac{\partial}{\partial x}(q_x) \tag{2.5}$$

The rate of change of enthalpy is given by

$$\dot{E}_{stored} = mc_p \frac{\partial T}{\partial t} \tag{2.6}$$

Substituting Eqs. (2.2)–(2.6) in Eq. (2.1), we have

$$(q_x + q_y + q_z)A + q_v \Delta x \Delta y \Delta z = (q_{x+dx} + q_{y+dy} + q_{z+dz})A + mc_p \frac{\partial T}{\partial t} \tag{2.7}$$

$$q_v \Delta x \Delta y \Delta z A = (q_{x+\Delta x} + q_{y+\Delta y} + q_{z+\Delta z} - q_x - q_y - q_z)A + mc_p \partial T / \partial t \tag{2.8}$$

According to Fourier's law, the heat flux in the x-direction is given by

$$q_x = -k_x \frac{\partial T}{\partial x} \tag{2.9}$$

The heat flux in the y-direction is

$$q_y = -k_y \frac{\partial T}{\partial y} \tag{2.10}$$

and the heat flux in the z-direction is

$$q_z = -k_z \frac{\partial T}{\partial z} \tag{2.11}$$

Eq. (2.8) then becomes,

$$q_v \Delta x \Delta y \Delta z = \Delta x \frac{\partial}{\partial x}\left(-k\frac{\partial T}{\partial x}\right)\Delta y \Delta z + \Delta y\left(-k\frac{\partial T}{\partial y}\right)\Delta x \Delta z$$
$$+\Delta z \frac{\partial}{\partial z}\left(-k\frac{\partial T}{\partial z}\right)\left(\Delta x \Delta y + \rho \Delta x \Delta y \Delta z c_p \frac{\partial T}{\partial t}\right) \tag{2.12}$$

For a homogenous material, the thermal conductivity is constant in space (i.e., $k_x = k_y = k_z = k$). In the view of this, Eq. (2.12) can be rewritten as

$$k\frac{\partial^2 T}{\partial x^2} + k\frac{\partial^2 T}{\partial y^2} + k\frac{\partial^2 T}{\partial z^2} + q_v = \rho c_p \frac{\partial T}{\partial t} \tag{2.13}$$

Equation 2.13 may be rewritten as

$$\frac{\partial^2 T}{\partial x^2} + \frac{\partial^2 T}{\partial y^2} + \frac{\partial^2 T}{\partial z^2} + \frac{q_v}{k} = \frac{1}{\alpha}\frac{\partial T}{\partial t} \tag{2.14}$$

Here α is thermal diffusivity of the material and is given by $k/\rho\, c_p$. It is instructive to obtain the units of α.

$$\alpha = \frac{W/mK}{kg/m^3\, J/kgK} = m^2/s$$

We see that α has the units of m^2/s. The s^{-1} clearly shows that it is related to "some rate." That "some" is nothing but the diffusion of heat. It is now evident that α gives us an idea of the rate of diffusion of heat in a medium.

In other words, the quantity α signifies how quickly heat penetrates a solid body. The α for steel is far higher compared to that of wood. A wooden ladle in a boiling soup would not be as hot to feel as a steel ladle when other conditions are the same (i.e. when felt at the same time).

Eq. (2.14) is called the three dimensional transient heat conduction equation with heat generation. The three terms on the left-hand side of Eq. (2.14) represent the net diffusion of heat in the three directions x, y, and z, respectively. The fourth term represents the uniform volumetric heat generation due to either nuclear fusion, metabolism (say cancer) or an exothermic chemical reaction. Eq. (2.14) has the units K/m^2. Mathematically, the equation is a second-order linear partial differential equation (PDE) in space, a first-order PDE in time, and supports six boundary conditions (two each on x, y, and z) and one initial condition (in time).

Mathematicians refer to such equations as IVBP problems, with the abbreviation denoting initial value boundary problems. Even k_x, k_y, and k_z need not be equal; as in the case of orthotropic materials where the governing equation is still linear. However if k_x, k_y, and k_z are functions of space (x, y, z) or temperature, the governing equation is no longer linear, and its solution is nontrivial.

For steady state heat transfer, $\partial T / \partial t = 0$, and Eq. (2.14) becomes

$$\frac{\partial^2 T}{\partial x^2} + \frac{\partial^2 T}{\partial y^2} + \frac{\partial^2 T}{\partial z^2} + \frac{q_v}{k} = 0 \tag{2.15}$$

Eq. (2.15) is called the Poisson's equation.

For steady $(\partial T / \partial t) = 0$ and no heat generation $(q_v = 0)$, Eq. (2.14) becomes

$$\frac{\partial^2 T}{\partial x^2} + \frac{\partial^2 T}{\partial y^2} + \frac{\partial^2 T}{\partial z^2} = 0 \tag{2.16}$$

Eq. (2.16) is called the Laplace equation and can be written compactly as $\nabla^2 T = 0$ where ∇^2 is frequently referred to as the Laplacian operator or simply Laplacian.

For one-dimensional, steady state heat conduction and without heat generation, we have

$$\frac{d^2 T}{dx^2} = 0 \tag{2.17}$$

By using a similar control volume approach, the heat conduction equation in cylindrical coordinates can be derived in the form shown in Eq. (2.20).

$$\frac{1}{r} \frac{\partial}{\partial r} \left(kr \frac{\partial T}{\partial r} \right) + \frac{1}{r^2} \frac{\partial}{\partial \phi} \left(k \frac{\partial T}{\partial \phi} \right) + \frac{\partial}{\partial z} \left(k \frac{\partial T}{\partial z} \right) + q_v = \rho c_p \frac{\partial T}{\partial t}. \tag{2.18}$$

In Eq. (2.18), r is the radius, ϕ is the azimuthal angle and z is the axial direction. (Please see Fig. 2.2A). In an analogous fashion, the heat conduction equation in spherical coordinates can be derived in the form shown in Eq. (2.19).

$$\frac{1}{r^2} \frac{\partial}{\partial r} \left(kr^2 \frac{\partial T}{\partial r} \right) + \frac{1}{r^2 \sin\theta} \frac{\partial}{\partial \theta} \left(k \sin\theta \frac{\partial T}{\partial \theta} \right) + \frac{1}{r^2 \sin^2\theta} \frac{\partial}{\partial \phi} \left(k \frac{\partial T}{\partial \phi} \right) + q_v = \rho c_p \frac{\partial T}{\partial t} \tag{2.19}$$

In Eq. (2.19), r is the radius, θ is the zenith angle and ϕ is the azimuthal angle. (Please see Fig. 2.2)

In compact, coordinate-free form, the governing equations (Eqns. 2.14, 2.18 and 2.19) can be written as

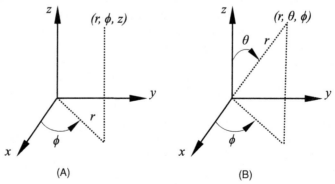

(A)   (B)

FIGURE 2.2

Schematic of coordinate systems (A) cylindrical (B) spherical.

$$\nabla^2 T + \frac{q_v}{k} = \frac{1}{\alpha} \partial T / \partial t \qquad (2.20)$$

for constant thermal conductivity, k.

2.2.1 Boundary conditions

For any given problem, a complete mathematical description requires imposing the appropriate boundary conditions. Some possibilities are:

1. **Dirichlet boundary condition:** A known value is imposed on boundary for temperature as shown in Fig. 2.3A. Also known as boundary condition of first kind

The conditions for Fig. 2.3A are as follows

$$at\ x = 0; T = T_1 \qquad (2.21)$$
$$a\ x = L; T = T_2 \qquad (2.22)$$

2. **Neumann boundary condition:** Gives a condition for the first derivative of temperature and is shown in Fig. 2.3B. Also known as boundary condition of second kind

$$at\ x = 0; q = -k\frac{\partial T}{\partial x}\ (\text{Neumann}) \qquad (2.23)$$
$$at\ x = L; T = T_2\ (\text{Dirichlet}) \qquad (2.24)$$

3. **Robin boundary condition (or) mixed condition (or) third kind of boundary condition:**

This concerns convection or radiation or both at the surfaces in question. In Fig. 2.3C, the key point is that both T_1 and T_2 are unknown.

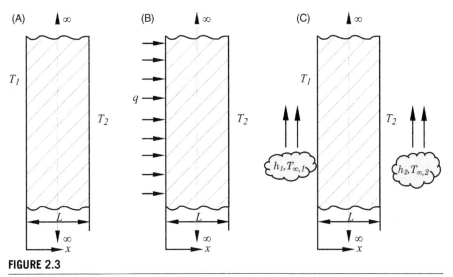

FIGURE 2.3

Schematic representation of different boundary conditions in heat conduction (A) Dirichlet (B) Neumann and (C) Robin or mixed.

$$at\ x = 0; q = -k\frac{\partial T}{\partial x} = h_1(T_{\infty,1} - T_1) \tag{2.25}$$

$$at\ x = L; q = -k\frac{\partial T}{\partial x} = h_2(T_2 - T_{\infty,2}) \tag{2.26}$$

From Eqs. 2.25 and 2.26, it is intuitively apparent that a heat conduction equation with Robin conditions will be a lot harder to solve.

2.3 Steady state, one-dimensional conduction in a few commonly encountered systems

Let us now look at simple yet potent solutions to one-dimensional, steady conduction in three representative geometries.

1. Plane wall
2. Cylinder
3. Sphere

They are not only amenable to a "clean" mathematical analysis, but offer considerable insights into the engineering of several heat transfer systems.

2.3.1 Heat transfer in a plane wall

Consider a simple solid plane wall (or slab) that is infinitely long with a thickness L, with Dirichlet boundary conditions, as shown in Fig. 2.4.

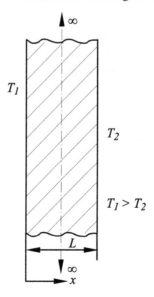

FIGURE 2.4

Schematic representation of the plane wall under consideration in section 2.3.1.

The following assumptions are made:

a. $T = f(x)$ alone.
b. Steady state prevails.
c. No heat generation in the solid.
d. Constant thermophysical properties for the solid.

The governing Eq. (2.14) reduces to the following.

$$\frac{d^2T}{dx^2} = 0 \qquad (2.27)$$

Integrating Eq. (2.27) twice, we have

$$\frac{dT}{dx} = c_1 \qquad (2.28)$$

$$T = c_1 x + c_2 \qquad (2.29)$$

In Eq. (2.29), c_1 and c_2 are constants.
Boundary condition 1: At $x = 0$, $T = T_1$
From Eq. (2.29), we have

$$c_2 = T_1 \qquad (2.30)$$

Boundary condition 2: At $x = L$; $T = T_2$

$$T_2 = c_1 L + T_1 \qquad (2.31)$$

$$c_1 = \frac{T_2 - T_1}{L} \qquad (2.32)$$

Substituting for c_1 and c_2 in Eq. (2.29), we obtain the following expression for temperature, T, across the slab.

$$T = \frac{T_2 - T_1}{L} x + T_1 \qquad (2.33)$$

Eq. (2.33) tells us that the temperature profile is linear across the slab. This is sometimes referred to as LTP (Linear Temperature Profile). This will not be the case when the thermal conductivity is varying (see Problem 2.2 at the end of the chapter). Differentiating Eq. (2.33) with respect to x gives the following expression.

$$\frac{dT}{dx} = \frac{T_2 - T_1}{L} \qquad (2.34)$$

The heat transfer through the plane wall is calculated by the Fourier's law of heat conduction.

$$Q = -kA\frac{dT}{dx} \qquad (2.35)$$

$$Q = -kA \frac{T_2 - T_1}{L} \tag{2.36}$$

Eq. (2.36) can also be written, on purpose (we will see why in the ensuing section), as

$$Q = \frac{T_1 - T_2}{L / kA} \tag{2.37}$$

2.4 Electrical analogy and thermal resistance

A class of heat transfer problems can be eminently analyzed using an electrical analogy. In this approach, we consider the flow of heat to be analogous to the flow of electrical current, and the following equivalences apply: $Q \Leftrightarrow I$, $R_{th} \Leftrightarrow R$, $\Delta T \Leftrightarrow \Delta V$. R_{th} here refers to thermal resistance.

A typical resistance circuit for the problem considered above is shown in Fig. 2.5.

From Fourier's law and also from the solution to conduction in a plane wall that we just saw Eq. (2.37), we know that

$$Q = \frac{\Delta T}{L / kA} \tag{2.38}$$

From Ohm's law, we know that because of the existence of potential difference (ΔV) between the two ends of an electrical conductor, a current I passes through the conductor against the resistance R, and the relation between the three is

$$I = \frac{\Delta V}{R} \tag{2.39}$$

A comparison of Eqs. (2.38) and (2.39) clearly shows that the thermal resistance offered by the plane wall is given by

$$R_{thermal} = L / kA \tag{2.40}$$

From Eq. (2.40), it is clear that thicker the wall more is the resistance and higher the thermal conductivity of a material, lower is the resistance. The equation also confirms that $R_{thermal}$ varies inversely with the cross sectional area A.

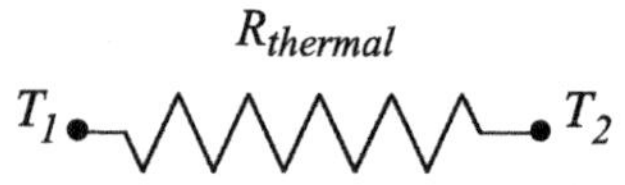

FIGURE 2.5

Schematic representation of an electrical resistance network for solving a heat conduction problem.

2.5 Heat transfer in cylindrical coordinates

Conduction problems in cylindrical geometries are commonly encountered in pipes, tubes, current-carrying conductors, and nuclear fuel rods, to name a few.

The governing given by Eq. (2.18) is reproduced here for the sake of completeness.

$$\frac{1}{r}\frac{\partial}{\partial r}\left(kr\frac{\partial T}{\partial r}\right)+\frac{1}{r^2}\frac{\partial}{\partial \phi}\left(k\frac{\partial T}{\partial \phi}\right)+\frac{\partial}{\partial z}\left(k\frac{\partial T}{\partial z}\right)+q_v = \rho c_p \frac{\partial T}{\partial t} \tag{2.41}$$

For one-dimensional, steady state heat transfer without heat generation, where temperature varies only with radius, r Eq. (2.41) reduces to

$$\frac{1}{r}\frac{d}{dr}\left(kr\frac{dT}{dr}\right)=0 \tag{2.42}$$

For constant thermal conductivity, Eq. (2.42) can be written as

$$\frac{1}{r}\frac{d}{dr}\left(r\frac{dT}{dr}\right)=0 \tag{2.43}$$

Consider a cylindrical annulus as shown in Fig. 2.6. Let us assume that $T_1 > T_2$. The boundary conditions for the problem shown in the figure are as follows

At $r = r_1; T = T_1$

At $r = r_2; T = T_2$

Integrating Eq. (2.44) once, we get

$$\frac{d}{dr}\left(r\frac{dT}{dr}\right)=0 \tag{2.44}$$

$$r\frac{dT}{dr}=c_1 \tag{2.45}$$

$$\frac{dT}{dr}=\frac{c_1}{r} \tag{2.46}$$

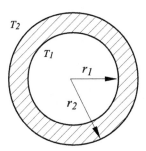

FIGURE 2.6

Schematic representation of a cross-sectional view of a cylindrical annulus undergoing one dimensional steady state conduction.

Integrating again, we obtain an expression for T or $T(r)$ as

$$T = c_1 \ln r + c_2 \tag{2.47}$$

Applying the two boundary conditions to Eq. (2.47), we have

$$T_1 = c_1 \ln r_1 + c_2 \tag{2.48}$$

$$T_2 = c_1 \ln r_2 + c_2 \tag{2.49}$$

On subtracting Eq. (2.49) from Eq. (2.48), we obtain the expression for c_1 as

$$c_1 = \frac{T_1 - T_2}{\ln\left(\dfrac{r_1}{r_2}\right)} \tag{2.50}$$

Substituting for c_1 in Eq. (2.48); we obtain the expression for c_1 as

$$c_2 = T_1 - \frac{T_1 - T_2}{\ln\left(\dfrac{r_1}{r_2}\right)} \ln(r_1) \tag{2.51}$$

Substituting for both c_1 and c_2 in Eq. (2.48), we now arrive at the final "usable" form of Eq. (2.47).

$$T = T_1 + \frac{T_1 - T_2}{\ln\left(\dfrac{r_1}{r_2}\right)} \ln\left(\frac{r}{r_1}\right) \tag{2.52}$$

Therefore from Eq. (2.52), one can clearly see that the temperature distribution in a cylinder or rather a cylindrical annulus is logarithmic.

The heat transfer through the cylinder can be calculated as

$$Q = -kA_n \frac{dT}{dr}\Big|_{r=r_1} \tag{2.53}$$

$$Q = -k2\pi r_1 L \frac{T_1 - T_2}{\ln\left(\dfrac{r_1}{r_2}\right)} \times \frac{1}{r_1} \tag{2.54}$$

$$Q = \frac{\Delta T}{\dfrac{\ln(r_2 / r_1)}{2\pi k L}} \tag{2.55}$$

Please note that $Q(\text{at } r = r_1) = Q(\text{at } r = r_2)$. In the above equation, L is the length in the direction perpendicular to the plane of the paper.

Invoking the concept of electrical analogy by looking at Eq. (2.55), we obtain the expression for $R_{cond,cyl}$ as

$$R_{condn,cyl} = \frac{\ln\left(\dfrac{r_2}{r_1}\right)}{2\pi k L} \tag{2.56}$$

2.5.1 Critical radius of insulation for cylinder

Consider a cylindrical pipe of radius r and length L in the direction perpendicular to the plane of the paper. It is necessary to add insulation with thermal conductivity of k to this pipe to reduce heat transfer. In the face of it, the proposal looks like a perfect engineering solution for the problem at hand. However, as r_2 increases conduction becomes "more difficult" but because the area for convection given by $2\pi r_2 L$ increases, the increase in conduction resistance will be offset by a decrease in convection resistance. Hence, there is an interplay of these two competing phenomena, which also suggests that there would be a particular r_2 at which the total resistance is an extremum (Fig. 2.7).

The heat transfer through the cylinder is given by

$$Q = \frac{T_1 - T_\infty}{\dfrac{\ln(r_2/r_1)}{2\pi kL} + \dfrac{1}{h2\pi r_2 L}} \tag{2.57}$$

In order to maximize or minimize the heat transfer for the specified $(T_1 - T_\infty)$, we need to differentiate the denominator (R_{total}) in Eq. (2.57) with respect to r_2 and set it to zero to determine the value of r_2 at which R_{total} becomes stationary.

$$\frac{dR_{total}}{dr} = 0 \tag{2.58}$$

$$\frac{1}{2\pi k_2 L}\left(\frac{1}{r_2}\right) + \frac{1}{h2\pi L}\left(\frac{-1}{r_2^2}\right) = 0 \tag{2.59}$$

$$r_2 = \frac{k}{h} = r_c \tag{2.60}$$

In Eq. (2.60), r_c is known as the critical radius of insulation. Second order conditions are required to check if R_{total} is maximum or minimum at $r_2 = r_c$

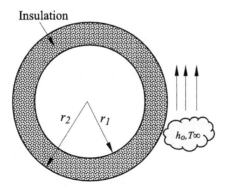

Insulation

r_2 r_1

$h_0, T\infty$

FIGURE 2.7

Schematic representation of conduction in a cylinder with insulation, and convection on the outside.

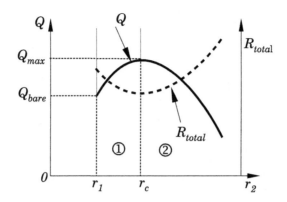

FIGURE 2.8

Variation of thermal resistance and heat transfer with thickness of insulation.

Alternatively we can plot Q and R_{total} against r_2 and examine what happens. A qualitative variation of thermal resistance and heat transfer with the thickness of insulation is shown in Fig. 2.8.

From r_1 to r_c, the total resistance keeps decreasing. This is due to the dominance of the decreasing convection resistance compared to conduction resistance in Eq. (2.57). Beyond r_{c1}, R_{total} increases due to the dominance of the increased conduction resistance over the decreasing convection resistance in Eq. (2.57). The result is counterintuitive in the sense that the total resistance is a minimum at $r_2 = r_c$, and so to have an insulating effect, r_2 must be much greater than r_c.

2.6 Steady state conduction in a spherical shell

A spherical shell or simply a sphere is an essential geometry in heat transfer engineering that is used in a variety of applications, such as nuclear reactor waste disposal, ball bearings (solid sphere), and rocket nozzles (nose cone can have a hemispherical shape) to name a few.

Consider a spherical shell of which a cross-sectional view is shown in Fig. 2.9, along with the geometrical details and the two boundary conditions.

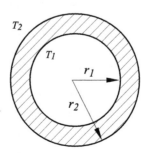

FIGURE 2.9

Schematic representation of a cross-sectional view of a sphere, for studying one dimensional steady conduction.

We make the following assumptions:

1. Steady state prevails
2. $T = f(r)$ alone
3. $q_v = 0$ and thermo physical and transport properties (ρ, c_p, k) are constant

The governing equation for this situation can be obtained by setting to zero, the terms in the general conduction equation that are not contributing to the heat transfer process in the geometry.

$$\frac{1}{r^2}\frac{d}{dr}\left(r^2\frac{dT}{dr}\right) = 0 \tag{2.61}$$

$$\frac{d}{dr}\left(r^2\frac{dT}{dr}\right) = 0 \tag{2.62}$$

Integrating once

$$r^2\frac{dT}{dr} = c_1 \tag{2.63}$$

$$\frac{dT}{dr} = \frac{c_1}{r^2} \tag{2.64}$$

Integrating again

$$T = \frac{-c_1}{r} + c_2 \tag{2.65}$$

Boundary condition 1: At $r = r_1$; $T = T_1$
Boundary condition 2: At $r = r_2$; $T = T_2$

$$T_1 = \frac{-c_1}{r_1} + c_2 \tag{2.66}$$

$$T_2 = \frac{-c_1}{r_2} + c_2 \tag{2.67}$$

By solving Eq. (2.66) and (2.67), we obtain the expression for c_1 and c_2 as

$$c_1 = \frac{T_1 - T_2}{\left(\dfrac{1}{r_2} - \dfrac{1}{r_1}\right)} \tag{2.68}$$

$$c_2 = T_1 + \frac{T_1 - T_2}{\left(\dfrac{1}{r_2} - \dfrac{1}{r_1}\right)}\frac{1}{r_1} \tag{2.69}$$

Finally, the temperature distribution in the spherical shell is given by

$$T = T_1 - \frac{T_1 - T_2}{\left(\dfrac{1}{r_2} - \dfrac{1}{r_1}\right)}\left(\frac{1}{r} - \frac{1}{r_1}\right) \tag{2.70}$$

The heat transfer through the sphere can be calculated by using Fourier's law as follows (Please recognize that $A = f(r)$ and so we need to be cautious in evaluating Q)

$$Q = -kA\frac{dT}{dr}\Big|_{r=r_1} \tag{2.71}$$

$$Q = -k4\pi r_1^2 \frac{T_1 - T_2}{\left(\dfrac{1}{r_2} - \dfrac{1}{r_1}\right)}\left(\dfrac{1}{r_1^2}\right) \tag{2.72}$$

Invoking the electrical analogy, we can obtain an expression for the conduction resistance across the shell, $R_{cond,sphere}$

$$Q = \frac{T_1 - T_2}{\dfrac{(r_2 - r_1)}{4\pi k r_1 r_2}} \tag{2.73}$$

$$R_{cond,sphere} = \frac{(r_2 - r_1)}{4\pi k r_1 r_2} \tag{2.74}$$

It is instructive to mention that $Q\ (At\ r = r_1) = Q\ (At\ r = r_2)$.

2.7 Steady state conduction in a composite wall, cylinder and sphere

2.7.1 Composite wall

A simple composite plane wall made of three materials with thermal conductivities k_a, k_b, with convection boundary conditions at the end wall k_c and three thicknesses L_a, L_b, and L_c respectively is shown in Fig. 2.10, is undergoing one dimensional steady state conduction. A resistance circuit can be drawn, as shown in Fig. 2.11. Under steady state, with no heat generation we have the following:

$$Q = Q_{conv,left} = Q_{conduction,a} = Q_{conduction,b} = Q_{conduction,c} = Q_{conv,right} \tag{2.75}$$

This may be alternatively written as (Refer to Fig. 2.7)

$$Q = \frac{T_{\infty_1} - T_1}{R_{conv1}} = \frac{T_1 - T_2}{R_a} = \frac{T_2 - T_3}{R_b} = \frac{T_3 - T_4}{R_c} = \frac{T_4 - T_{\infty_2}}{R_{conv2}} \tag{2.76}$$

$$= \frac{T_{\infty,1} - T_1 + T_1 \ldots + \ldots - T_{\infty,2}}{R_{conv1} + \ldots + R_{conv2}} \tag{2.76a}$$

Using the Componendo-Divedendo rule, i.e., adding all the numerators and denominators respectively and equating the result to Q, we have

$$Q = \frac{T_{\infty,1} - T_{\infty,2}}{R_{total}} \tag{2.76b}$$

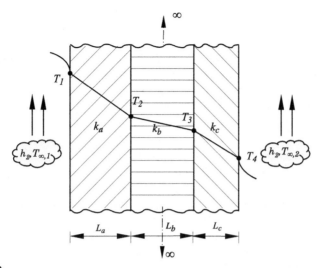

FIGURE 2.10

Schematic representation of a composite wall undergoing steady state conduction.

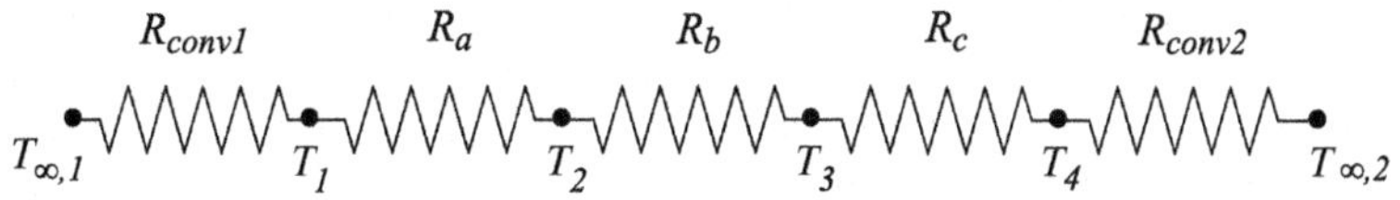

$$R_{conv1} \qquad R_a \qquad R_b \qquad R_c \qquad R_{conv2}$$

$$T_{\infty,1} \quad T_1 \quad T_2 \quad T_3 \quad T_4 \quad T_{\infty,2}$$

FIGURE 2.11

Schematic representation of a resistance network for the composite wall under consideration, under heat conduction.

In Eq. (2.76b), R_{total} can be calculated as

$$R_{total} = R_{conv1} + R_a + R_b + R_c + R_{conv2} \qquad (2.77)$$

Eq. (2.77) is for a series connection and is much more powerful than what it appears to be. Without solving for conduction equation in three materials with two Robin conditions, we straight away solved the problem. Whenever two or more materials are kept side by side along the direction of heat transfer, their resistances are considered to be in series as shown in Fig. 2.11. The individual resistances in the composite wall are given by

$$R_{conv1} = \frac{1}{h_1 A} \qquad (2.78)$$

$$R_a = \frac{L_a}{k_a A} \qquad (2.79)$$

$$R_b = \frac{L_b}{k_b A} \qquad (2.80)$$

$$R_c = \frac{L_c}{k_c A} \qquad (2.81)$$

$$R_{conv2} = \frac{1}{h_2 A} \tag{2.82}$$

An overall heat transfer coefficient, U, may now be defined as follows:

$$Q = UA(T_{\infty_1} - T_{\infty_2}) = \frac{(T_{\infty,1} - T_{\infty,2})}{R_{total}} \tag{2.83}$$

From Eq. (2.83), it follows that

$$\frac{1}{UA} = R_{total} \tag{2.84}$$

In Eq. (2.84), $1/UA$ in general is given by,

$$\frac{1}{UA} = \frac{1}{h_1 A} + \frac{L_a}{k_a A} + \frac{L_b}{k_b A} + \frac{L_c}{k_c A} + \frac{1}{h_2 A} \tag{2.85}$$

Upon simplification Eqn. (2.85) becomes

$$\frac{1}{U} = \frac{1}{h_1} + \frac{L_a}{k_a} + \frac{L_b}{k_b} + \frac{L_c}{k_c} + \frac{1}{h_2} \tag{2.86}$$

In a practical problem, all the terms on the right hand side of Eq. (2.86) are known. Using these, the overall heat transfer coefficient, U can be determined which inturn can be used to calculate Q, which is of primary interest. Now we can get back to Eq. (2.76) and pull out all the intermediate temperatures. This approach may be termed as "Divide and Conquer" and is clean, neat and smart.

2.7.1.1 Parallel connection
A typical case of heat transfer through a parallel connection is shown in Fig. 2.12. Along the normal to the direction of heat transfer, if the materials are kept one over

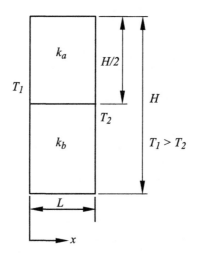

FIGURE 2.12

Schematic representation of a plane wall with parallel resistances.

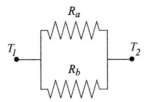

FIGURE 2.13

Schematic representation of an electrical resistances network for solving a heat conduction problem with resistances in parallel.

the other, their thermal resistances are considered to be in parallel and are shown in Fig. 2.13. Let the boundary temperatures be T_1 and T_2 with $T_1 > T_2$, as before. The total heat transfer through the slab can be calculated as

$$Q = \frac{T_1 - T_2}{R_{ab}} \tag{2.87}$$

In Eq. (2.87), R_{ab} is calculated as

$$\frac{1}{R_{ab}} = \frac{1}{R_a} + \frac{1}{R_b} \tag{2.88}$$

2.7.1.2 Series-parallel connection

A series-parallel connection is shown in Fig. 2.14, and the corresponding resistance network is shown in Fig. 2.15. Let the temperatures at x = 0 and x = $(L_1 + L_2 + L_3)$ be T_1 and T_4 respectively with $T_1 > T_4$.

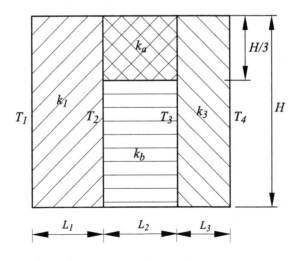

FIGURE 2.14

Schematic representation of a series parallel combination of thermal resistances.

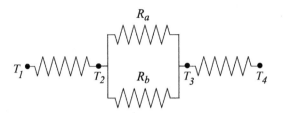

FIGURE 2.15

Schematic representation of a composite resistance network involving series and parallel resistances.

The heat transfer through the slab is given by

$$Q = \frac{T_1 - T_4}{R_{total}}$$ (2.89)

$$R_{total} = R_1 + R_{ab} + R_3$$ (2.90)

In Eq. (2.90), R_{ab} can be calculated as follows

$$\frac{1}{R_{ab}} = \frac{1}{R_a} + \frac{1}{R_b}$$ (2.91)

Here, $R_a = \dfrac{L_2}{k_a(A/3)}$ and $R_b = \dfrac{L_2}{k_b(2A/3)}$.

R_1 (i.e., between T_1 and T_2) and R_3 (i.e., between T_3 and T_4) can be calculated, as explained before.

Example 2.1: *Consider a heat-generating chip that is 12 × 12 mm² in cross-section. The chip is very thin and can be considered to be spatially isothermal. It is mounted on a substrate made of aluminum (k = 205 W/mK) that is 10 mm thick. The dimensions of the substrate are the same as the chip. A joint made of resin with a thickness of 0.04 mm holds the chip and the substrate together. The joint has a thermal conductivity of 2 W/mK on both ends there is a convective heat transfer for coefficient of 100 W/m²K with an ambient at 30 °C.*

1. *If the heat flux from the chip is 10,000 W/m², what is the maximum temperature in the assembly, and where does it occur?*
2. *If the joint offers an additional resistance (known as thermal contact resistance) of 0.8 × 10⁻⁴ m² K/W, what is the maximum temperature in the assembly?*
3. *If a designer says that one can power the chip all the way up to a maximum of 85 °C, what is the maximum possible heat flux or power?*

Solution:

Schematic representation, as given in the problem is shown in Fig. 2.16

1. When there is no thermal contact resistance, the heat flux q_1 shown in the figure is given by

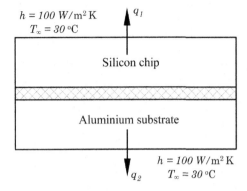

FIGURE 2.16

Schematic representation of heat conduction in a chip that is under consideration in Example 2.1. Please note that this figure is not to scale.

$$q_1 = \frac{T_{chip} - 30}{\dfrac{1}{100}}$$

$$\text{Similarly, } q_2 = \frac{T_{chip} - 30}{\dfrac{10 \times 10^{-3}}{205} + \dfrac{0.04 \times 10^{-3}}{2} + \dfrac{1}{100}} \approx \frac{T_{chip} - 30}{0.01}$$

$$q = q_1 + q_2 = \frac{T_{chip} - 30}{0.01} + \frac{T_{chip} - 30}{0.01}$$

$$10000 = 2 \times \left(\frac{T_{chip} - 30}{0.01} \right)$$

$$T_{chip} = \left(\frac{10000}{2} \right) \times 0.01 + 30 = 80 \, ^\circ\text{C}$$

$$q_2 = \frac{80 - 30}{0.01} = 5000 = \frac{T_{chip} - 30}{\dfrac{0.04 \times 10^{-3}}{2}}$$

$$T_{int} = 79.9 \, ^\circ\text{C}$$

Thus the maximum temperature in the assembly is 80 °C at the interface between chip and joint.

2. When considering a finite thermal contact resistance

$$q_2 = \frac{T_{chip} - 30}{\dfrac{10 \times 10^{-3}}{205} + \dfrac{0.04 \times 10^{-3}}{2} + 0.8 \times 10^{-4} + \dfrac{1}{100}}$$

$$q_2 = \frac{T_{chip} - 30}{0.0101}$$

$$q = q_1 + q_2 = \frac{T_{chip} - 30}{0.01} + \frac{T_{chip} - 30}{0.0101}$$

$$10000 = 10T_{chip} - 3000 + 99T_{chip} - 2970.3$$

$$T_{chip} = 80.25 \,°C$$

As expected, the addition of contact resistance increases the temperature, but its effect in this case is negligibly small.

3. Maximum allowable heat flux

$$q_1 = \frac{T_{chip} - 30}{\dfrac{1}{100}}$$

$$q_1 = 100 \times (85 - 20) = 5500 \text{ W/m}^2$$

$$q_2 = \frac{T_{chip} - 30}{0.0101} = \frac{85 - 30}{0.0101} = 5445.54 \text{ W/m}^2$$

Maximum allowable heat flux, $q = q_1 + q_2 = 10945$ W/m^2.

$$q \approx 10.95 \text{ } kW/m^2.$$

Example 2.2: *Consider a material with constant properties and no heat generation. Temperatures have been measured at a certain instant of time and have been regressed to a quadratic form as* $T(x,y,z) = 2x^2 + 3y^2 - 5z^2 + 2xy - 3yz + 6xz$. *Using the general heat conduction equation, determine if there is any region in the body where* $T = f(t)$.
Solution:
The governing equation is

$$\frac{\partial^2 T}{\partial x^2} + \frac{\partial^2 T}{\partial y^2} + \frac{\partial^2 T}{\partial z^2} = \frac{1}{\alpha} \frac{\partial T}{\partial t}$$

$$4 + 6 - 10 = \frac{1}{\alpha} \frac{dT}{dt}$$

$$\frac{\partial T}{\partial t} = 0$$

Since $\partial T / \partial t = 0$ everywhere, it is seen that transients exist nowhere. Hence, steady state prevails in the material.

2.7.1.3 Thermal contact resistance

When two surfaces are kept together, because of the presence of a surface roughness at the contacting surfaces, the gaps are invariably filled with air, and this introduces convection resistance. This is called thermal contact resistance, which reduces the heat transfer through the composite wall.

Needless to say, for maximizing the heat transfer through the composite wall, the thermal contact resistance at the interface has to be reduced.

Methods of reducing thermal contact resistance:

1. By applying high thermal conductivity gel at the interface. A gel is typically a semisolid material. When the gel is exposed to a higher temperature, it gets converted into liquid and slowly drops out and can be used for low-temperature applications.
2. Placing thin soft and high thermal conductivity metallic sheets at the interface during the assembly.
3. By increasing the contact area, by applying more pressure during assembly.

Thermal contact resistance is quite a difficult quantity to measure and, in practice, can often be a "sore thumb" in thermal system design.

2.7.2 Composite cylinder

Consider one-dimensional steady-state conduction in a composite cylinder, schematic representation of which is in Fig. 2.17. The corresponding resistance network is presented in Fig. 2.18. Geometrical details of thermophysical properties of the materials making up composite cylinder are also shown in Fig. 2.17.

Invoking, the resistance analogy, presented earlier, the total heat transfer from the inside to the outside through the cylindrical shell is given by

$$Q = \frac{T_{\infty_1} - T_{\infty_2}}{R_{total}} \qquad (2.92)$$

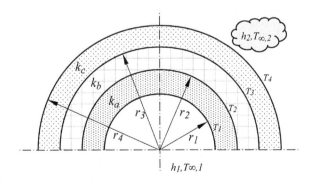

FIGURE 2.17

Schematic representation of a composite cylinder undergoing one-dimensional steady-state conduction.

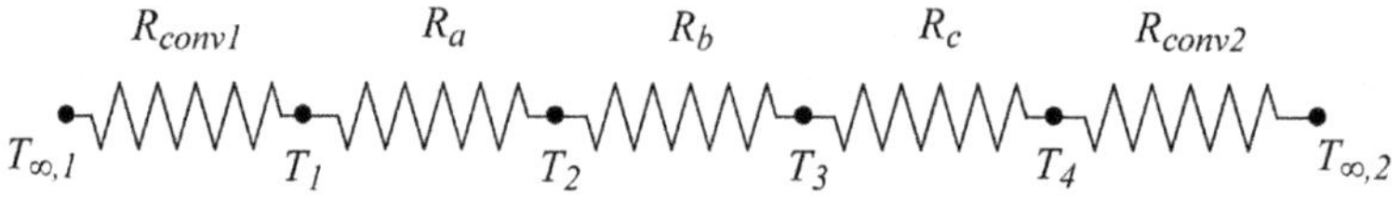

FIGURE 2.18

Schematic representation of a resistance network one dimensional steady state conduction in a composite cylinder.

The total resistance R_{total} is given by

$$R_{total} = R_{conv1} + R_a + R_b + R_c + R_{conv2} \tag{2.93}$$

The individual resistances in Eq. (2.93) can be calculated as follows:

$$R_{conv1} = \frac{1}{h_1 A_1} \tag{2.94}$$

$$R_a = \frac{\ln \frac{r_2}{r_1}}{2\pi k_a L} \tag{2.95}$$

$$R_b = \frac{\ln \frac{r_3}{r_2}}{2\pi k_b L} \tag{2.96}$$

$$R_c = \frac{\ln \frac{r_4}{r_3}}{2\pi k_c L} \tag{2.97}$$

$$R_{conv2} = \frac{1}{h_2 A_1} \tag{2.98}$$

Please note that $A_1 = 2\pi r_1 L_1$ and $A_4 = 2\pi r_4 L_4$. An overall heat transfer coefficient can now be defined for the cylindrical shell as follows.

$$Q = \frac{T_{\infty_1} - T_{\infty_2}}{R_{total}} = U_1 A_1 (T_{\infty_1} - T_{\infty_2}) \tag{2.99}$$

Here U_1 is the overall heat transfer coefficient, based on inner area A_1, and is given by

$$U_1 = \frac{1}{A_1 R_{total}} \tag{2.100}$$

$$\frac{1}{U_1} = A_1 \left(\frac{1}{2\pi r_1 L h_1} + \frac{\ln\left(\frac{r_2}{r_1}\right)}{2\pi k_a L} + \frac{\ln\left(\frac{r_3}{r_2}\right)}{2\pi k_b L} + \frac{\ln\left(\frac{r_4}{r_3}\right)}{2\pi k_c L} + \frac{1}{h_2 2\pi r_4 L} \right) \tag{2.101}$$

Please note, as opposed to a plane wall, the overall heat transfer coefficient varies across the cylinder, and its specification is incomplete without a declaration of the area on which it is based.

Example 2.3: *Consider a very long[†] cylindrical rod of 100 mm radius that consists of a nuclear-reacting material (k = 0.05 W/mK), generating 24,000 W/m³ uniformly throughout the volume (q_g). This rod is enclosed within a tube having an outer radius of 200 mm, whose thermal conductivity is 4 W/mK. The outer surface is surrounded by a fluid at 100 °C and h = 20 W/m²K. Find the temperature at the interface of the two cylinders and the outer surface. A schematic representation of the given problem is shown in Fig. 2.19.*

Solution:

From energy balance we have

$$q_g \times volume = hA(T_2 - T_{\infty_2}) \tag{2.102}$$

$$24000 \times \frac{\pi}{4} \times 0.2^2 \times L = 20 \times 2\pi \times 0.2L(T_2 - 100) \tag{2.103}$$

$$T_2 = 130\,°C$$

Under steady state

$$q_g \times volume = \frac{T_1 - T_2}{\dfrac{\ln \dfrac{r_2}{r_1}}{2\pi kL}} \tag{2.104}$$

$$24000 \times \frac{\pi}{4} \times 0.2^2 \times L = \frac{T_1 - 130}{\dfrac{\ln \dfrac{200}{100}}{2\pi \times 4 \times 1}} \tag{2.105}$$

$$T_1 = 150.8\,°C$$

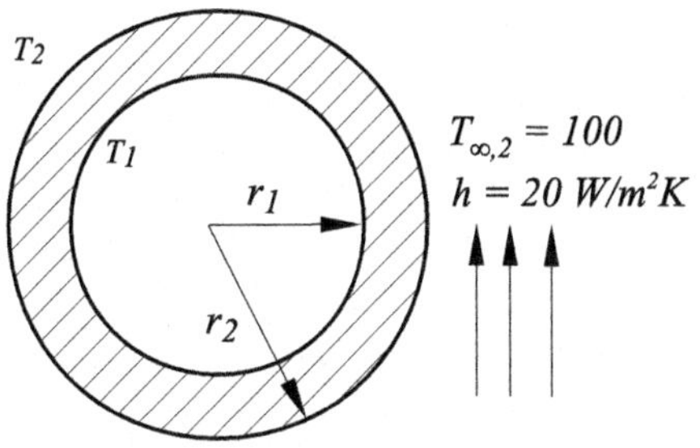

FIGURE 2.19

Schematic representation of the cylindrical rod under consideration in Example 2.3.

[†]In the direction perpendicular to the plane of the paper.

Example 2.4: *Revisit Example 2.3 and calculate the center temperature.*
Solution:

$$\frac{1}{r}\frac{d}{dr}\left(r\frac{dT}{dr}\right)+\frac{q_g}{k}=0 \tag{2.106}$$

$$\frac{d}{dr}\left(r\frac{dT}{dr}\right)=-\frac{q_g}{k}\times r \tag{2.107}$$

$$r\frac{dT}{dr}=-\frac{q_g}{k}\times\frac{r^2}{2}+c_1 \tag{2.108}$$

$$\frac{dT}{dr}=-\frac{q_g}{k}\times\frac{r}{2}+\frac{c_1}{r} \tag{2.109}$$

Boundary condition 1: At $r=0;\dfrac{dT}{dr}=0$

Therefore

$$c_1=0 \tag{2.110}$$

Integrating Eq. (2.128) with respect r on both sides, we have

$$T=-\frac{q_g}{k}\times\frac{r^2}{4}+c_2 \tag{2.111}$$

Boundary condition 2: At $r=0.1\,m;T=150.8\,°C$
Therefore, c_2 can be determined as follows

$$150.8=-\frac{24000}{0.5}\times\frac{0.1^2}{4}+c_2$$
$$c_2=270.8\,°C$$

At the center $r=0$, the temperature is given by

$$T=T_{center}=c_2=270.8\,°C$$

This example shows how we can use a "divide-and-conquer approach" to solve problems involving the third kind of boundary condition and a volumetric heat generation. Such problems are called conjugate problems, particularly when "*h*" at the surface also needs to be determined from, say, first-principles modeling.

2.7.3 Composite sphere

The next logical extension, as done before, is to work out the heat transfer rate from a composite spherical shell.

The cross-sectional view of a composite spherical shell is shown in Fig. 2.20 along with all the necessary details for us to be able to compute the heat transfer rate from the composite spherical shell.

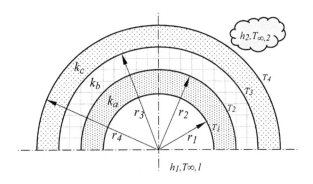

FIGURE 2.20

Schematic representation of a cross-sectional view of a composite sphere.

Under steady state conditions, and by invoking an electrical analogy, we can write out an expression for the heat transfer rate, Q, across the shell as

$$Q = \frac{T_{\infty_1} - T_1}{\dfrac{1}{h_1 4\pi r_1^2}} = \frac{T_1 - T_2}{\dfrac{r_2 - r_1}{4\pi k_a r_1 r_2}} = \frac{T_2 - T_3}{\dfrac{r_3 - r_2}{4\pi k_b r_2 r_3}}$$

$$= \frac{T_3 - T_4}{\dfrac{r_4 - r_3}{4\pi k_c r_3 r_4}} = \frac{T_4 - T_{\infty_2}}{\dfrac{1}{h_4 4\pi r_4^2}} \tag{2.112}$$

Using the Dividendo componendo rule, Eq. (2.112) may be written as

$$Q = \frac{T_{\infty_1} - T_{\infty_2}}{R_{total}} \tag{2.113}$$

$$\text{Where,} \quad R_{total} = \frac{1}{h_1 4\pi r_1^2} + \frac{r_2 - r_1}{4\pi k_a r_1 r_2} + \frac{r_3 - r_2}{4\pi k_b r_2 r_3} \tag{2.114}$$
$$+ \frac{r_4 - r_3}{4\pi k_c r_3 r_4} + \frac{1}{h_4 4\pi r_4^2}$$

By using Eq. (2.112), one can calculate any interface temperature.

Example 2.5: *Consider a composite spherical shell of inner diameter 50 cm, made of lead and stainless steel with outer diameters 60 and 62 cm respectively (Fig. 2.21). The cavity contains radioactive waste that generates heat volumetrically (q_v) and uniformly at the rate of 4.8×10^5 W/m³. The melting point of lead is 600 K; $k_{lead} = 35.3$ W/mK and $k_{ss} = 15.1$ W/mK. The shell is kept in ambient at a temperature of 30 °C, with a convection coefficient of $h = 500$ W/m²K. What is the maximum temperature in the lead? Will the lead melt?*

Solution:

Given $r_1 = 25$ cm, $r_2 = 30$ cm, $r_3 = 31$ cm

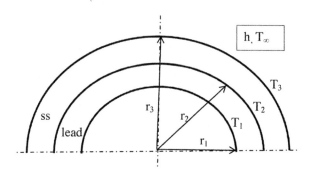

FIGURE 2.21

Schematic representation of a cross-sectional view of the composite sphere under consideration in Example 2.5.

Energy balance equation:

$$q_v \times volume = hA_3(T_3 - T_\infty)$$

$$4.8 \times 10^5 \times \frac{4\pi}{3} \times 0.25^3 = 500 \times 4\pi \times 0.31^2(T_3 - 30) \tag{2.115}$$

$$T_3 = 82\,°C$$

T_3 is the outermost surface temperature of the composite shell
Under steady state

$$\frac{T_2 - T_3}{\dfrac{r_3 - r_2}{4\pi k_{ss} r_2 r_3}} = \frac{T_3 - T_\infty}{\dfrac{1}{h 4\pi r_3^2}} \tag{2.116}$$

$$\frac{T_2 - 82}{\dfrac{0.31 - 0.3}{4\pi \times 15.1 \times 0.31 \times 0.3}} = \frac{82 - 30}{\dfrac{1}{500 \times 4\pi \times 0.31^2}} \tag{2.117}$$

$$T_2 = 99.8\,°C$$

Under steady state

$$\frac{T_2 - T_3}{\dfrac{r_3 - r_2}{4\pi k_{ss} r_2 r_3}} = \frac{T_1 - T_2}{\dfrac{r_2 - r_1}{4\pi k_{lead} r_1 r_2}} \tag{2.118}$$

$$\frac{99.8 - 82}{\dfrac{0.31 - 0.3}{4\pi \times 15.1 \times 0.3 \times 0.31}} = \frac{T_1 - 99.8}{\dfrac{0.3 - 0.25}{4\pi \times 30 \times 0.3 \times 0.25}} \tag{2.119}$$

$$T_1 = 155.3\,°C$$

T_1 is the temperature at $r_1 = 0.25$.m.

The melting point of lead is given as 600 K. Since all the temperatures inside the composite sphere are less than 600 K, there is no chance of the lead melting.

Example 2.6: *Revisit Example 2.5 and take k for radioactive waste as 7 W/mK. Calculate the maximum temperature in the cavity.*

$$\frac{1}{r^2}\frac{d}{dr}\left(r^2\frac{dT}{dr}\right)+\frac{q_v}{k}=0 \tag{2.120}$$

$$\frac{d}{dr}\left(r^2\frac{dT}{dr}\right)=-\frac{q_v}{k}r^2 \tag{2.121}$$

$$r^2\frac{dT}{dr}=-\frac{q_v}{k}\frac{r^3}{3}+c_1 \tag{2.122}$$

$$\frac{dT}{dr}=-\frac{q_v}{k}\frac{r}{3}+\frac{c_1}{r^2} \tag{2.123}$$

Boundary condition 1: At $r = 0$; $dT/dr = 0$

$$c_1 = 0 \tag{2.124}$$

$$T=-\frac{q_v}{k}\frac{r^2}{6}+c_2 \tag{2.125}$$

Boundary condition 2: At $r = r_1$; $T = T_1$

$$T_1=-\frac{q_v}{k}\frac{r^2}{6}+c_2 \tag{2.126}$$

$$155.3=-\frac{4.8\times10^5}{7}\frac{0.25^2}{6}+c_2 \tag{2.127}$$
$$c_2 = 869.6\,°C$$

At $r = 0$;
$$T=T_{center}=T_{max}=c_2=869.6\,°C$$

Please note that we again used "divide and conquer" to solve the problem. We tried to use every bit of information that was presented in advance to simplify the problem as much as possible. As a general rule, we have to examine first if we can employ energy balance to simplify the boundary conditions. The revelation here is that without solving the original governing equation for the shell, we are not only able to obtain the overall heat transfer but are also able to get estimates of the intermediate temperatures. This is classic heat transfer engineering practice and is very quick!

2.8 One-dimensional, steady state heat conduction with heat generation

2.8.1 Plane wall with heat generation

Consider the plane wall, shown in Fig. 2.22 with a uniform volumetric heat generation, given by q_v W/m^3 with Dirichlet boundary condition at $x = \pm L$. Such a situation arises in nuclear fuel rods or when chemical reactions (exothermic) take place in a medium.

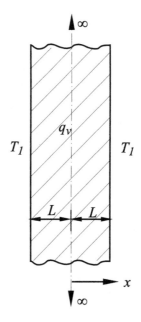

FIGURE 2.22

Schematic representation of a plane wall with heat generation.

We apply the following simplifying assumptions

1. One dimensional conduction
2. Steady state
3. Constant properties

The governing equation for the problem under consideration becomes

$$\frac{d}{dx}\left(k\frac{dT}{dx}\right)+q_v=0 \tag{2.128}$$

$$\frac{d^2T}{dx^2}+\frac{q_v}{k}=0 \tag{2.129}$$

$$\frac{d^2T}{dx^2}=-\frac{q_v}{k} \tag{2.130}$$

On integrating once, we get

$$\frac{dT}{dx}=-\frac{q_v}{k}x+c_1 \tag{2.131}$$

Boundary condition 1: At $x=0; \frac{dT}{dx}=0$
From Eq. (2.131),

$$c_1=0 \tag{2.132}$$

Integrating Eq. (2.131), we have

$$T = -\frac{q_v}{k}\frac{x^2}{2} + c_2 \tag{2.133}$$

Boundary condition 2: At $x = \pm L; T = T_1$
From Eq. (2.133), we solve for c_2

$$c_2 = T_1 + \frac{q_v}{k}\frac{L^2}{2} \tag{2.134}$$

Substituting for c_2 in Eq. (2.133), we have

$$T = -\frac{q_v}{k}\frac{x^2}{2} + T_1 + \frac{q_v}{k}\frac{L^2}{2} \tag{2.135}$$

$$T = T_1 + \frac{q_v}{2k}(L^2 - x^2) \tag{2.136}$$

$$T - T_1 = \frac{q_v}{2k}(L^2 - x^2) \tag{2.137}$$

Here $(T - T_1)$ is excess temperature due to heat generation inside the slab.

Eq. (2.137) tells us that the temperature distribution is quadratic. A slightly more involved version of the above problem is when we have a situation where there is a plane wall with heat generation and with convection boundary conditions, as shown in Fig. 2.23.

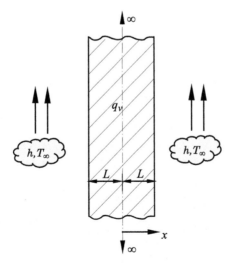

FIGURE 2.23

Schematic representation of a plane wall with heat generation with convection boundary conditions on both the sides.

The governing equation remains the same as before. The boundary conditions, though, are different.

Boundary condition 1: At $x = 0$; $dT/dx = 0$

From Eq. (2.131)

$$c_1 = 0 \tag{2.138}$$

Integrating Eq. (2.131), we have

$$T = -\frac{q_v}{k}\frac{x^2}{2} + c_2 \tag{2.139}$$

Boundary condition 2: At $x = L$; $q_{conduction} = q_{convection}$

$$-k\frac{dT}{dx} = h(T_1 - T_\infty) \tag{2.140}$$

$$-k\left(\frac{-q_v}{k}L\right) = h(T_1 - T_\infty) \tag{2.141}$$

$$T_1 = T_\infty + \frac{q_v L}{h} \tag{2.142}$$

we know

$$c_2 = T_1 + \frac{q_v}{k}\frac{L^2}{2} \tag{2.143}$$

Therefore,

$$T = -\frac{q_v}{k}\frac{x^2}{2} + T_1 + \frac{q_v}{k}\frac{L^2}{2} \tag{2.144}$$

$$T = T_\infty + \frac{q_v L}{h} + \frac{q_v L^2}{2k}\left(1 - \frac{x^2}{L^2}\right) \tag{2.145}$$

Similarly, the heat transfer in a cylinder and sphere with internal heat generation can be derived.

Example 2.7: *Consider a one-dimensional wall that uniformly generates heat at the rate of q_v W/m^2 whose thickness is $2L = 50$ mm. The wall is made of a material whose thermal conductivity is 50 W/mK. Temperature measurements show the following relation, $T(°C) = 200 - 2000x^2$. In the above expression, x is in meters. (1) Determine the volumetric heat generation rate q_v in the wall. (2) Determine the heat fluxes at the two walls. (3) How are these related to the q_v?*

Solution:

Fig. 2.22 can be used for this problem.

Temperature symmetry about the mid-plane is clear from the given temperature distribution

The governing equation is

$$\frac{d^2T}{dx^2} + \frac{q_v}{k} = 0 \quad . \tag{2.146}$$

$$\frac{d^2T}{dx^2} = -\frac{q_v}{k} \tag{2.147}$$

$$\frac{dT}{dx} = -\frac{q_v}{k}x + c_1 \tag{2.148}$$

$$T = -\frac{q_v}{k}\frac{x^2}{2} + c_1 x + c_2 \tag{2.149}$$

Boundary condition 1: At $x = 0$; $dT/dx = 0$

$$c_1 = 0 \tag{2.150}$$

Boundary condition 2: At $x = L = 0.025$m; $T = T_1$

$$T_1 = 200 - 2000[0.025^2]$$

$$T_1 = 198.7\,°C$$

$$\frac{d^2T}{dx^2} = -\frac{q_v}{k} = 2b \tag{2.151}$$

$$q_v = 200\ kW/m^3$$

$$q\Big|_{x=+L} = -k\frac{dT}{dx}\Big|_{x=+L}$$
$$= -50(2(-2000)0.025) = 5000\ W/m^2 \tag{2.152}$$

$$q\Big|_{x=-L} = k\frac{dT}{dx}\Big|_{x=-L}$$
$$= 50(2(-2000)(-0.025)) = 5000\ W/m^2 \tag{2.153}$$

The total heat generation is

$$q = q_v \times 2L \times A = 200 \times 10^3 \times 2 \times 25 \times 10^{-3}$$
$$q = 10000\ W/m^2$$

[†]From Eqs. (2.152) and (2.153)

$$q_{left} + q_{right} = 10000\ W/m^2$$

Therefore, Heat generated = Heat lost

[†]We still use q instead of Q as we are assuming A=1m^2.

Example 2.8: *Consider steady state conduction in a composite slab with heat generation in the middle portion of the slab. The problem description is shown in Fig. 2.24. Given* $T_2 = 270\,°C$, $T_3 = 220\,°C$, *determine the thermal conductivity of the middle portion of the slab and the volumetric heat generation in the middle portion of the slab, for steady state one dimensional conduction with constant properties of all the materials.*

Solution:

Under steady sate

$$Q = \frac{T_2 - 30}{\dfrac{1}{hA} + \dfrac{L_a}{k_a A}}$$

$$q_{left} = \frac{270 - 30}{\dfrac{1}{1000} + \dfrac{0.03}{16}}$$

$$q_{left} = 83478 \text{ W/m}^2$$

Similarly from the right side, we have

$$q_{right} = \frac{220 - 30}{\dfrac{1}{1000} + \dfrac{0.02}{50}}$$

$$q_{right} = 135714 \text{ W/m}^2$$

$$q_{total} = 83478 + 135714 = 219192 \text{ W/m}^2$$

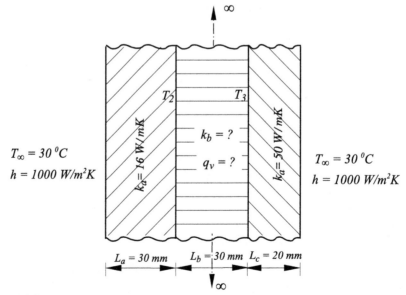

FIGURE 2.24

Schematic representation of a composite plane wall with heat generation in the middle and convection boundary conditions at the two ends.

In order to obtain the volumetric heat generation rate, q_v, we use energy balance.

$$\text{Total heat transferred} = \text{Total heat generated}$$
$$q_{total} A = q_v AL$$
$$q_v = \frac{219192}{0.03} = 7306.4 \; kW/m^3 \tag{2.154}$$

Now, for the thermal conductivity calculation, we consider only the middle wall. The governing equation for the middle wall is

$$\frac{d^2 T}{dx^2} + \frac{q_v}{k_b} = 0 \tag{2.155}$$

$$\frac{dT}{dx} = -\frac{q_v}{k_b} x + c_1 \tag{2.156}$$

$$T = -\frac{q_v}{k_b} \frac{x^2}{2} + c_1 x + c_2 \tag{2.157}$$

At the left of the middle wall: $x = 0$; $T = T_2$; $q_{left} = -k\dfrac{dT}{dx}$

At the right of middle wall: $x = L = 0.03\,\text{m}$; $T = T_3$; $q_{right} = -k\dfrac{dT}{dx}$

At $x = 0$; $T = T_2$

$$c_2 = T_2 = 270 \; °C$$

At $x = L = 0.03\,\text{m}$; $T = T_3$

$$220 = \frac{-7.306 \times 10^6 \times 0.03^2}{k_b} + c_1 \times 0.03 + 270 \tag{2.158}$$

At the left of the middle wall: $x = 0$; $T = T_2$; $q_{left} = -k\dfrac{dT}{dx}$

$$\therefore \; k\frac{dT}{dx}\bigg|_{x=0} = 83478 = c_1 k_b \tag{2.159}$$

$$k_b = \frac{83478}{c_1} \tag{2.160}$$

Substituting for k_b in Eq. (2.158), we have

$$c_1 = 1026.7 \; °C/m$$

$$k_b = 81.3 \; \text{W/mK}$$

What we have solved above is a typical "inverse" problem. A direct problem is one in which all causes are given and the effects are determined. An inverse problem is one in which the effects are known (typically temperature measurements in heat transfer problems), and the causes (like thermal conductivity) have to be estimated.

2.9 Fin heat transfer

For maintaining the temperatures of equipment as low as possible, it is necessary to maximize the heat transfer from them, subject to various constraints such as cost, pressure drop, and space. One practical way of increasing heat transfer from surfaces is by providing more surface area, and these extended surfaces are called fins.
The convection heat transfer from a surface is given by

$$Q = hA_s(T_w - T_\infty) \tag{2.161}$$

Here the goal is to increase "Q."

Methods to increase Q:
1. Increase heat transfer coefficient (h): when velocity increases, h will increase. However, the pressure drop also increases.
2. Increase surface area (A): the passive method.
3. Increase ($T_w - T_\infty$): restrictions may be there on the capabilities of the material or legislation or safety in respect of the maximum temperatures allowed.

Maximizing the heat transfer with an increase in surface area is an effective method. We accomplish this by using extended surfaces or fins on the base surface (Kraus et al., 2002). Addition of fins may marginally decrease the heat transfer coefficient. However, (hA) with fins is usually much greater than (hA) without fins, if they are engineered based on sound scientific principles to be discussed here.

Examples: Electrical transformers, condensers, economizers, motor cycle, IC engines, etc.

Desirable properties of fin material:
1. High thermal conductivity to maximize heat transfer.
2. Density of fin material is as low as possible; otherwise, the weight of the system increases significantly, outweighing the benefits.
3. Reasonable cost.

2.10 Analysis of fin heat transfer

Fig. 2.25 shows a schematic representation of a typical variable area fin with a rectangular cross-section.

The following assumptions are in order:

1. One-dimensional, steady state heat transfer, i.e., T = f(x) alone
2. Constant properties

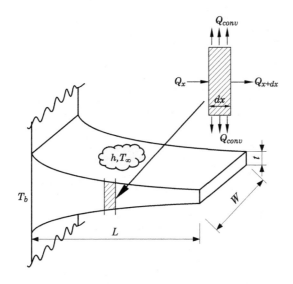

FIGURE 2.25

Schematic representation of a rectangular cross-section fin with variable area.

3. No heat generation in the fin
4. (h, T_∞) are constant
5. Radiation is negligible

Consider the energy balance across a control volume of a fin, as shown in Fig. 2.25

$$Q_x = Q_{x+dx} + Q_{conv} \tag{2.162}$$

$$Q_x = Q_x + \Delta x \frac{d}{dx} Q_x + Q_{conv} \tag{2.163}$$

$$\Delta x \frac{d}{dx}\left(-kA_c \frac{dT}{dx}\right) + hpdx(T - T_\infty) = 0 \tag{2.164}$$

$$k \frac{d}{dx}\left(A_c \frac{dT}{dx}\right) - hp(T - T_\infty) = 0 \tag{2.165}$$

Let $\theta = T - T_\infty$

$$\frac{d}{dx}\left(A_c \frac{d\theta}{dx}\right) - \frac{hp}{k}\theta = 0 \tag{2.166}$$

Eq. (2.166) is the fin equation and is a second-order differential equation. It supports two boundary conditions.

1. Boundary condition 1: at $x = 0$; $T = T_b$ or $\theta = \theta_b$
2. Boundary condition 2: Three possibilities exist

a. Insulated condition: at $x = L$; $\dfrac{d\theta}{dx} = 0$ (adiabatic fin tip)

b. Convection boundary: at $x = L$; $-k \left.\dfrac{d\theta}{dx}\right|_{x=L} = h\theta \left.\right|_{x=L}$

c. Long fin: at $x = L$; (rather $x \rightarrow \infty$) $\theta = 0$

For a constant area fin, Eq. (2.166) reduces to

$$\frac{d^2\theta}{dx^2} - \frac{hp}{kA_c}\theta = 0 \qquad (2.167)$$

$$\frac{d^2\theta}{dx^2} - m^2\theta = 0 \qquad (2.168)$$

$$m^2 = \frac{hp}{kA_c} \qquad (2.169)$$

In Eq. (2.169), m is the fin parameter and is known for any fin *a priori*.

For a rectangular fin of constant cross sectional area shown in Fig. 2.26, the perimeter, p and the cross sectional area A_c of the fin are given by,

$$p = 2(W + t)$$

$$A_c = W \cdot t$$

$$m = \sqrt{\frac{2h(W + t)}{k \cdot W \cdot t}} \qquad (2.170)$$

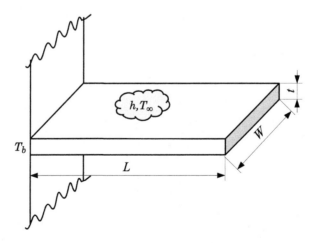

FIGURE 2.26

Schematic representation of a rectangular fin with constant cross section area.

If $w \gg t$, the expression for m for a constant area rectangular fin reduces to

$$m = \sqrt{\frac{2h}{kt}} \qquad (2.171)$$

2.10.1 Case 1: Insulated tip

Assumption: $T_b > T_\infty$

Boundary condition 1: At $x = 0$; $\theta = \theta_b$

Boundary condition 2: At $x = L$; $\dfrac{d\theta}{dx} = 0$

$$\text{Let} \frac{d}{dx} = D$$

Therefore Eq. (2.168) becomes

$$[D^2 - m^2]\theta = 0 \qquad (2.172)$$

The solution to the above equation is given by

$$\theta = c_1 \cosh(mx) + c_2 \sinh(mx) \qquad (2.173)$$

Boundary condition 1: At $x = 0$; $\theta = \theta_b$

$$\theta_b = c_1 + 0 \qquad (2.174)$$

$$c_1 = \theta_b \qquad (2.175)$$

$$\frac{d\theta}{dx} = c_1 (\sinh(mx))m + c_2 (\cosh(mx))m \qquad (2.176)$$

Boundary condition 2: At $x = L$; $\dfrac{d\theta}{dx} = 0$

$$0 = m[\theta_b \sinh(mL) + c_2 \cosh(mL)] \qquad (2.177)$$

$$c_2 = -\theta_b \tanh(mL) \qquad (2.178)$$

$$\theta = \theta_b \cosh(mx) - \theta_b \tanh(mL) \sinh(mx) \qquad (2.179)$$

$$\theta = \theta_b \left(\frac{\cosh(mx)\cosh(mL) - \sinh(mL)\sinh(mx)}{\cosh(mL)} \right) \qquad (2.180)$$

$$\theta = \theta_b \left(\frac{\cosh(m(L-x))}{\cosh(mL)} \right) \qquad (2.181)$$

$$\frac{\theta}{\theta_b} = \frac{\cosh(m(L-x))}{\cosh(mL)} \qquad (2.182)$$

Eq. (2.182) indicates that the temperature distribution across the fin is exponential.

The heat transfer from the fin can now be calculated as

$$Q = -kA_c \frac{d\theta}{dx}\Big|_{x=0} \tag{2.183}$$

$$Q = -kA_c\theta_b \frac{1}{cosh(mL)} sinh(m(L-x))(-m)\,|_{x=0} \tag{2.184}$$

$$Q = kA_c\theta_b \sqrt{\frac{hp}{kA_c}} tanh(mL) \tag{2.185}$$

$$Q = \sqrt{hpkA_c}\,\theta_b\, tanh(mL) \tag{2.186}$$

Fig. 2.27A represents the temperature variation along the length of the fin for the ideal case. In this case, the fin temperature is equal to the base temperature everywhere.

However for a typical fin, the temperature varies along the length, exponentially, and is shown in Fig. 2.27B.

As seen in Fig. 2.27C, the unhatched portion indicates the penalty we pay for the fin not being isothermal. If the fin is not maintained isothermally throughout the length, its efficiency to transfer heat (we will formally introduce the term efficiency in the next sub-section) will get reduced and is shown with the unhatched portion in Fig. 2.27C. Therefore, we try to make the exponential variation as close to the horizontal line as possible in order to get maximum efficiency (close to ideal efficiency).

$$\theta = \theta_b \left(\frac{cosh(m(L-x))}{cosh(mL)}\right) \tag{2.187}$$

The fin tip temperature can be obtained by evaluating θ at $x = L$.

$$\theta = \theta_b \left(\frac{1}{cosh(mL)}\right) \tag{2.188}$$

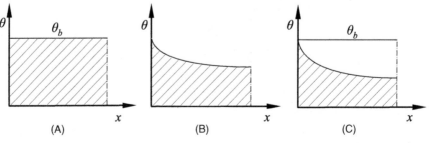

FIGURE 2.27

Graphical representation of fin performance. (A) ideal case, (B) typical fin case, and (C) reduction in θ for a fin not being isothermal.

The fin tip temperature is a kind of 'bell-weather' to capture the non-isothermality of the fin and represents the lowest temperature in the fin.

Fin efficiency

The fin efficiency is defined as the ratio of actual heat transfer to the maximum possible heat transfer (happens when the fin is isothermal) and is mathematically given by

$$\eta = \frac{Q_{actual}}{Q_{max}} \tag{2.189}$$

For the insulated tip fin considered above

$$\eta = \frac{\sqrt{hpkA_c}\,\theta_b \tanh(mL)}{h(pL)\theta_b} \tag{2.190}$$

$$\eta = \frac{\tanh(mL)}{mL}, \left(\text{Since } m = \sqrt{\frac{hp}{kA_c}} \right) \tag{2.191}$$

Effectiveness of the fin

The effectiveness of a fin is defined as the ratio of heat transfer with fins to the heat transfer without fins. The effectiveness is a key engineering metric for making a decision as to whether fins can or need to be used in a particular situaration or not.

$$\varepsilon = \frac{Q_{fin}}{Q_{withoutfin}} \tag{2.192}$$

As a general, it is advisable to use fins, only if $\varepsilon \geq 2$.

For a general case with n fins, each of which has an adiabatic tip and with a fin parameter m, as shown in Fig. 2.28, the overall effectiveness is given by

$$\varepsilon = \frac{(\text{Heat transfer from fins} + \text{Heat transfer from the unfinned area})}{\text{Heat transfer from the base area}} \tag{2.193}$$

We obtain an expression for ε as

$$\varepsilon = \frac{n\sqrt{hpkA_c}\,\theta_b \tanh(mL) + h[A - nA_c]\theta_b}{hA\theta_b} \tag{2.194}$$

$$\varepsilon = \frac{n\sqrt{hpkA_c}\,\theta_b \tanh mL + h[A - nA_c]\theta_b}{nhA_c\theta_b + h[A - nA_c]\theta_b} \tag{2.195}$$

For a single fin with an insulated tip and no unfinned area, Eq. (2.195) reduces to

$$\varepsilon = \frac{\sqrt{hpkA_c}\,\theta_b \tanh(mL)}{hA_c\theta_b} \tag{2.196}$$

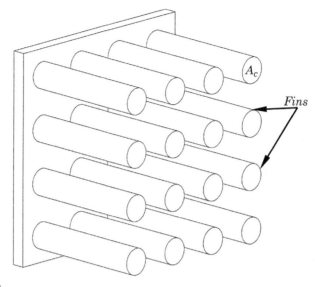

FIGURE 2.28

Schematic representation of n number of fins over a base plate.

$$\varepsilon = \sqrt{\frac{pk}{hA_c}} \tanh mL \qquad (2.197)$$

Eq. (2.197) confirms that it is more advantageous to use high thermal conductivity fins in situations where the heat transfer coefficient "h" is low. In so far as the geometry of the fin is concerned, lower the fin cross section and higher the perimeter, higher is the effectiveness. We can now work out the effectiveness relation for a rectangular fin.

Rectangular fin

For a rectangular fin, the effectiveness which turns out to be the key engineering metric, as already discussed is

$$\varepsilon = \sqrt{\frac{2k}{ht}} \tanh mL \qquad (2.198)$$

We would like to have an effectiveness as high as possible. Hence, we choose a material with a high thermal conductivity and we try to make a fin thin, as already discussed. A surprising and counterintuitive result is that as h is increasing while Q_{fin} keeps increasing, ε comes down. So, we are confronted with a situation where a fin appears to be good but not effective. In simple English, effficiency tells us if a fin is thermally the best of many competing fin designs, while effectiveness adresses the fundamental question as to whether we need to use a fin, however efficient it may be in an engineering problem.

2.10.2 Case 2: Long fin

The fin equation, for this case, is given by

$$\frac{d^2\theta}{dx^2} - m^2\theta = 0 \qquad (2.199)$$

The general solution is

$$\theta = c_1 e^{mx} + c_2 e^{-mx} \qquad (2.200)$$

Boundary condition 1: At $x = 0$; $T = T_b$, $\theta = \theta_b$
From Eq. (2.200) $c_1 = 0$; $c_2 = \theta_b$

$$\theta = \theta_b e^{-mx} \qquad (2.201)$$

The heat transfer from a long fin is given by

$$Q = -kA_c \left.\frac{d\theta}{dx}\right|_{x=0} \qquad (2.202)$$

$$Q = \sqrt{hpkA_c}\,\theta_b \qquad (2.203)$$

The efficiency of a long fin is given by

$$\eta = \sqrt{\frac{hpkA_c}{h(pL)\theta_b}} \qquad (2.204)$$

$$\eta = \frac{1}{mL} \qquad (2.205)$$

The effectiveness of the fin is given by

$$\varepsilon = \sqrt{\frac{pk}{hA_c}} \qquad (2.206)$$

Eq. (2.205) can be reduced for the specific case of a rectangular long fin to be $\varepsilon = \sqrt{(2k/ht)}$.

2.10.3 Case 3: Convecting tip

The more general case is; where the fin tip is convective.
The governing equation is

$$\frac{d^2\theta}{dx^2} - m^2\theta = 0 \qquad (2.207)$$

The general solution is

$$\theta = c_1 \cosh(mx) + c_2 \sinh(mx) \qquad (2.208)$$

Boundary condition 1: At $x = 0$; $\theta = \theta_b$
From Eq. (2.208)

$$c_1 = \theta_b \qquad (2.209)$$

$$\theta = \theta_b cosh(mx) + c_2 sinh(mx) \tag{2.210}$$

Boundary condition 2: $-k\dfrac{d\theta}{dx}\Big|_{x=L} = h_{tip}\theta\,|_{x=L}$

$$\frac{d\theta}{dx} = \theta_b m sinh(mx) + c_2 m cosh(mx) \tag{2.211}$$

From boundary condition 2, we get

$$-km\left(\theta_b sinh(mL) + c_2 cosh(mL)\right) = h_{tip}\left(\theta_b cosh(mL) + c_2 sinh(mL)\right) \tag{2.212}$$

$$c_2\left(\frac{h_{tip}}{mk} sinh mL + cosh(mL)\right) = -\theta_b sinh(mL) - \frac{h_{tip}}{mk}\theta_b cosh(mL) \tag{2.213}$$

$$c_2 = -\theta_b \frac{\left(sinh(mL) + \dfrac{h_{tip}}{mk} cosh(mL)\right)}{cosh(mL) + \left(\dfrac{h_{tip}}{mk}\right) sinh(mL)} \tag{2.214}$$

The temperature distribution in the fin is given by

$$\theta = \theta_b cosh(mx) - \theta_b\left(\frac{\left(sinh(mL) + \dfrac{h_{tip}}{mk} cosh(mL)\right)}{cosh(mL) + \dfrac{h_{tip}}{mk} sinh(mL)}\right) sinh(mx) \tag{2.215}$$

$$\theta = \theta_b \frac{cosh(m(L-x)) + \dfrac{h_{tip}}{mk} sinh(m(L-x))}{cosh(mL) + \dfrac{h_{tip}}{mk} sinh mL} \tag{2.216}$$

It is instructive to see that the expression reduces to the following for an adiabatic fin tip case, when we set h_{tip} to be 0.

$$\frac{\theta}{\theta_0} = \frac{cosh(m(L-x))}{cosh(mL)} \tag{2.217}$$

Eqn. (2.217) is the same as the equation obtained for the insulated fin tip case (Eqn. 2.182)

The heat transfer from the fin with convecting fin tip is given by

$$Q = -kA_c \frac{d\theta}{dx}\Big|_{x=0} \tag{2.218}$$

$$Q = kA_c\theta_b m \frac{sinh(mL) + \dfrac{h_{tip}}{mk} cosh(mL)}{cosh(mL) + \dfrac{h_{tip}}{mk} sinh(mL)} \tag{2.219}$$

$$Q = \sqrt{hpkA_c}\,\theta_b \, \frac{\tanh(mL) + \dfrac{h_{tip}}{mk}}{\left(1 + \dfrac{h_{tip}}{mk}\tanh(mL)\right)} \tag{2.220}$$

Again when $h_{tip} \to 0$ Eq. (2.220) reduces to Eq. (2.186); i.e., it reduces to the case of a fin with adiabatic tip.

We now look at an engineering approach to get rid of the mathematical tedium associated with this case

Fig. 2.29 represents the idea behind the use of a corrected length in the analysis of fin heat transfer. Fig. 2.29 shows that we are looking for an equivalent length L_c of a similar fin as the original, which transfers the same heat but has an adiabatic tip. Assuming $(L_c - L)$ to be at temperature T_L, we equate the heat transfer at the tip from Fig. 2.29A to the convection across $(L_c - L)$ in Fig. 2.29B.

$$hA_c\theta_L = hp(L_c - L)\theta_L \tag{2.221}$$

In the above equation θ_L is the temperature excess at the tip given by $(T_L - T_\infty)$.

$$L_c = L + \frac{A}{p} \tag{2.222}$$

Here L_c is the corrected length.
The heat transfer from the fin is given by

$$Q = \sqrt{hpkA_c}\,\theta_b \tanh mL_c \tag{2.223}$$

The efficiency of the fin is given by

$$\eta = \frac{\tanh mL_c}{mL_c} \tag{2.224}$$

The above-corrected length was conceptualized in Harper and Brown (1923) and must be considered as a smart simplification, considering the fact that it is nearly 100 years since the original work was published, is still frequently used.

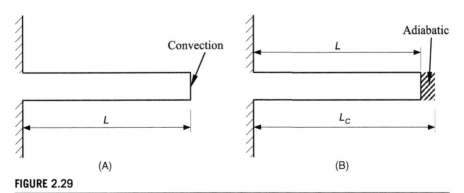

(A)   (B)

FIGURE 2.29

Schematic representation of a fin to demonstrate the corrected fin length concept. (A) Actual length with convection at the tip; (B) increased length with an adiabatic tip.

Example 2.9: *Consider a long rod that is placed vertically. The rod cools in ambient air at 30 °C. Thermocouples at 30 mm and 60 mm from the base indicate tempera-tures of 90 °C and 74 °C respectively. Determine the base temperature.*

Solution:

At $x = x_1 = 30$ mm, $T_1 = 90$ °C

At $x = x_2 = 60$ mm, $T_2 = 74$ °C

$$\theta_1 = T_1 - T_\infty = 60 \text{ °C}$$

$$\theta_2 = T_2 - T_\infty = 44 \text{ °C}$$

For a long fin:

$$\frac{\theta}{\theta_b} = e^{-mx} \tag{2.225}$$

$$\ln\left(\frac{\theta}{\theta_b}\right) = -mx$$

$$\ln\left(\frac{\theta_1}{\theta_b}\right) = -mx_1 \tag{2.226}$$

$$\ln\left(\frac{\theta_2}{\theta_b}\right) = -mx_2 \tag{2.227}$$

Dividing Eq. (2.226) by Eq. (2.227) results in Eq. (2.228).

$$\frac{\ln(\theta_1/\theta_b)}{\ln(\theta_2/\theta_b)} = \frac{x_1}{x_2} = \frac{30}{60} = \frac{1}{2} \tag{2.228}$$

$$\ln\left(\frac{\theta_1}{\theta_b}\right) = \left(\frac{1}{2}\right)\ln\left(\frac{\theta_2}{\theta_b}\right) \tag{2.229}$$

$$\frac{\theta_1}{\theta_b} = \left(\frac{\theta_2}{\theta_b}\right)^{1/2} \tag{2.230}$$

$$\theta_b = 81.82 \text{ °C}$$

$$T_b = 111.82 \text{ °C}$$

Please note that the above example is again an inverse problem, wherein from temperatures at two locations we were able to obtain the base temperature. In all of the preceding discussions, θ_b was given or known. Hence this problem repre-sents a 'surprising' example. One can go one step further and obtain m, from either Eq. (2.225) or Eq. (2.226). Proceeding even further, if the geometrical details and the thermal conductivity are known, making use of the definition of m as $\sqrt{(hp/kA_c)}$, we can actually "pull out" the heat transfer coefficient in the situation. If more measurements are available, one can find the "best" value of the heat transfer coefficient, h that minimizes the difference between the measure temperatures and those obtained with given values of 'h.' This minimization is invariably done in a least square sense and opens up new vistas in the application of heat transfer principles in the solution of practical inverse problems in thermal science and engineering.

2.10.4 Variable area fins

For variable area fins with trapezoidal or parabolic profile, the solution leads to Bessel functions. These can be concisely presented as a chart for fin efficiency as a function of the fin parameter and are shown in Fig. 2.30. A similar chart can also be made for radial fins across a cylinder. These charts can be used to estimate the heat transfer from such variable area fins quickly. Analytical solutions to these are available in many advanced texts on conduction (see, for example, Poulikakos, 1994).

Fig. 2.30 shows the fin efficiency chart of rectangular, triangular, and parabolic fins with and rectangular cross-section as a function of fin parameters. Please note that the corrected length should be used here along with specific expressions for the profile area, A_p and length parameter ξ which are embedded in the figure.

Fig. 2.31 shows efficiency curves for annular fins placed circumferentially over a pipe or tube. Please note that here too corrected fin length should be used.

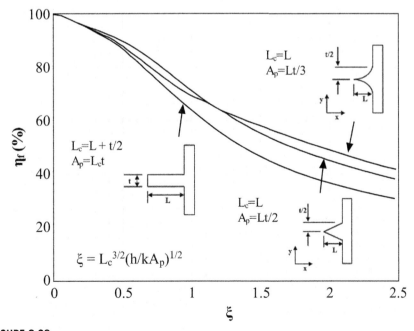

FIGURE 2.30

Fin efficiency of rectangular, triangular, and parabolic profiles as a function of key parameters.

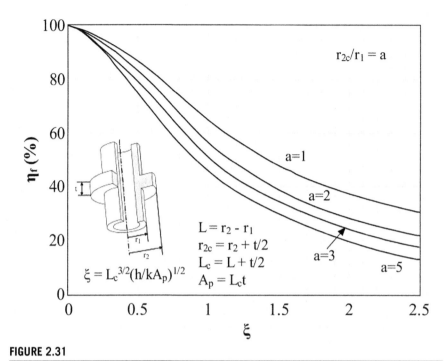

FIGURE 2.31

Efficiency of an annular fin as a function of key parameters.

For a detailed presentation of the analytical results for these, please refer to Poulikakos (1994).

Problems

2.1 A composite wall is made of a 30 mm stainless steel plate with a thermal conductivity of 16 W/mK, 50 mm of fiberglass insulation with a thermal conductivity of 0.04 W/mK, and 20 mm of plastic insulation with thermal conductivity of 0.03 W/mK. The temperature on the stainless steel side of the wall is 600 °C. The right side is exposed to ambient air with a heat transfer coefficient of 6 W/m²K and $T_\infty = 30$ °C.

    a. Draw a resistance diagram for this problem.

    b. Determine the heat transfer across the composite wall.

    c. Calculate all the intermediate temperatures.

2.2 Consider a plane wall, 15 cm in thickness, made of stainless steel (SS304) and very long such that the heat transfer across the wall is one dimensional (axial direction-x, with the origin at the left wall). The left end of the wall is maintained at 573 K, and the right end is at 373 K. The thermal conductivity of the material is known to vary as $k = 16.5 + 0.0175(T - 373)$, where T is the temperature in Kelvin and k is in W/mK. Steady state conditions prevail, and there is no heat generation in the material.

   a. Write down the equation governing the temperature distribution in the plane wall.
   b. Solve (a) for the boundary conditions given.
   c. Determine the value of dT/dx at both ends of the wall.
   d. Determine the heat transfer rate across the wall.

2.3 A long heat-generating wall is bathed by a cold fluid on both sides. The thickness of the wall is 12 cm, and the thermal conductivity of the wall material is 48 W/mK. A thermocouple located in the midplane shows a temperature of 88 °C. The cold fluid is at a temperature of 30 °C. The heat transfer coefficient on both sides of the wall is 200 W/m²K.
   a. Identify the governing equation for the problem, recognizing that a heat generation (which may be assumed to be uniform) is present in the problem.
   b. Solve the governing equation assuming that the temperatures are symmetric about the midplane.
   c. From the data given and the solution obtained, determine the volumetric heat generation in the slab.
   d. Determine the temperature at the two ends of the slab.

2.4 A very long copper wire of specification AWG 4/0 (AWG, American wire gauge) has a diameter of 11.68 mm. It is capable of carrying a current of 200 A. The thermal conductivity of copper is 400 W/mK, and its electrical resistivity is 1.72×10^{-8} Ωm. The surroundings are at 30 °C with a free convection heat transfer coefficient of 6 W/m²K.
   a. Write down the governing equation for temperature distribution in the wire under the conditions of steady state, together with boundary conditions.
   b. Solve (a) to obtain the temperature distribution in the wire.
   c. Determine the maximum temperature in the wire.
   d. Determine the minimum temperature in the wire. Where does it occur?
   e. From the solution to this problem, reason out why a maximum current-carrying capacity is specified on a wire.

2.5 Revisit Problem 2.4. However, now the goal is to insulate the wire to reduce losses. It is preferable to use Bakelite ($k = 0.2$ W/mK) to insulate the wire.
   a. Determine the critical radius of insulation.
   b. Determine the total heat rate with the Bakelite insulation for a thickness equal to the critical thickness of insulation.
   c. Is the result obtained in (b) baffling to you? Why?
   d. Determine the temperature at the interface between the copper wire and Bakelite.
   e. Determine the maximum temperature inside the copper wire with the insulation.
   f. Has the insulation helped or hurt the heat transfer from the wire?

2.6 One possibility of having a nuclear fuel element is to have a spherical capsule of fissionable material and a spherical shell of aluminum cladding. Coolant flows outside of the cladding and removes the heat generated due to nuclear fission. Consider a homogeneous fissionable material of radius $r_1 = 5$ cm, with uniform heat generation of $q_v = 4.5 \times 10^7$ W/m^3. The thermal conductivity of the fissionable material is 10 W/m/K. The thermal conductivity of the aluminum cladding is 200 W/mK and has a thickness of 1 cm. The heat transfer coefficient on the coolant side is 4000 W/m^{2}K, and the bulk temperature of the coolant is 90 °C. Assume steady state, one-dimensional heat transfer.

   a. Write down the governing equations for the temperature distribution within the fissionable material along with the boundary conditions.

   b. Solve the governing equation and obtain an expression for temperature distribution in the sphere.

   c. From the data given, determine the temperature at the outer end of the aluminum cladding.

   d. Determine the temperature at the interface between the cladding and the fuel.

   e. Determine the temperature at the center of the sphere.

2.7 Consider two long rods (or long fins) made of aluminum ($k = 205$ W/mK) that have diameter $d = 12$ mm. They are soldered together end to end with the solder having a melting point of 700 °C. The rods are kept in quiescent air at 30 °C, with a convection coefficient of 10 W/m^{2}K. Determine the minimum power input that is required to accomplish the soldering.

2.8 An inverse problem in heat transfer concerns the estimation of properties like k, C_p, and α, to name a few, from heat flux or temperature measurements. A long fin is a heat transfer device that can be ingeniously used to estimate the thermal conductivity of material if simple temperature measurements are made. One approach would be to use two identical, very long rods (or long fins) that are geometrically the same in all aspects and are subject to the same environment and base temperatures but are made of different materials, k_1 and k_2. Consider, a situation where thermocouple measurements with a fin of thermal conductivity k_1 show a temperature of 72 °C at an axial location x, when $T_b = 100$ °C and $T_\infty = 25$ °C. For the fin of thermal conductivity k_2, the temperature is measured to be 85 °C at the same location x. If $k_2 = 205$ W/mK corresponding to aluminum, what is k_1?

2.9 Consider an aluminum rectangular fin ($k = 205$ W/m K) of length $L = 10$ mm, thickness $t = 1$ mm, and width $w \gg t$. The base temperature of the fin is $T_b = 100$ °C, and the fin is exposed to a fluid of temperature = 30 °C. (a) Assuming a uniform convection coefficient of h = 100 W/m^{2}K over the entire fin surface, determine the fin heat transfer rate per unit width, efficiency, effectiveness, thermal resistance per unit width, and tip temperature for the following variants of the problem:

   a. Convection at the tip

b. Adiabatic tip
c. Adiabatic tip with length corrected for tip convection and compare these with case (a)
d. Infinite fin (long fin) approximation

2.10 Consider a horizontally oriented triangular fin made of aluminum (k = 205 W/mK) having a rectangular cross-section placed on a base at $T = 100\,°C$. The thickness of the fin is 3.5 mm at the base, and the length of the fin is 16 mm. The fin is exposed to the ambient at $T_\infty = 30\,°C$ that affords a convection heat transfer coefficient of $h = 70\ W/m^2K$. For unit width of the fin, determine the following:
a. Fin tip temperature
b. Heat transfer rate from the fin
c. Fin efficiency
d. Fin effectiveness

2.11 Revisit problem 2.10. Now consider two variants (1) rectangular fin with the same thickness of 3.5 mm throughout, and (2) trapezoidal fin with a tip thickness of 1.5 mm. All other conditions remain the same as before. For these two cases, determine the fin tip temperature, fin heat transfer rate, fin efficiency, and fin effectiveness, and compare their values along with those of the fin considered in problem 2.10. The tip may be assumed to be insulated for both the cases.

2.12 Annular aluminum fins of rectangular profile ($k = 205$ W/mK) are attached to a circular tube The outside diameter of the tube 50 mm and the outer surface temperature of the tube is 200 °C. The fins are 3 mm thick and 15 mm long (in the radial direction). The system is in ambient air at 30 °C, and the associated convection heat transfer coefficient is 40 W/m^2K.
a. Determine fin efficiency.
b. Determine the heat transfer from one fin.
c. Determine the effectiveness of one fin.
d. If there are 120 fins per meter length of the tube, what is the overall heat transfer rate per unit length of the system?
e. Determine the overall surface effectiveness of the fin array.

2.13 A heat sink consists of a horizontal base with an array of vertical aluminum fins of a rectangular cross-section. The base of the heat sink is 15 cm long and 8 cm deep. There are 20 equispaced (spaced equally along the length) fins, each of 3 mm thickness and 3 cm height. The heat transfer coefficient under these conditions for the finned and unfinned portions of the heat sink is 6.5 W/m^2K with a T_∞ of 30 °C. The heat sink is placed on heat-generating electronic equipment, and it can be assumed that the electronic equipment and the base are at the same temperature. If the electronic component dissipates 6W of heat and the heat transfer from the underside of the base and the equipment can be neglected, what is the steady state temperature of the electronic equipment? (Note: you may assume the tip of the fins to be insulated, and properties of aluminum are $k = 205$ W/mK, $\rho = 2700\ kg/m^3$, and $C_p = 900$ J/kg K.)

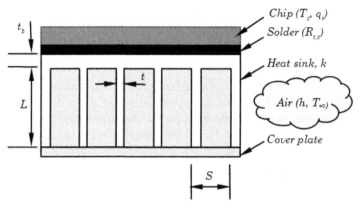

FIGURE 2.32

Schematic representation of heat sink with aluminium fins described in Problem 2.14.

2.14 A silicon chip of width $W = 24.2$ mm on a side is soldered to an aluminum heat sink ($k = 205$ W/m K) of the same width as shown in Fig. 2.32. The heat sink has a base thickness of $t_b = 3$ mm and consists of an array of vertical rectangular fins, each of length $L = 14$ mm. Air flows with $T_\infty = 30$ °C through channels formed between the fins and the cover plate. The convection coefficient h is 150 W/m²K, and a minimum fin spacing of 2 mm is required considering the restrictions on the pressure drop. The solder joint offers a thermal resistance of $R_{t,c} = 3 \times 10^{-6}$ m² K/W. The silicon chip may be assumed to be isothermal.

The array has 13 fins, the fin thickness $t = 0.2$ mm, and pitch $S = 2$ mm. If the maximum permissible temperature of the chip is $T_c = 80$ °C, determine the chip power. Assume adiabatic fin tip condition (adapted from Incropera et al., *Fundamentals of Heat Transfer*, 2013).

References

Harper, R.R., and Brown, W.B., 1923. Mathematical equations for heat conduction in the fins of air-cooled engines. 1923, NACA-TR-158, NACA Annual Report 8, pp. 677–708.

Incropera F.P., Lavine A.S., Bergman T.L., DeWitt D.P., 2013. Principles of Heat and Mass Transfer. John Wiley and sons, New York.

Kraus, A.D., Aziz, A., Welty, J., 2002. Extended Surface Heat Transfer. John Wiley & Sons Inc., New York.

Poulikakos, D., 1994. Conduction Heat Transfer. Prentice Hall, Englewood Cliffs, NJ.

Conduction: One-dimensional transient and two-dimensional steady state

3

3.1 Introduction

There are many situations and applications where temperature is a function of time alone, or time and space, for example, in the heat treatment of ball bearings, response of thermocouples, aerodynamic heating of rockets, transients in avionic cooling packages, and so on. In view of this, it is necessary to know the temperature at a location at a given instant of time. One way to do this is to simply measure it. The "analytically" inclined, though, always ask the question, "Can we obtain this information by theory?" The answer is yes, and in this chapter we see how to obtain this information in a few typical cases of engineering interest.

Please recall the general heat conduction equation derived in chapter 2, which is

$$\nabla^2 T + \frac{q_v}{k} = \frac{1}{\alpha}\frac{dT}{dt} \tag{3.1}$$

In Cartesian coordinates, Eq. (3.1) can be written as

$$\frac{\partial^2 T}{\partial x^2} + \frac{\partial^2 T}{\partial y^2} + \frac{\partial^2 T}{\partial z^2} + \frac{q_v}{k} = \frac{1}{\alpha}\frac{\partial T}{\partial t} \tag{3.2}$$

Now, let us consider a stationary plane wall of thickness $2L$ initially at T_i that is suddenly immersed in a convection environment with a heat transfer coefficient "h" and an ambient temperature of T_∞.

A schematic representation of the plane wall is shown in Fig. 3.1

The following assumptions hold:

1. $T_i > T_\infty$
2. One-dimensional, unsteady state heat transfer, that is, $T = f(x,t)$ alone
3. Constant thermophysical properties of the medium
4. No heat generation in the solid, that is, $q_v = 0$
5. (ρ, c_p, k) are constant
6. Radiation is negligible

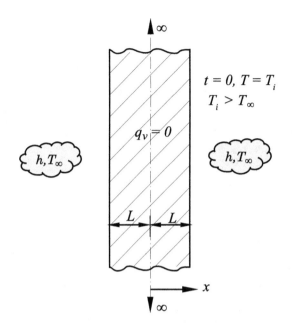

FIGURE 3.1

Schematic representation of a one-dimensional slab undergoing transient conduction $(T_i > T_\infty)$.

The general governing equation reduces to

$$\frac{\partial^2 T}{\partial x^2} = \frac{1}{\alpha}\frac{\partial T}{\partial t} \tag{3.3}$$

Eq. (3.3) supports two boundary conditions, as it is second order in x and one initial condition, as it is first order in time.

Boundary condition 1: At $x = 0$ $\dfrac{\partial T}{\partial x} = 0$ (symmetry condition)

Boundary condition 2: At $x = \pm L$; $-k\dfrac{\partial T}{\partial x} = h(T - T_\infty)$

Initial condition: At time $t = 0$ s, $T = T_i$ (throughout the slab)

A close look at the above boundary and initial conditions shows that the slab is initially hot and is getting convectively cooled at the two ends. After sufficient time elapses, the entire slab will be at T_∞. Hence, there is little interest in the steady state solution to this problem. The challenge is to determine $T(x,t)$ for any (x,t).

We can now carry out a non-dimensionalization of the governing equations and the boundary and initial conditions to get down to the bottom of the problem.

Let ξ, the nondimensional length, be defined as x/L, and ϕ, the nondimensional temperature excess, be defined as $\dfrac{T - T_\infty}{T_i - T_\infty}$. Then

$$\frac{\partial T}{\partial x} = (T_i - T_\infty)\frac{\partial \phi}{\partial x} \tag{3.4}$$

$$\frac{\partial^2 T}{\partial x^2} = (T_i - T_\infty)\frac{\partial^2 \phi}{\partial x^2} \tag{3.5}$$

Now, non-dimensionalizing x as $\xi = x/L$, we have

$$\frac{\partial^2 T}{\partial x^2} = \frac{(T_i - T_\infty)}{L^2}\frac{\partial^2 \phi}{\partial \xi^2} \tag{3.6}$$

Similarly,

$$\frac{\partial T}{\partial t} = (T_i - T_\infty)\frac{\partial \phi}{\partial t} \tag{3.7}$$

Substituting Eqs. (3.6) and (3.7) in Eq. (3.3), we have

$$\frac{(T_i - T_\infty)}{L^2}\frac{\partial^2 \phi}{\partial \xi^2} = \frac{1}{\alpha}(T_i - T_\infty)\frac{\partial \phi}{\partial t} \tag{3.8}$$

$$\frac{\partial^2 \phi}{\partial \xi^2} = \frac{L^2}{\alpha}\frac{\partial \phi}{\partial t} \tag{3.9}$$

Eq. (3.9) may also be written as

$$\frac{\partial^2 \phi}{\partial \xi^2} = \frac{\partial \phi}{\partial\left(\dfrac{\alpha t}{L^2}\right)} \tag{3.10}$$

From the boundary condition 2: At $x = \pm L$; $\xi = \pm 1$; $-k\dfrac{\partial T}{\partial x} = h(T - T_\infty)$

$$-k\frac{T_i - T_\infty}{L}\frac{\partial \phi}{\partial \xi} = h(T_i - T_\infty)\phi \tag{3.11}$$

$$\frac{\partial \phi}{\partial \xi} = -\frac{hL}{k}\phi \tag{3.12}$$

From the initial condition: At time $t = 0$ s or $\dfrac{\alpha t}{L^2} = 0$; $T = T_i$; $\phi = 1$, and from the boundary condition 1: At $x = 0$ or $\xi = 0$, $\dfrac{\partial T}{\partial x} = 0$ or $\dfrac{\partial \phi}{\partial \xi} = 0$

From the above set of equations, it is clear that we can immediately recognize that all quantities appearing in Eq. (3.10) and Eq. (3.12) are dimensionless.

From the above non-dimensionalization, it is evident that ϕ can be written as a function of the pertinent dimensionless parameters, as follows

$$\phi = f\left(\xi, \frac{\alpha t}{L^2}, \frac{hL}{k}\right) \tag{3.13}$$

Please note that the dependence of ϕ on ξ and $\dfrac{\alpha t}{L^2}$ follows Eq. (3.10), while its dependence on $\dfrac{hL}{k}$ follows the convective boundary condition on the sides, given by Eq. (3.12).

We now introduce, two key dimensionless numbers, namely, Biot number (Bi) and Fourier number (Fo) as:

$$\text{Biot number} = Bi = \frac{hL}{k} = \frac{L/kA}{1/hA} = \frac{R_{conduction}}{R_{convection}} \tag{3.14}$$

where A is the area of the slab perpendicular to the plane of the paper.

$$\text{Fourier number} = Fo = \frac{\alpha t}{L^2} \tag{3.15}$$

The Fourier number can be rewritten as $Fo = \dfrac{t}{(L^2/\alpha)}$ with $\dfrac{L^2}{\alpha}$ having the units of time, and let us denote it as t^*, a characteristic time in the problem. Therefore, $Fo = \dfrac{t}{t^*}$. In other words, the Fourier number is nothing but a scaled or normalized or dimensionless time.

The Biot number, as shown above, is the ratio of conduction to the convection resistance.

As $R_{conduction} \to 0$, the temperature variation within the solid body approaches zero. The resulting problem at hand is frequently referred to as one with negligible internal temperature gradients. In the parlance of heat transfer, this formulation or simplification is referred to as the lumped capacitance method with the word "lumped" implying that the object is assumed to be spatially isothermal. We now look at this variant of the transient heat conduction problem in a little more detail.

3.2 Lumped capacitance method

Consider a body of arbitrary shape shown in Fig. 3.2 with a mass of m, surface area A, volume V, specific heat c_p.

Let T_i be the initial temperature of the body. The body is losing heat to the surroundings, which are at a temperature of T_∞ ($T_i > T_\infty$). Performing an energy balance on the body, we get the following:

$$\text{Rate of change of internal energy} = \dot{E}_{in} - \dot{E}_{out} \tag{3.16}$$

$$mc\frac{dT}{dt} = -Q_{convection} \tag{3.17}$$

(with the understanding that $c_p = c_v = c$ here)

$$mc_p\frac{dT}{dt} = -hA(T - T_\infty) \tag{3.18}$$

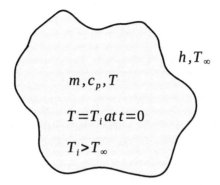

FIGURE 3.2

Schematic representation of an irregular body used for lumped capacitance method of solving a transient heat conduction problem.

Let the temperature excess be, $\theta = T - T_\infty$

$$mc_p \frac{d\theta}{dt} = -hA\theta \tag{3.19}$$

The initial condition at $t = 0$; is $T = T_i$ or $\theta = \theta_i$

$$\therefore \frac{d\theta}{dt} = \frac{-hA}{mc_p}\theta \tag{3.20}$$

$$\frac{d\theta}{\theta} = \frac{-hA}{mc_p}dt \tag{3.21}$$

Integrating both sides from $\theta = \theta_i$ to $\theta = \theta$, we have

$$\int_{\theta_i}^{\theta} \frac{d\theta}{\theta} = \frac{-hA}{mc_p}\int_0^t dt \tag{3.22}$$

$$\ln\left(\frac{\theta}{\theta_i}\right) = \frac{-hA}{mc_p}t \tag{3.23}$$

$$\frac{\theta}{\theta_i} = e^{\frac{-hA}{mc_p}t} \tag{3.24}$$

$$\frac{\theta}{\theta_i} = e^{\frac{-t}{mc_p/hA}} \tag{3.25}$$

$$\frac{\theta}{\theta_i} = e^{\frac{-t}{\tau}} \tag{3.26}$$

where τ is the time constant given by $\dfrac{mc_p}{hA}$.

Working further, we have

$$\frac{hA}{mc_p}t = \frac{hAt}{\rho Vc_p} = \frac{ht}{\rho c_p V/A} \tag{3.27}$$

The characteristic length, L_c, is given by V/A. The characteristic length, L_c for a transient conduction problem is, L for a plane slab of thickness $2L$, $R/2$ (or $D/4$) for a cylinder of radius R (or diameter D) and $R/3$ (or $D/6$) for a sphere of radius R (or diameter D). We can work on $\dfrac{ht}{\rho c_p V/A}$ a little more to pull out the two dimensionless numbers, Bi and Fo, of interest in this problem, as follows,

$$\frac{ht}{\rho c_p V/A} = \frac{ht}{\rho c_p L_c} \times \frac{L_c k}{L_c k} = \frac{hL_c}{k} \times \frac{\alpha t}{L_c^2} = BiFo \tag{3.28}$$

$$\frac{\theta}{\theta_i} = e^{-BiFo} \tag{3.29}$$

Eq. (3.29) reconfirms that the dimensionless temperature $\dfrac{\theta}{\theta_i} = f$ (Bi, Fo, dimensionless length scale), with the dependence on the third quantity on the right-hand side vanishing due to the spatial isothermality of the body under consideration.

The temperature time history of the body is an exponential curve and is qualitatively shown in Fig. 3.3.

It would be instructive to calculate the enthalpy lost by the body in the time interval 0 to t.

$$\text{Enthalpy lost} = \int_0^t Q\,dt \tag{3.30}$$

$$= \int_0^t hA\theta\,dt \tag{3.31}$$

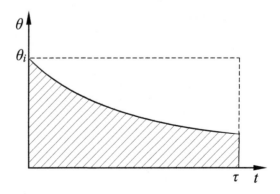

FIGURE 3.3

Schematic representation of the temperature time history of the spatially isothermal body losing heat to cooler surroundings.

$$= \int_0^t hA\theta_i e^{\frac{-t}{\tau}}\, dt \qquad (3.32)$$

$$= hA\theta_i \tau \left(1 - e^{\frac{-t}{\tau}}\right) \qquad (3.33)$$

$$= hA\theta_i \frac{mc_p}{hA}\left(1 - e^{\frac{-t}{\tau}}\right) \qquad (3.34)$$

We can work on eqn. 3.34 to come out with a form which is physically more intuitive to us.

$$\text{Enthalpy lost} = mc_p\theta_i \left(1 - e^{\frac{-t}{\tau}}\right)$$

$$= mc_p\theta_i \left(1 - \frac{\theta}{\theta_i}\right) \qquad (3.35)$$

$$= mc_p(\theta_i - \theta) = mc_p(T_i - T)$$

Eqns. 3.34 and 3.35 are essentially the same with the latter serving as confirmation.

The enthalpy lost in the time interval 0 to t is given by the difference between $mc_p\theta_i$ and the area under the curve in Fig. 3.3. A quick asymptotic check is in order.

At $t = 0$, Enthalpy lost is "0" (zero)

At $t \to \infty$, Enthalpy lost $\to mc_p\theta_i$

These are asymptotically correct and are consistent with the physics of the problem under consideration. The normally accepted criterion for considering a body to be spatially isothermal is that the Biot number (Bi), based on the characteristic length L_c given by the ratio of the volume (V) to the surface area (A), must be less than 0.1.

Example 3.1: *A copper sphere 12 mm in diameter is initially 100 °C. It is then immersed in cold surroundings with an ambient temperature of 30 °C, and a natural heat transfer coefficient of h = 7 W/m² K. The thermal conductivity of copper is 369 W/mk.*
1. For this transient conduction problem, determine the Biot number.
2. Can lumped capacitance method be used to solve this problem?
3. If the answer to (2) is yes, determine the temperature at the center of the sphere after 100 s.
4. What is the total enthalpy lost by the copper sphere in the first 50 s? (The thermophysical properties of copper are thermal conductivity = 396 W/mK density = 8960 kg/m³, and specific heat = 385 J/kg k).

Solution:
Given data

The diameter of sphere (D) = 12 mm; initial temperature of the sphere (T_i) = 100 °C; ambient temperature (T_∞) = 30 °C; heat transfer coefficient (h) = 7 W/m²K; thermal conductivity (k) = 396 W/mK.

$$Bi = \frac{hL_c}{k}$$

$$L_C = \frac{V_s}{A_s} = \frac{4/3\pi r^3}{4\pi r^2}$$

$$L_C = \frac{D}{6}$$

This is a key result. The characteristic length in transient conduction is L for a plain slab of thickness 2L, R/2 or D/4 for a cylinder of radius R and R/3 or D/6 for a sphere of a radius of R, as already mentioned.

1. $Bi = \dfrac{7 \times 12 \times 10^{-3}}{6 \times 396} = 35.35 \times 10^{-6}$

2. Since Bi < 0.1, the lumped capacitance method is applicable.
3. From the solution for lumped capacitance heat transfer, the temperature excess is given by

$$\frac{T - T_\infty}{T_i - T_\infty} = exp\left(\frac{-hA_s}{\rho V c_p} \times t\right)$$

The temperature of the sphere at the center after 100 s (same everywhere in the sphere under the lumped capacitance assumption)

$$\frac{T_{100s} - 30}{100 - 30} = exp\left(\frac{-7 \times 6 \times 10^3}{8960 \times 12 \times 385} \times 100\right)$$

$$T = 30 + 70 \times exp\left(\frac{-42 \times 10^3 \times 10^2}{8960 \times 12 \times 385}\right)$$

$$T_{100s} = 93.25 \,°C.$$

4. Enthalpy lost $= mc_p(T_i - T_{50s}) = \rho V c_p(T_i - T_{50s})$

To be able to calculate this, first we need to obtain the temperature of the sphere at 50 s

$$\frac{T_{50s} - 30}{100 - 30} = exp\left(\frac{-7 \times 6 \times 10^3}{8960 \times 12 \times 385} \times 50\right)$$

$$T = 30 + 70 \times exp\left(\frac{-2.1 \times 10^6}{8960 \times 12 \times 385}\right)$$

$$T_{50s} = 96.54 \,°C.$$

Enthalpy lost $= 8960 \times 9.04 \times 10^{-7} \times 385 \times (100 - 96.54) = 10.79$ J.

3.3 Semi-infinite approximation

Let us revisit our earlier discussion on transient conduction in an infinitely long one-dimensional slab of thickness 2L. However, now we are looking at a situation where the Bi > 0.1 and, in view of this, the assumption of spatial isothermality no longer holds.

Consider a plane wall that is semi-infinite in character, that is, the left end is specified ($x = 0$) while the right end is infinite in extent at an initial temperature of T_i.

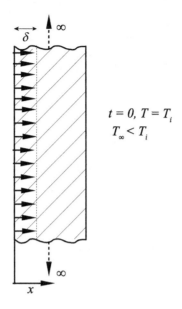

FIGURE 3.4

Schematic representation of a semi-infinite plane wall undergoing transient conduction.

Now, at $x = 0$ (left end) suddenly, the temperature is lowered to T_∞ ($T_\infty < T_i$). A schematic is shown in Fig. 3.4. It is intuitive to imagine that the effect of the thermal disturbance at the wall will be initially felt only near the wall, and this region of disturbance, denoted by δ, slowly grows with time along x. Now consider a time interval for which

$$\frac{\delta}{L} \ll 1$$

This assumption involves a situation where we do not know the value of δ a priori. However, there will be no flaw in the analysis so long as $\frac{\delta}{L} \ll 1$. "L" is the characteristic length scale for the problem. For times when $\delta \ll L$, large portions of the slab would simply not know that a thermal disturbance happened at $x = 0$. They would still be at $T = T_i$. We call this the "early regime". This is the reason why we stated earlier that the body under consideration, a slab in this case, is semi-infinite in extent. In what follows, we will see that the 'L' itself does not appear in the final solution, as it should be for a truly semi-infinite approximation.

The δ referred to above is the thermal penetration depth or thermal penetration thickness.

The governing equation for the problem under consideration is

$$\frac{\partial^2 T}{\partial x^2} = \frac{1}{\alpha}\frac{\partial T}{\partial t} \tag{3.36}$$

The initial condition is for all x: At $t = 0\,s$, $T = T_i$.

Boundary conditions: At $x = 0$, $T = T_\infty$ for $t > 0$.

It is intuitively apparent that when $x \to \infty$; $T = T_i$.

We now perform a scale analysis on this problem, wherein we try to reason out scales or orders of magnitude of the key quantities for the problem under consideration.

Let $\theta = T - T_\infty$. Eq. (3.36) turns out to be

$$\frac{\partial^2 \theta}{\partial x^2} = \frac{1}{\alpha} \frac{\partial \theta}{\partial t} \tag{3.37}$$

The scale for θ is θ_i, while the scale for x is δ. Substituting for these in Eq. (3.37), we have

$$\frac{\theta_i}{\delta^2} \sim \frac{\theta_i}{t} \tag{3.38}$$

$$\delta^2 \sim \alpha t \tag{3.39}$$

$$\delta \sim \sqrt{\alpha t} \tag{3.40}$$

As already mentioned,

$$\frac{\delta}{L} \ll 1$$

substituting δ from Eq. (3.40), we have

$$\frac{\sqrt{\alpha t}}{L} \ll 1$$

Hence, $\sqrt{\alpha t} \ll L$ for the semi-infinite model to be valid. The key question to be asked now is the motivation to do all this. The answer to this question lies in the fact that not only for $t \to 0$, $T = T_i$ everywhere, but also for $x \to \infty$; that is, when $\frac{\delta}{L} \ll 1$, large portions of the slab remain at only "T_i" at all times for which $\frac{\delta}{L} \ll 1$." In view of this, one initial and one boundary condition will fuse into just one condition if a new variable, η, is introduced that takes care of both x and t. If the introduction of η reduces the original partial differential equation (PDE) into an ordinary differential equation (ODE), with the understanding that the latter is a lot easier to solve than the former, then its introduction is well worth the effort. In this effort, the fusing of one initial and one boundary condition will ensure that there is no degeneracy, when the reduction is done. Let us now see if this reduction is possible!

Let $\eta = \dfrac{x}{2\sqrt{\alpha t}}$, then Eq. (3.37) can be expressed as

$$\frac{\partial \theta}{\partial x} = \frac{d\theta}{d\eta} \frac{\partial \eta}{\partial x} = \frac{d\theta}{d\eta} \frac{1}{2\sqrt{\alpha t}} \tag{3.41}$$

$$\frac{\partial^2 \theta}{\partial x^2} = \frac{d^2 \theta}{d\eta^2} \frac{1}{4\alpha t} \tag{3.42}$$

$$\frac{\partial\theta}{\partial t} = \frac{d\theta}{d\eta}\frac{\partial\eta}{\partial t} = \frac{d\theta}{d\eta}\frac{x}{2}\times-\frac{1}{2}\times(\alpha t)^{-3/2}\times\alpha \tag{3.43}$$

Substituting in Eq. (3.37) we get the following

$$\therefore \frac{d^2\theta}{\partial\eta^2} + 2\eta\frac{d\theta}{d\eta} = 0 \tag{3.44}$$

Please note that Eq. (3.44) is an ordinary differential equation (ODE), as opposed to Eq. (3.37), which is a partial differential equation (PDE). Eq. (3.44) is a second-order differential equation that supports two conditions. These are

$$\eta = 0;\ \theta = 0$$

$$\eta \to \infty;\ \theta = \theta_i$$

The second condition is quite interesting. This actually means that, $x \to \infty$ as well at $t = 0$, $\theta = \theta_i$. Hence, our proposal to introduce a new variable η, frequently referred to as a similarity variable has helped us to maintain the integrity of the original governing equation, as now we have only two conditions for θ, as opposed to three in the original PDE.

Let $\dfrac{d\theta}{d\eta} = p$

$$\frac{dp}{d\eta} + 2\eta p = 0 \tag{3.45}$$

$$\frac{dp}{p} = -2\eta d\eta \tag{3.46}$$

Integrating both sides, we have

$$\ln p = -\eta^2 + C \tag{3.47}$$

$$p = Ae^{-\eta^2} \tag{3.48}$$

$$\frac{d\theta}{d\eta} = Ae^{-\eta^2} \tag{3.49}$$

$$d\theta = Ae^{-\eta^2}d\eta \tag{3.50}$$

Integrating Eq. (3.50) between limits $(0,\eta)$ in η and $(0, \theta)$ in θ, we have

$$\int_0^\theta d\theta = A\int_0^\eta e^{-\eta^2}\,d\eta \tag{3.51}$$

In order to obtain the constant A, We need to integrate Eq. (3.50) from 0 to ∞ and the corresponding limits for θ are 0 and θ_i.

$$\int_0^{\theta_i} d\theta = A\int_0^\infty e^{-\eta^2}\,d\eta \tag{3.52}$$

$$A = \int_0^\infty e^{-\eta^2} d\eta \tag{3.53}$$

Combining Eqs. (3.51) and (3.52), we have

$$\frac{\theta}{\theta_i} = \frac{\int_0^\eta e^{-\eta^2} d\eta}{\int_0^\infty e^{-\eta^2} d\eta} \tag{3.54}$$

At this point in time, we can invoke the use of error function, $erf(\eta)$ as follows,

$$erf(\eta) = \frac{2}{\sqrt{\pi}} \int_0^\eta e^{-\eta^2} d\eta \tag{3.55}$$

Consequent upon the invoking of $erf(\eta)$, Eq. (3.54) may be rewritten as,

$$\therefore \frac{\theta}{\theta_i} = \frac{erf(\eta)}{erf(\infty)} = erf(\eta) \tag{3.56}$$

Eq. (3.56) uses the information that $erf(\infty) = 1$.

The error function can be tabulated in a form that is easy to use. Values of $erf(\eta)$ against η are given in Table 3.1.

Table 3.1 Error function table.

x	erf(x)	x	erf(x)	x	erf(x)
0.00	0.00000	0.36	0.38933	1.04	0.85865
0.02	0.02256	0.38	0.40901	1.08	0.87333
0.04	0.04511	0.40	0.42839	1.12	0.88679
0.06	0.06762	0.44	0.46623	1.16	0.89910
0.08	0.09008	0.48	0.50275	1.20	0.91031
0.10	0.11246	0.52	0.53790	1.30	0.93401
0.12	0.13476	0.56	0.57162	1.40	0.95229
0.14	0.15695	0.60	0.60386	1.50	0.96611
0.16	0.17901	0.64	0.63459	1.60	0.97635
0.18	0.20094	0.68	0.66378	1.70	0.98379
0.20	0.22270	0.72	0.69143	1.80	0.98909
0.22	0.24430	0.76	0.71754	1.90	0.99279
0.24	0.26570	0.80	0.74210	2.00	0.99532
0.26	0.28690	0.84	0.76514	2.20	0.99814
0.28	0.30788	0.88	0.78669	2.40	0.99931
0.30	0.32863	0.92	0.80677	2.60	0.99976
0.32	0.34913	0.96	0.82542	2.80	0.99992
0.34	0.36936	1.00	0.84270	3.00	0.99998
				∞	1.00000

$$\therefore \frac{\theta}{\theta_i} = erf\left(\frac{x}{2\sqrt{\alpha t}}\right) \tag{3.57}$$

Eq. (3.57) can be used to determine the temperature at any point "x" at any time "t" in the slab. Please note that nowhere does the length scale L of the problem appear in the final solution.

3.4 The method of separation of variables

Let us now move on to the more general and less restrictive case where the dimensionless temperature $\phi = \frac{\theta}{\theta_i} = f(\xi, Bi, Fo)$. Consider a one-dimensional slab of thickness $2L$. Let ρ, c_p and k be the density, specific heat capacity, and thermal conductivity of the body respectively. The body is initially at a temperature of T_i throughout. At time $t = 0$, the two ends at $x = \pm L$ are suddenly brought to T_∞ with $T_\infty < T_i$. The problem of transient conduction begins with the challenge being our ability to get $T(x,t)$ in the domain for $-L \le x \le L$ and for all $t \ge 0$.

Let us now try to sketch qualitatively the variation of ϕ across the slab at various times, as shown in Fig. 3.5

From the figure it is seen that at "early" times a substantial part of the slab is still at T_i, and it is in this regime that the semi-infinite approximation is valid. Additionally,

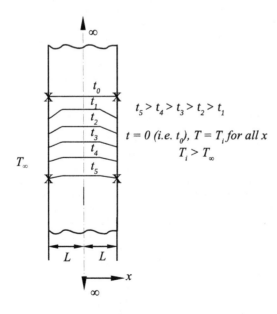

FIGURE 3.5

Qualitative variation of temperature time history for various time instances in an infinitely long slab of thickness $2L$.

for $t \gg 0$, known as the late regime, all temperatures within the slab are nearly the same, and in this late regime, the lumped capacitance method is valid regardless of the Biot number. At time instants not covered by either of these two asymptotic limits and the consequent simplifications thereof, one has to necessarily solve the governing equation in its full strength to obtain $T(x,t)$. Assume $T_i > T_\infty$.

Let $\phi = \dfrac{T - T_\infty}{T_i - T_\infty}$

$\therefore 0 \leq \phi \leq 1$ The governing Eq. (3.3) reduces to

$$\frac{\partial^2 \phi}{\partial x^2} = \frac{1}{\alpha} \frac{\partial \phi}{\partial t} \tag{3.58}$$

The initial condition is: $\phi = \phi_i = 1 \; for \; all \; x, \; for \; t = 0$

Boundary condition 1: At $x = \pm L; \phi = 0 \; for \; t > 0$

Boundary condition 2: At $x = 0; \dfrac{\partial \phi}{\partial x} = 0 \; for \; t \geq 0$

We now assume a product solution as follows.

$$\phi(x,t) = X(x)T(t) \tag{3.59}$$

The objective is to see whether, by doing this, the governing equation, which is a partial differential equation (PDE), can be reduced to an ordinary differential equation (ODE). The motivation for doing this is the fact that it is a lot easier to solve an ODE rather than a PDE.

$$\frac{\partial^2 \phi}{\partial x^2} = \frac{d^2 X}{dx^2} T \tag{3.60}$$

$$\frac{\partial \phi}{\partial t} = \frac{dT}{dt} X \tag{3.61}$$

$$\frac{d^2 X}{dx^2} T = \frac{1}{\alpha} \frac{dT}{dt} X \tag{3.62}$$

$$\frac{1}{X} \frac{d^2 X}{dx^2} = \frac{1}{\alpha T} \frac{dT}{dt} = -\lambda^2 \tag{3.63}$$

In Eq. (3.63), λ^2 is a constant and the negative sign ensures that the resulting ODEs do not result in solutions inconsistent with the physics of the problem. The above is a consequence of the fact that the left hand side of eqn. 3.63 is only a function of x while the right hand side is only a function of t. For this to happen, both must be equal to a constant.

The two ordinary differential equations are

$$\frac{dT}{dt} + \alpha \lambda^2 T = 0 \tag{3.64}$$

$$\frac{d^2 X}{dx^2} + \lambda^2 X = 0 \tag{3.65}$$

The general solution to Eq. (3.64) is

$$T = Ae^{-\alpha\lambda^2 t} \tag{3.66}$$

The general solution to Eq. (3.65) is

$$X = B\cos\lambda x + C\sin\lambda x \tag{3.67}$$

The general solution to Eq. (3.58) then becomes

$$\phi = Ae^{-\alpha\lambda^2 t}(B\cos\lambda x + C\sin\lambda x)^\dagger \tag{3.68}$$

We now evaluate A, B, C, and λ using the initial and boundary conditions.

$$\frac{\partial\phi}{\partial x} = Ae^{-\alpha\lambda^2 t}(-B\lambda\sin\lambda x + C\lambda\cos\lambda x) \tag{3.69}$$

At $x = 0$; $\dfrac{\partial\phi}{\partial x} = 0$

Therefore

$$C = 0$$

The general solution then becomes

$$\phi = Ae^{-\alpha\lambda^2 t}B\cos\lambda x \tag{3.70}$$

Let $AB = A'$, and Eq. (3.70) becomes

$$\phi = A'e^{-\alpha\lambda^2 t}\cos\lambda x \tag{3.71}$$

At $x = \pm L$; $\phi = 0$

$$0 = A'e^{-\alpha\lambda^2 t}\cos\lambda L \tag{3.72}$$

From Eq. (3.72), it is clear that

$$\cos\lambda L = 0 \tag{3.73}$$

Eq. (3.73) is satisfied by a succession of λ's that satisfy the following equation.

$$\lambda_n = \frac{n\pi}{2L} \tag{3.74}$$

where $n = 1, 3, 5.....$
Hence, the solution to Eq. (3.70) becomes

$$\phi = \sum_{n=1}^{\infty} A_n \cos(\lambda_n x)e^{-\lambda_n^2 t} \tag{3.75}$$

---

†An incorrect choice of λ^2 instead of $-\lambda^2$ would have resulted in a solution for temperature that is exponential in x and periodic in t, which completely goes against the physics of the problem.

In Eq. (3.75), A_n is as yet unknown. To obtain A_n we invoke the initial condition. At $t = 0$ s; $\phi = \phi_i = 1$

$$1 = \sum_{n=1}^{\infty} A_n \cos(\lambda_n x) e^0 \tag{3.76}$$

$$\therefore 1 = \sum_{n=1}^{\infty} A_n \cos(\lambda_n x) \tag{3.77}$$

We need to use the property of a set of functions called orthogonal functions to obtain A_n.

Orthogonal functions

An infinite set of functions $g_1(x), g_2(x), g_3(x).....g_n(x)$ is said to be orthogonal in the closed interval $a \le x \le b$, if $\int_a^b g_m(x) g_n(x) = 0$; for $m \ne n$.

Many functions like $\sin x$ and $\cos x$ exhibit orthogonality. Hence if we multiply Eq. (3.77) by $\cos(\lambda_n x)$ on both sides and integrate from 0 to 1, we can exploit the property of orthogonal functions to evaluate A_n.

$$\int_0^L \cos(\lambda_n x) dx = \int_0^L A_n \cos^2(\lambda_n x) dx \tag{3.78}$$

Please note that the right hand side of the equation 3.78 was originally a summation. However, in the view of the orthogonality property of $\cos \lambda_n x$, only the term with $\cos^2 \lambda_n x$ will remain in the series, as all other terms will be zero.

$$A_n = \frac{\int_0^L \cos(\lambda_n x) dx}{\int_0^L \cos^2(\lambda_n x) dx} \tag{3.79}$$

$$A_n = \frac{\left[\dfrac{\sin(\lambda_n x)}{\lambda_n} \right]_0^L}{\int_0^L \dfrac{1 + 2\cos(2\lambda_n x)}{2} dx} \tag{3.80}$$

$$A_n = \frac{\left[\dfrac{\sin(\lambda_n x)}{\lambda_n} \right]_0^L}{\left[\dfrac{x}{2} \right]_0^L + \left[\dfrac{\sin(2\lambda_n x)}{4\lambda_n} \right]_0^L} \tag{3.81}$$

$$A_n = \frac{\dfrac{1}{\lambda_n} \sin(\lambda_n L)}{\dfrac{L}{2} + \dfrac{2\sin(\lambda_n L)\cos(\lambda_n L)}{4\lambda_n}} \tag{3.82}$$

$$A_n = \frac{\dfrac{\phi_i}{\lambda_n} \sin(\lambda_n L)}{\dfrac{L}{2} + \dfrac{\sin(\lambda_n L)\cos(\lambda_n L)}{2\lambda_n}} \tag{3.83}$$

Substituting for A_n, in Eq. (3.75) and rewriting $\phi = (T\text{-}T_\infty)/(T_i\text{-}T_\infty)$ as θ/θ_i. We get the final expression for θ (or ϕ) as,

$$\frac{\theta}{\theta_i} = \sum_{n=1}^{\infty} \frac{1}{\lambda_n} \frac{\sin(\lambda_n L)\cos(\lambda_n x)e^{-\alpha\lambda_n^2 t}}{\dfrac{L}{2} + \dfrac{\sin(\lambda_n L)\cos(\lambda_n L)}{2\lambda_n}} \tag{3.84}$$

A tougher variant of this problem arises when the sides are exposed to convection with a heat transfer coefficient of "h" and free stream temperature "T_∞". All the conditions are the same as the previous case except that at the convection boundaries, we have

At $x = \pm L; \; -k\dfrac{\partial\phi}{\partial x} = h\phi$

$$\phi = Ae^{-\alpha\lambda_n^2 t}\cos\lambda_n x \tag{3.85}$$

$$-kAe^{-\alpha\lambda_n^2 t}(-\sin\lambda_n L)\lambda_n = hAe^{-\alpha\lambda_n^2 t}\cos\lambda_n L \tag{3.86}$$

$$\lambda_n Lk\sin\lambda_n L = hL\cos\lambda_n L \tag{3.87}$$

$$\xi\tan\xi = Bi \quad (\text{Where } \xi = \lambda L) \tag{3.88}$$

Again eq. 3.88 is satisfied by a succession of ξs that have to be obtained by numerically solving the transcendental equation 3.88
From Eq. (3.84)

$$\frac{\theta}{\theta_i} = \sum_{n=1}^{\infty} \frac{2L}{n\pi} \frac{\sin(\xi_n)\cos(\lambda_n x)e^{-\alpha\lambda_n^2 t}}{\dfrac{L}{2} + \dfrac{\sin(2\lambda_n L)}{4\lambda_n}} \tag{3.89}$$

$$\frac{\theta}{\theta_i} = \sum_{n=1}^{\infty} \frac{2L}{n\pi} \frac{\sin(\xi_n)\cos(\lambda_n x)e^{-\alpha\lambda_n^2 t}}{\dfrac{2L\lambda_n + \sin 2\lambda_n L}{4\lambda_n}} \tag{3.90}$$

$$\frac{\theta}{\theta_i} = \sum_{n=1}^{\infty} \frac{2L}{n\pi} 4\frac{n\pi}{2L} \frac{\sin(\xi_n)\cos(\lambda_n x)e^{-\alpha\lambda_n^2 t}}{2\xi_n + \sin(2\xi_n)} \tag{3.91}$$

$$\frac{\theta}{\theta_i} = \sum_{n=1}^{\infty} \frac{4\sin(\xi_n)}{2\xi_n + \sin(2\xi_n)} e^{-\xi_n^2 Fo}\cos(\lambda_n x) \tag{3.92}$$

The first-term approximation of the solution is given as

$$\Phi = \frac{\theta}{\theta_i} = A_1 e^{-\xi_1^2 Fo}\cos(\lambda_1 x) \tag{3.93}$$

The location $x = 0$, corresponds to the mid-plane temperature. Eq. (3.93) can be simplified at $x = 0$ as.

$$\frac{\theta_0}{\theta_i} = A_1 e^{-\xi_1^2 Fo} \tag{3.94}$$

where $\dfrac{\theta_0}{\theta_i}$ is the dimensionless mid-plane temperature of the plate.

We can now calculate the ratio of enthalpy transfer to the maximum enthalpy transfer possible in the one dimensional slab.

$$\frac{Q}{Q_{max}} = \frac{\int_0^V \rho c_p (T - T_i) dV}{\rho c_p (T_\infty - T_i) V} \tag{3.95}$$

$$= \frac{\int_0^V \rho c_p (T - T_\infty + T_\infty - T_i) dV}{\rho c_p (T_\infty - T_i) V} \tag{3.96}$$

$$\frac{Q}{Q_{max}} = \frac{1}{V} \int_0^V (1 - \phi) dV \tag{3.97}$$

$$= \frac{1}{AL} \int_0^L (1 - \phi) A \, dx \tag{3.98}$$

$$\frac{Q}{Q_{max}} = 1 - \frac{1}{AL} \int_0^L \phi A \, dx \tag{3.99}$$

$$= 1 - \frac{1}{L} \int_0^L \phi \, dx \tag{3.100}$$

Invoking eqn. 3.93 for ϕ, we have

$$= 1 - \frac{1}{L} \int_0^L A_1 \cos\left(\xi \frac{x}{L}\right) e^{-\xi^2 F_o} dx \tag{3.101}$$

$$= 1 - \frac{1}{L} A_1 e^{-\xi^2 F_o} \int_0^L \cos\left(\xi \frac{x}{L}\right) dx \tag{3.102}$$

$$\frac{Q}{Q_{max}} = 1 - \frac{1}{L} A_1 e^{-\xi^2 F_o} \left[\frac{\sin\left(\xi \frac{x}{L}\right)}{\frac{\xi}{L}} \right]_0^L \tag{3.103}$$

$$\therefore \frac{Q}{Q_{max}} = 1 - A_1 e^{-\xi^2 F_o} \frac{\sin(\xi)}{\xi} \tag{3.104}$$

$$\text{where } A_1 = \frac{4\sin(\xi_1)}{2\xi_1 + \sin(2\xi_1)}$$

Though we are able to obtain the solution analytically, Eq. (3.92) is quite tedious to use. Heisler (1947) used the first term in the series Eq. (3.94) and presented charts to solve the problem swiftly; these have now come to be known as "Heisler's charts." and are valid for Fo > 0.2.

Let us use the first term of Eq. (3.92) to redo these charts for the problem under consideration for convenience. We can get the mid-plane temperature first and then develop a chart for $\dfrac{\phi}{\phi_{center}}$ for any plane. This is mainly for convenience.

Analytical solutions are also possible for a cylinder and a sphere. There are formidable and involve Bessel's functions and Legendre polynomials respectively. However, these are quite involved. Charts based on the first term in the series solution for the plane wall are given in Figs. 3.6–3.7. Figs. 3.9–3.10, as

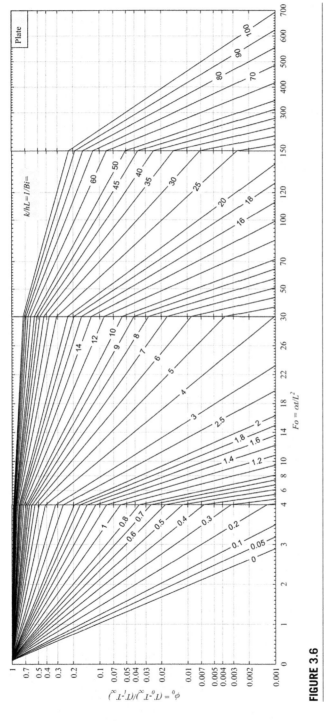

FIGURE 3.6

Mid-plane temperatures charts for a plane wall with thickness $2L$ initially at a uniform temperature of T_i subject to convection on both sides with a heat transfer coefficient of h and an ambient temperature of T_∞.

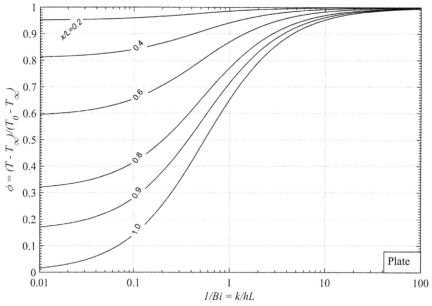

FIGURE 3.7

Temperature distribution chart for a plane wall with thickness $2L$ initially at a uniform temperature of T_i subjected to convection on both sides with a heat transfer coefficient of h and an ambient temperature of T_∞.

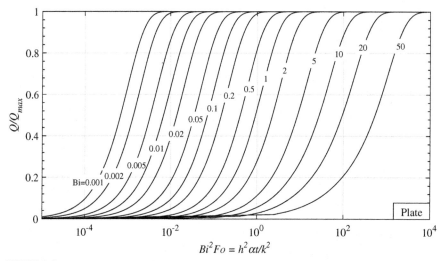

FIGURE 3.8

Enthalpy chart for a plane wall with thickness $2L$ initially at a uniform temperature of T_i subjected to convection on both sides with a heat transfer coefficient of h and an ambient temperature of T_∞.

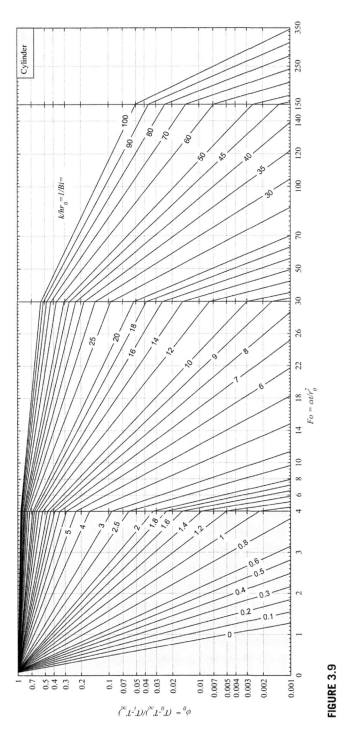

FIGURE 3.9

Center-line temperature charts for a long cylinder of radius r_0 initially at a uniform temperature of T_i, subject to convection on both sides with a heat transfer coefficient of h and an ambient temperature of T_∞.

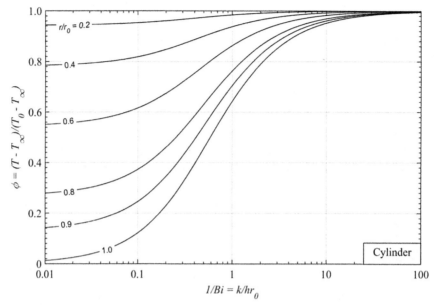

FIGURE 3.10

Temperature distribution chart for a long cylinder of radius r_0 initially at a uniform temperature of T_i, subjected to convection on both sides with a heat transfer coefficient of h and an ambient temperature of T_∞.

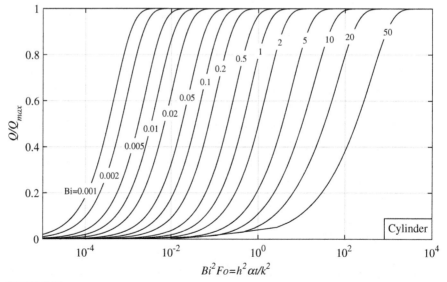

FIGURE 3.11

Enthalpy chart for a long cylinder of radius r_0 initially at a uniform temperature of T_i, subjected to convection on both sides with a heat transfer coefficient of h and an ambient temperature of T_∞.

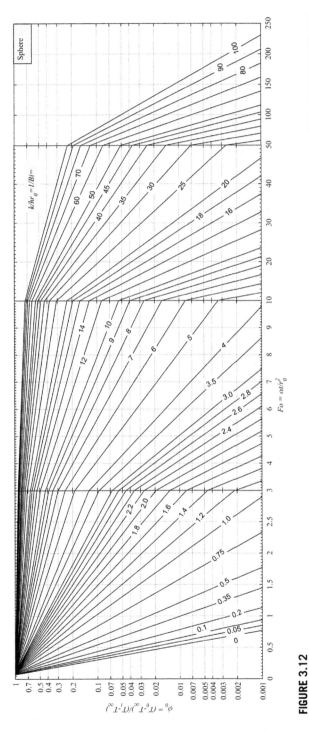

FIGURE 3.12

Midpoint temperature charts for a sphere of radius r_0 initially at a uniform temperature of T_i, subject to convection on both sides with a heat transfer coefficient of h and an ambient temperature of T_∞.

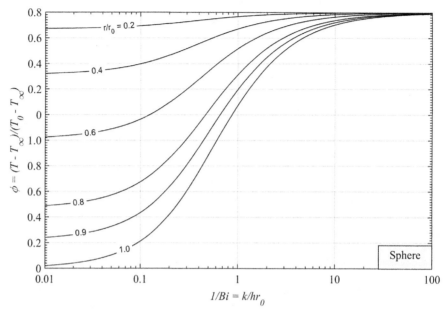

FIGURE 3.13

Temperature distribution chart for a sphere of radius r_0 initially at a uniform temperature of T_i subjected to convection on both sides with a heat transfer coefficient of h and an ambient temperature of T_∞.

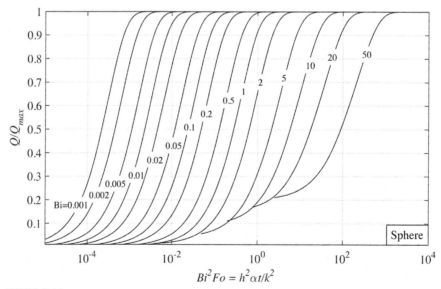

FIGURE 3.14

Enthalpy chart for a sphere of radius r_0 initially at a uniform temperature of T_i subjected to convection on both sides with a heat transfer coefficient of h and an ambient temperature of T_∞.

well as, Figs. 3.12–3.13, are Heisler's charts for the plane wall (or plate), cylinder and sphere respectively. For each of the three charts, the first two concern the temperature at the mid-plane followed by the temperature at any other plane. A third chart is usually added to these charts, for all the geometries, and this is known as Grober's chart. These give the ratio of enthalpy to the initial enthalpy excess of the body in question. In what follows, we present fully worked examples for plane wall. Computer codes based on MATLAB have been developed as a part of this book project to generate Heisler's charts for the plane wall, cylinder and sphere. These are available through the online support for this book. End of chapter problems 3.7 to 3.9 deal with transient conduction in a cylinder and sphere.

Example 3.2: *Consider a thick slab made of a material whose thermal conductivity "k" is unknown. The slab is initially at 100 °C, and one end of the slab is suddenly brought to 30 °C by the use of a temperature bath. A thermocouple, placed 8 mm from the end, exposed to the cold fluid shows 63 °C at 3 min, after the surface is exposed to the cold fluid. Determine the thermal conductivity of the material. The density and specific heat of the material are respectively 1920 kg/m³ and 835 J/kg.K, respectively.*

Solution:
We use the semi-infinite approximation to solve this problem.

$$\frac{\phi}{\phi_i} = erf(\eta) \tag{3.105}$$

$$\frac{63-30}{100-30} = erf(\eta) \tag{3.106}$$

$$0.4714 = erf(\eta) \tag{3.107}$$

From the error function table, $\eta = 0.44$ for $erf(\eta) = 0.4714$

$$\eta = \frac{x}{2\sqrt{\alpha\tau}} = 0.44 \tag{3.108}$$

$$\frac{0.008}{2\sqrt{\alpha t}} = 0.44 \tag{3.109}$$

$$\alpha t = 8.26 \times 10^{-5} \tag{3.110}$$

$$\frac{k}{1920 \times 835} \times 180 = 8.26 \times 10^{-5} \tag{3.111}$$

$$k = 0.735 \text{ W/mK} \tag{3.112}$$

Example 3.3: *A 1.6 cm thick slab of carbon steel is initially at $T_i = 610\ °C$. This slab is suddenly immersed into a bath of water at $T_\infty = 25\ °C$. The heat transfer coefficient h is $10^4\ W/m^2\ K$. The properties of a carbon steel are k = 40 W/m K, and $\alpha = 1 \times 10^5\ cm^2/s$.*

1. Determine the time t at which the temperature in the mid-plane of the slab drops to $T_C = 100\ °C$. Also determine the average temperature of the slab corresponding to this time.
2. Calculate the corresponding temperature in a plane situated 0.2 cm from one of the ends.

Solution:

The problem is solved in both the methods, i.e., through the analytical method and by using the Heisler's charts.

$$Bi = \frac{hL_c}{k}$$

$$L_c = t/2 \text{ for a plate} = \frac{1.6 \times 10^{-2}}{2} = 0.008\ \text{m}$$

$$Bi = \frac{10^4 \times 0.008}{40} = 2 \gg 0.1$$

Since $Bi \gg 0.1$ the lumped capacitance method is not applicable.

Solution through the analytical method:

From the analytical solution, the first term approximation for Eq. (3.92) is given as

$$\frac{\phi}{\phi_i} = \frac{4\sin(\xi_n)}{2\xi_n + \sin(2\xi_n)} e^{\xi_n^2 Fo} \cos(\lambda_n x)$$

a. For mid-plane $x = 0$, so the first term approximation for Eq. (3.92) becomes

$$\frac{\phi_{x=0}}{\phi_i} = \frac{4\sin(\xi_n)}{2\xi_n + \sin(2\xi_n)} e^{-\xi_n^2 Fo}$$

From Eq. (3.88)

$$\xi \tan \xi = Bi$$

$$\xi \tan \xi = 2$$

$$\tan \xi = \frac{2}{\xi}$$

$$\xi = \tan^{-1}\left(\frac{2}{\xi}\right)$$

Solving for ξ using the successive substitution method[‡](i.e., iteratively),

$$\xi = 1.076$$

$$\frac{\phi_{t,x=0}}{\phi_i} = \frac{T_{(t,x=0)} - T_\infty}{T_i - T_\infty} = \frac{100 - 25}{610 - 25} = 0.1282$$

$$\Rightarrow 0.1282 = \frac{4 \times \sin\left(1.076 \times \dfrac{180}{\pi}\right)}{2 \times 1.076 + \sin\left(2 \times 1.076 \times \dfrac{180}{\pi}\right)} \times e^{-\xi^2 Fo}$$

$$\Rightarrow 0.1282 = 1.178 \times e^{-\xi^2 F_0}$$

$$\Rightarrow e^{-\xi^2 F_0} = \frac{0.1282}{1.178} = 0.1086$$

$$\Rightarrow -\xi^2 Fo = -2.22$$

$$\Rightarrow Fo = 1.917$$

$$\Rightarrow \frac{\alpha t}{L_c^2} = 1.917$$

$$\Rightarrow t = \frac{1.917 \times (0.008)^2}{0.1 \times 10^{-4}}$$

$$\therefore t = 12.27\, s$$

From eq. 3.103 we know that

$$\frac{Q}{Q_{max}} = 1 - \frac{1}{L} A_1\, e^{-\xi^2 Fo} \left[\frac{\sin\left(\xi\dfrac{x}{L}\right)}{\dfrac{\xi}{L}} \right]_0^L$$

$$\therefore \frac{Q}{Q_{max}} = 1 - A_1 e^{-\xi^2 Fo}\, \frac{\sin(\xi)}{\xi}$$

[‡]In this method we start with a guess of ξ. Use this value to evaluate 'ξ' again using the relation $\xi = \tan^{-1}(2/\xi)$. We get a new value of ξ. Use this again to get the next iteration of ξ and continue this till convergence.

$$\frac{T_{avg} - T_{\infty}}{T_i - T_{\infty}} = 1 - \frac{Q}{Q_{max}} = A e^{-\xi^2 Fo} \frac{\sin(\xi)}{\xi}$$

$$\frac{T_{avg} - T_{\infty}}{T_i - T_{\infty}} = 0.1272 \times \frac{\sin\left(1.076 \times \dfrac{180}{pi}\right)}{1.076}$$

$$T_{avg} = T_{\infty} + 0.1041 \times (T_i - T_{\infty})$$
$$= 25 + 0.1041 \times (610 - 25) = 85.59 \,^{\circ}\text{C}$$

$$\therefore T_{avg} = 85.89 \,^{\circ}\text{C}$$

b. Temperature in a plane situated 0.2 cm from the cooled surface of the plate (i.e., temperature in a plane at $x = 0.006$ m).

$$\frac{\phi_{t,x}}{\phi_i} = \frac{4 \sin(\xi_n)}{2\xi_n + \sin(2\xi_n)} \cos\left(\xi \times \frac{x}{L}\right) e^{-\xi_n^2 Fo}$$

$$F_o = 1.917, \ \xi = 1.076, \ x = 0.006, \ t = 12.27\,s \ \frac{x}{L} = \frac{0.006}{0.008} = 0.75$$

$$\frac{\phi_{x,t}}{\phi_i} = \frac{4 \sin\left(1.076 \times \dfrac{180}{pi}\right)}{2 \times 1.076 + \sin\left(2 \times 1.076 \times \dfrac{180}{pi}\right)}$$
$$\times \cos\left(1.076 \times \frac{180}{pi} \times 0.75\right) e^{-1.076^2 \times 1.917}$$
$$= 1.178 \times 0.691 \times 0.108 = 0.0879$$

$$\frac{\phi_{x,t}}{\phi_i} = \frac{T(x,t) - T_{\infty}}{T_i - T_{\infty}} = 0.0879$$

$$\therefore T(x,t) = 76.42 \,^{\circ}\text{C}$$

Solution using Heisler charts:

$$\phi = \frac{T - T_{\infty}}{T_i - T_{\infty}}$$

$$\phi = \frac{100 - 25}{610 - 25} = 0.1282$$

$$\frac{1}{Bi} = \frac{k}{hL_c} = 0.5$$

From the Heisler charts the centerline temperature chart for the case of a plane wall (or plate) is given by

$$\frac{\alpha t}{L^2} = 2$$

$$\therefore t = \frac{2 \times 0.008^2}{0.1 \times 10^{-4}}$$

$$t = 12.8\,\text{s}.$$

$$Bi^2 \times Fo = \frac{h^2 \alpha t}{k^2}$$
$$= \frac{(10^4)^2 \times 10^{-4} \times 12.8}{40^2} = 8$$

$$\therefore \phi_{avg} = \frac{T_{avg} - T_\infty}{T_i - T_\infty} = 1 - \frac{Q}{Q_{max}}$$

From Grober charts:

$$\frac{Q}{Q_{max}} = 0.875$$

$$\phi_{avg} = 1 - 0.875 = 0.125$$

$$\therefore T_{avg} = 0.125 \times (610 - 25) + 25$$
$$= 98.125\,°\text{C}$$

The temperature at 0.2 cm from one of the cooled surfaces of the plate is determined as

$$\frac{x}{L} = \frac{0.008 - 0.002}{0.002}$$

$$\therefore \phi = \frac{T_{(0.006,t)} - T_\infty}{T_o - T_\infty} = 0.65$$

$$\therefore T_{(0.006,t)} = 0.65 \times (T_o - T_\infty) + T_\infty$$
$$= 0.65 \times (100 - 25) + 25$$
$$= 73.75\,°\text{C}$$

We see that the analytical and graphical solutions are reasonably close. Confirmation of the average temperature of the slab obtained from the analytical solution with that from the Grober's chart is left as an exercise to the student.

3.5 Analysis of two-dimensional, steady state systems

Consider a two-dimensional slab of height "W" and length "L." The depth in the direction perpendicular to the plane of the paper is so large that T varies only with x and y.

Assumptions:

1. Steady state prevails.
2. No internal heat generation ($q_v = 0$).
3. The thermophysical properties such as k, ρ, c_p, etc., are constant (Fig. 3.15).

The governing equation for two-dimensional steady state conduction is given by

$$\frac{\partial^2 T}{\partial x^2} + \frac{\partial^2 T}{\partial y^2} = 0 \tag{3.113}$$

Let us introduce a dimensionless temperature, ϕ, as follows

$$\phi = \frac{T - T_C}{T_H - T_C} \tag{3.114}$$

Therefore, eqn. 3.113 becomes

$$\frac{\partial^2 \phi}{\partial x^2} + \frac{\partial^2 \phi}{\partial y^2} = 0 \tag{3.115}$$

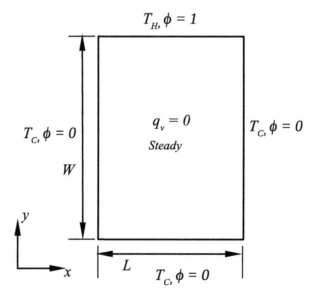

FIGURE 3.15

Schematic representation of a two-dimensional plane wall with the boundary condition $\left(\phi = \frac{T - T_C}{T_H - T_C}\right)$. T_H is the hot wall temperature (top) and T_c is the cold wall temperature on the remaining three walls.

Assume a product solution in the form given in Eq. (3.116).

$$\phi(x,y) = X(x)Y(y) \tag{3.116}$$

$$\frac{\partial^2 \phi}{\partial x^2} = \frac{d^2 X}{dx^2} Y \tag{3.117}$$

$$\frac{\partial^2 \phi}{\partial y^2} = \frac{d^2 Y}{dy^2} X \tag{3.118}$$

Using Eqs. (3.117) and (3.118), Eq. (3.115) can be expressed as

$$\frac{d^2 X}{dx^2} Y + \frac{d^2 Y}{dy^2} X = 0 \tag{3.119}$$

As before, we introduce a constant $-\lambda$, as follows.

$$\frac{1}{X}\frac{d^2 X}{dx^2} = \frac{-1}{Y}\frac{d^2 Y}{dy^2} = -\lambda^2 \tag{3.120}$$

$$\frac{d^2 X}{dx^2} + \lambda^2 X = 0 \tag{3.121}$$

The roots of the above equation are

$$D = \pm i\lambda \tag{3.122}$$

The general solution to Eq. (3.121) is given by

$$X = A\cos \lambda x + B\sin \lambda x \tag{3.123}$$

Consider, Eq. (3.120) which is rewritten below.

$$\frac{d^2 Y}{dy^2} - \lambda^2 Y = 0 \tag{3.124}$$

The roots for the above equation are

$$D = \pm \lambda \tag{3.125}$$

The general solution to Eq. (3.124) is given by

$$Y = C\cosh \lambda y + D\sinh \lambda y \tag{3.126}$$

The general solution to Eq. (3.115) can be written as

$$\phi = (A\cos \lambda x + B\sin \lambda x)(C\cosh \lambda y + D\sinh \lambda y) \tag{3.127}$$

The boundary condition for the left wall is at $x = 0$; $\phi = 0$.
From Eq. (3.127), we have

$$A = 0 \tag{3.128}$$

The boundary condition for the bottom wall is $y = 0$; $\phi = 0$.

Therefore, from Eq. (3.127), we have

$$C = 0 \tag{3.129}$$

Hence, the general solution reduces to

$$\phi = B\sin(\lambda x)D\sinh(\lambda y) \tag{3.130}$$

$$\phi = E\sin(\lambda x)\sinh(\lambda y) \tag{3.131}$$

where $E = BD$

The boundary condition at the right wall is $x = L$; $\phi = 0$.

$$E\sin\lambda L\sinh\lambda y = 0 \tag{3.132}$$

$$\sin\lambda L = 0 \tag{3.133}$$

Eq. (3.133) is satisfied by succession of λ's given by

$$\lambda = \frac{n\pi}{L} \tag{3.134}$$

Here $n = 0,1,2,3....$

$$\phi = \sum_{n=0}^{\infty} E_n \sin\frac{n\pi x}{L}\sinh\frac{n\pi y}{L} \tag{3.135}$$

The boundary condition at the top wall is given by $y = W$; $\phi = 1$. Using Eq. (3.135) and starting the expansion from $n = 1$, we have

$$1 = \sum_{n=1}^{\infty} E_n \sin(\lambda_n x)\sinh(\lambda_n W) \tag{3.136}$$

We can now exploit the property of orthogonal functions which we discussed earlier. We have for a general set of orthogonal functions given by $g_1(x)$, $g_2(x)$, ... $g_m(x)$ and $g_n(x)$ to represent a function $f(x)$ as an infinite series with c_n's being the constants of the individual terms in the series.

$$\int_a^b f(x)g_n(x)\,dx = \int_a^b c_n g_n^2(x)\,dx \tag{3.137}$$

Now invoking Eq. (3.137) and recognising that $f(x) = 1$, $g_n(x) = \sin(\lambda_n x)$ and $c_n = E_n \sinh(\lambda_n x)$. We have,

$$\int_0^L \sin(\lambda_n x)\,dx = \int_0^L E_n \sinh(\lambda_n W)\sin^2(\lambda_n x)\,dx \tag{3.138}$$

$$E_n \sinh(\lambda_n W) = \frac{\int_0^L \sin(\lambda_n x)\,dx}{\int_0^L \sin^2(\lambda_n x)\,dx} \tag{3.139}$$

$$E_n n\pi \sinh(\lambda_n W) = \frac{\left[\dfrac{-\cos(\lambda_n x)}{\lambda_n}\right]_0^L}{\int_0^L \dfrac{1-\cos 2\lambda_n x}{2}\,dx} \tag{3.140}$$

$$E_n n\pi \sinh(\lambda_n W) = \frac{\dfrac{-1}{\lambda_n}((-1)^n - 1)}{\dfrac{L}{2}} \tag{3.141}$$

$$E_n n\pi \sinh(\lambda_n W) = \frac{2}{L\dfrac{n\pi}{L}}(1 - (-1)^n) \tag{3.142}$$

$$E_n n\pi \sinh(\lambda_n W) = \left(\frac{2}{n\pi}\right)(1 - (-1)^n) \tag{3.143}$$

$$\therefore E_n = \left(\frac{2}{n\pi}\right)\frac{(1 - (-1)^n)}{\sinh(\lambda_n W)} \tag{3.144}$$

Substituting for E_n in Eq. (3.135), we have the final form of solution for the dimensionless temperature ϕ.

$$\phi = \sum_{n=1}^{\infty} \frac{2}{n\pi} \frac{(1 - (-1)^n)}{\sinh(\frac{n\pi W}{L})} \sin\left(\frac{n\pi x}{L}\right) \sinh\left(\frac{n\pi y}{L}\right) \tag{3.145}$$

For example consider a square slab where width (W) = length (L), as shown in Fig.3.16A. Please note that now the height is L and the width, W, just to let the readers know that there is no hard and fast rule about the nomenclature. It is important, though, to remember that the periodic part of the solution is along x and the exponential part of the solution is along y. Let the top temperature be 100 °C (403.15 K) and the other three walls be at 30 °C (303.15 K). Isotherms for this example would look like what are shown in Fig. 3.16B. Temperature symmetry around $x = \dfrac{W}{2}$ can be clearly seen. The exponential nature of temperature across y is also evident. This

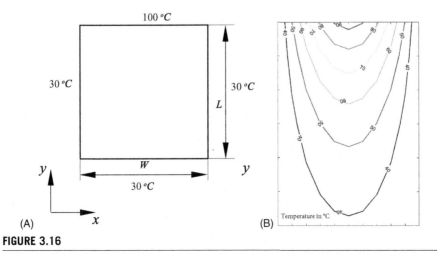

FIGURE 3.16

(A) Schematic representation of a two dimensional plane wall and (B) temperature contours.

explains the reason why we chose "$-\lambda^2$" in Eq. (3.120) Had we chosen "λ^2" instead, we would have landed up with an exponential temperature distribution in the x-direction and a symmetric distribution in the y-direction both of which violate the physics of the problem.

Example 3.4: *Consider steady, two-dimensional conduction heat transfer in a square metal plate of a constant thermophysical properties as shown in Fig. 3.17. The temperatures at the boundaries are prescribed to be 273.15 K on all the sides except at the top where the temperature is maintained at 373.15 K. Obtain the temperature from the analytical solution at nodes 1, 2, 3, and 4 with the first two nonzero terms of the series solution.*

Solution:

The analytical solution for two-dimensional heat transfer in a plate is given as

$$\phi = \sum_{n=1}^{\infty} \frac{2}{n\pi} \frac{(1-(-1)^n)}{\sinh(\frac{n\pi W}{L})} \sin\left(\frac{n\pi x}{L}\right) \sinh\left(\frac{n\pi y}{L}\right)$$

Temperature at node 1:

$W = 0.3$ m; $L = 0.3$ m; $x = 0.1$ m; $y = 0.2$ m

$$\phi = \frac{T_{(x,y)} - T_C}{T_H - T_C} = 0.63662 \times (0.5997 + 0 + 3.53 \times 10^{-10})$$

$$= 0.3818$$

$$\frac{T_{(x,y)} - 273.15}{373.15 - 273.15} = 0.3818$$

$$T_{(x,y)} = T_1 = 311.33 \text{ K}$$

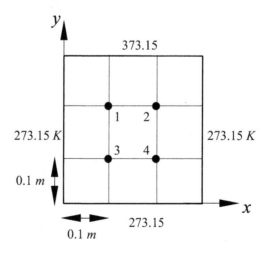

FIGURE 3.17

Schematic representation of a two-dimensional plane wall with boundary temperatures considered in example 3.4.

Temperature at node 2:
$W = 0.3$ m; $L = 0.3$ m; $x = 0.2$ m; $y = 0.2$ m

$$\phi = \frac{T_{(x,y)} - T_C}{T_H - T_C} = 0.63662 \times (0.5997 + 0 - 7.1 \times 10^{-18})$$

$$= 0.3818$$

$$\frac{T_{(x,y)} - 273.15}{373.15 - 273.15} = 0.3818$$

$$T_{(x,y)} = T_2 = 311.33 \text{ K}$$

Temperature at node 3:
$W = 0.3$ m; $L = 0.3$ m; $x = 0.1$ m; $y = 0.1$ m

$$\phi = \frac{T_{(x,y)} - T_C}{T_H - T_C} = 0.63662 \times (0.1874 + 0 + 1.52 \times 10^{-19})$$
$$= 0.1193$$

$$\frac{T_{(x,y)} - 273.15}{373.15 - 273.15} = 0.1193$$

$$T_{(x,y)} = T_3 = 285.08 \text{ K}$$

Temperature at node 4:
$W = 0.3$ m; $L = 0.3$ m; $x = 0.2$ m; $y = 0.2$ m

$$\phi = \frac{T_{(x,y)} - T_C}{T_H - T_C} = 0.63662 \times (0.1874 + 0 - 3 \times 10^{-19})$$
$$= 0.1193$$

$$\frac{T_{(x,y)} - 273.15}{373.15 - 273.15} = 0.1193$$

$$T_{(x,y)} = T_4 = 285.08 \text{ K}$$

The horizontal symmetry of the temperature distribution is confirmed with $T_1 = T_2$ and $T_3 = T_4$.

Problems

3.1 An electronic device "(like the processor of a desktop computer) weighing 0.28 kg generates 60 W of heat and reaches a temperature of 100 °C in ambient air at 30 °C under steady state conditions. The device is initially at 30 °C. Determine the temperature it will reach 6 min after the power is switched on? Assume the device to be spatially isothermal. Thermo physical properties of electronic chip are $c_p = 700$ J/kg.K, $\rho = 2329$ kg/m3, k = 130 W/m.K (Adapted and modified from Incropera et al. (2013)).

3.2 The time constant of a K-type (Chromel-Alumel) thermocouple of diameter 0.70 mm has been determined to be 1 s, and the temperature indicated by the thermocouple for a time t that equals its time constant τ value is 74.24 °C and the temperature indicated at 0.5 s is 57.54 °C. The specific heat and density of Chromel-Alumel are 420 J/kg-K and 8600 kg/m^3, respectively. Using the lumped heat capacity approach, determine

a. the temperature of the medium whose temperature needs to be measured.

b. the initial temperature of the thermocouple.

c. the rate of initial temperature change (at time $t = 0$ s) of the thermocouple.

3.3 A cylindrical stainless steel rod 1 cm in diameter and 20 cm in length is initially at 750 °C. It is then submerged in a bath of water at 90 °C. The heat transfer coefficient can be taken as 250 W/m^2 K. The density, specific heat, and thermal conductivity of the steel are $\rho = 7801 \, \text{kg/m}^3$, $c_p = 473 \, \text{J/kgK}$, and $k = 43 \, \text{W/mK}$, respectively. Determine the time required for the center of the rod to reach 300 °C. What is the key assumption/approximation required to solve the problem with the techniques presented in this chapter?

3.4 In a chocolate industry chocolates are made into spherical balls with a diameter of 2 cm and temperature of 20 °C, and they are kept in a freezer at 2 °C before packing. The chocolates have approximately the same thermophysical properties as that of water (Provided in Chapter 5), and the heat transfer coefficient is around 12 W/m^2 K. What will be the temperature of the center of the chocolate after 30 min? What is the time required to bring the center temperature of the chocolate to 6 °C. Also determine the total enthalpy removed by the freezer to bring the chocolate center temperature to 10 °C.

3.5 A wall 15 cm thick, made of clay brick, is initially 100 °C. The surface temperatures of both sides of the brick are suddenly reduced to 27 °C. Find the temperature at a plane 5 cm from the surface after 90 min have passed. How much enthalpy has been lost from the brick wall during that time? (Use Heisler's charts.)

Properties of the brick are $\rho = 1625 \, \text{kg/m}^3$, $c_p = 840 \, \text{J/kgK}$ and $k = 0.7 \, \text{W/mK}$, $\alpha = 5.25 \times 10^{-7} \, \text{m}^2/\text{s}$.

3.6 Revisit problem 3.5, in which the two sides of the brick are suddenly exposed to a medium that is at 27 °C with a heat transfer coefficient of 80 W/m^2 K. Find the temperature at a point 1.5 cm from the surface after 7 hours have passed. Determine (1) the enthalpy loss from the wall during that time and (2) the average temperature within the wall at the end of 7 hours (Use Heisler's charts.)

3.7 In a plastic welding process, a cylindrical polypropylene rod (filler material with $k = 0.5 \, \text{W/mK}$) of diameter 3 mm is used and is initially 25 °C. The rod is suddenly exposed to a hot air jet of diameter 2 cm at 560 °C. The effective convective and radiative heat transfer coefficient (total) is 60 W/m^2 K. How much time will it take for the polypropylene rod to reach its melting temperature of 160 °C? The density and specific heat of polypropylene are 920 kg/m^3 and 1800 J/kg K respectively.

3.8 Stainless steel bearings ($\rho = 7900 \, \text{kg/m}^3$, $c_p = 477 \, \text{J/kgK}$, and $k = 15 \, \text{W/mK}$) that have been uniformly heated to 840 °C are hardened by quenching them in an oil bath that is maintained at 40 °C. The ball diameter is 20 mm, and the associated convection coefficient is 1000 W/m^2 K. If quenching is to occur until the surface temperature of the balls reaches 100 °C, how long must the balls be kept in oil? What is the center temperature at the end of the cooling period?

3.9 An iron sphere of diameter 80 mm is initially at a uniform temperature of 200 °C. It is suddenly exposed to ambient air at 30 °C with a convection coefficient of 510 W/m^2 K.

    a. Determine the temperature at the center of the sphere and at a depth of 5 mm from the surface at $t = 1$ min after the sphere is exposed to air. Also determine the average temperature of the slab at $t = 1$ min.

    b. Calculate the enthalpy removed from the sphere in this duration.

    Assume the density, specific heat, thermal conductivity, and thermal diffusivity for the sphere as 8000 kg/m^3, 460 J/kgK, 60 W/mK, and 1.6×10^{-5} m^2/s, respectively.

3.10 In northern India, the highest temperature of a summer day can go up to 45 °C. In places where refrigeration facilities are not available, drinking such warm water or taking baths is very unpleasant. What minimum burial depth would you recommend to the company laying water pipelines so that even in summer one can get water at a temperature not exceeding 25 °C? Assume that initially the soil is at 20 °C, and then it is subjected to a constant surface temperature of 40 °C for 60 days. The thermal diffusivity of soil at 20 °C is 0.138×10^{-6} m^2/s.

3.11 A semi-infinite slab is initially at a temperature of 100 °C. Suddenly one end of the slab is exposed to boiling water at 100 °C. An arrangement is made with a thermocouple to measure the temperature at a location 12 mm from one end. For the slab under consideration, the temperature at this location at $t = 2$ min is 65 °C. Determine the thermal conductivity of the material. The density and specific heat of the solid are known to be 2300 kg/m^3 and 750 J/kg K, respectively.

3.12 A square slab of dimensions 10 cm $\times$ 10 cm is very deep in the direction perpendicular to the plane of the paper. The slab is made of material with a thermal conductivity of $k = 15$ W/mK. Steady state prevails in the slab, there is no heat generation, and all the properties are assumed to be constant. The boundary conditions are given in accompanying Fig. 3.18.

    a. Suggest a strategy of making use of the analytical solution presented in this chapter to determine the temperature distribution in the slab.

    b. Using the strategy obtained in (a), determine the center temperature, that is, the temperature at (0.05, 0.05) using the first two non-zero terms in the series solution.

3.13 Consider the problem of conduction in a square slab as given in problem 3.12. However, the boundary conditions are now different and are given in accompanying Fig. 3.19. The slab is made of material with a thermal conductivity of $k = 15$ W/mK. Steady state prevails in the slab, there is no heat generation, and all the properties are assumed to be constant.

    a. Develop a strategy to obtain a solution to the problem of determination of temperature distribution for this problem using the analytical solution presented earlier in this chapter.

    b. Hence, determine the temperature at (0.05, 0.05) using the first two nonzero terms in the series solution.

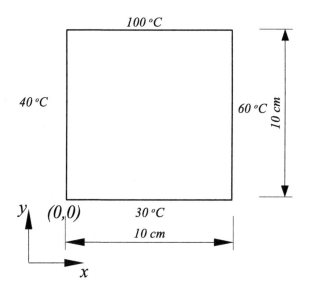

FIGURE 3.18

Two-dimensional plane wall with boundary temperatures considered in problem 3.12.

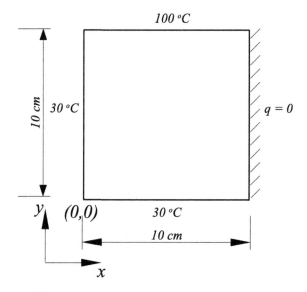

FIGURE 3.19

Two-dimensional plane wall with boundary conditions considered in problem 3.13.

References

Bergman, T.L., Incropera, F.P., DeWitt, D.P. and Lavine, A.S., 2011. Fundamentals of Heat and Mass Transfer. John Wiley and Sons, New York.

Fundamentals of convection

4

4.1 Introduction

In the previous chapters, we saw, in detail, how heat transfer is affected by conduction. Throughout our discussions, we considered heat transfer within a *stationary* medium. Even so, in these discussions, we also encountered several situations where this stationary medium (like a solid slab) shared an interface with a fluid, such as air or water. In these situations, we accounted for this interface with the fluid by applying convective boundary conditions. These boundary conditions were specified *a priori* with an assumed value of the heat transfer coefficient h. We will now see the origins of this "*h*".

So, what is convection? It is a mode of heat transfer or energy exchange that takes place between a (usually solid) surface and a fluid moving relative to it. You may ask: "Why only a moving fluid? Can convection not take place at the interface of two solids that move relative to each other?" The short answer is that there is no convection possible without a fluid. The key here, as discussed in Chapter 1, is the existence of a boundary layer in the fluid near solid-fluid interfaces. We will see this in greater detail in the coming chapters.

Convection is a complex topic of immense practical importance in heat transfer. Practical problems in convective heat transfer involve a mix of exact science, approximate solutions, and empirical correlations. This combination of approaches will be, therefore, present in the coming chapters as well. Because of the practical relevance of this topic, we have dedicated three chapters to it.

The current chapter sets the physical and mathematical fundamentals of convection. We will introduce the fundamentals of the heat transfer coefficient and follow up with the full equations of fluid motion. The following two chapters, after the current one, apply these fundamentals to various applications.

4.2 Fundamentals of convective heat transfer

Imagine that you have just touched a hot pan. What would be your instinctive reaction? It would be either shaking your palm vigorously or blowing on your fingers. Why is blowing on your fingers a useful response? Why not keep your hand still and let the surrounding air cool your hand? (You might also like to think of the typical temperature of the surrounding air versus the typical temperature of the air inside

Heat Transfer Engineering. http://dx.doi.org/10.1016/B978-0-12-818503-2.00004-6

your mouth). The difference between the action of still air and moving air in cooling your hand is the difference between conduction and convection.

4.2.1 Conduction, advection, and convection

In still air, there is a microscopic motion of the air molecules, which results in heat transfer due to macroscopic conduction. When we blow air over our fingers, there is an additional bulk motion of the air. This bulk motion is called advection. Advection has a transporting effect on heat transfer. The combined effect of diffusion due to conduction and transport due to advection is called convection. That is,

$$\text{Convection} = \textbf{Cond}\text{uction} + \text{Ad}\textbf{vection}$$

A note of caution—the above equation is only heuristically true. That is, the physical effect of convection is due to the combination of the physical effects of conduction and advection. This does not mean that one can mathematically add or superpose these effects linearly. The two effects often interact nonlinearly, and the addition is only in terms of physics, not mathematics.

When the medium is at rest, convection devolves into conduction. Conduction, in this situation, may be thought of as a special case of convection. Purists, however, will frown at this statement, because this is a lot like saying that statics is a special case of dynamics!

In general, however, there is bulk motion (advection). There are two possible sources of this bulk motion. The first type of motion is obvious; motion might arise from an external pumping source such as a fan, pump, or a suction device. This type of heat transfer due to an external, forced motion of the fluid is known as forced convection. Fig. 4.1 shows a simple example of forced convection over a flat plate.

The second type of source is subtler. Consider a hot plate or a hot cup of coffee kept in still air. Is there convection in these cases? At first look, it seems like there is none, because the air outside is stationary. However, at a closer look, we notice that there is a current of rising air near the hot surfaces. This motion is set up naturally by the density difference between the hot air near the surface and the cooler air farther away. This bulk motion, caused not by external sources but by natural density differences occurring freely within the flow, causes convection as well. For obvious reasons, this type of convection is called *free* or **natural convection**. Fig. 4.2 shows

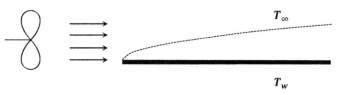

FIGURE 4.1

Forced convection over a flat plate.

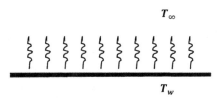

T_∞

T_w

FIGURE 4.2

Natural convection over heated plate facing upward with cooler quiescent fluid.

an example of natural convection over a horizontal heated plate kept in a cooler quiescent fluid.

Natural convection is typically weaker in magnitude than forced convection. When both effects are present in a situation, it is called **mixed convection**. Whether forced, natural, or mixed, the macroscopic picture of convection remains the same—advection in addition to conduction. But what happens microscopically at the molecular level?

4.2.2 The microscopic picture

Microscopically, we still have a picture very much like that in conduction. Molecules exchange energy due to their motion, and this energy causes convective heat transfer. The primary difference between conduction and convection is that, in a macroscopically quiescent medium, molecules are all moving in a purely random fashion with a mean velocity of zero, whereas, in a macroscopically moving medium, molecules have a non-zero mean velocity due to bulk motion in addition to random motion.

4.2.3 Fundamental definition of convection

This discussion leads us to the following understanding of convection:

1. Convection happens due to the combined physical effects of conduction and advection.
2. If we have an interface between a solid and a moving fluid, then there is no slip at the interface; therefore, due to the lack of relative motion, there is zero advection at the interface. We can conclude that, at the interface, convective heat transfer is equal to conduction.
3. From Fourier's law of conduction, we can therefore conclude that, at the wall/interface,

$$q_{convection} = -k_f \frac{\partial T}{\partial y}\bigg|_{wall} \tag{4.1}$$

This is assuming we have a flat interface that is at $y = 0$. For an interface of general shape, we have

$$q_{convection} = -k_f \left. \frac{\partial T}{\partial n} \right|_{interface}$$

(4.2)

where n is the local normal direction at the interface and k_f stands for the thermal conductivity of the fluid.

It is essential to understand that, for continuum flows, Eq. (4.2) is the fundamental way of defining convective heat transfer from first principles. The reason for emphasizing this is that, starting from the very next section, you will be flooded with a variety of expressions, methods, and formulae for calculating convective heat transfer. Every single one of these alternates is a secondary or tertiary derivation—usually approximate or empirical. If and when you are confused about how to compute convective heat transfer in a situation, the definition given in Eq. (4.2) is where you should return for solid conceptual ground. This is because the equation remains true in all continuum cases, forced or natural convection, laminar or turbulent flow, etc.

4.3 The heat transfer coefficient

The assertion that Eq. (4.2) is fundamental might make you wonder about the role of Newton's law of cooling and why we were using it in the earlier chapters. Recall that, according to Newton's law of cooling,

$$Q = hA_s \left(T_w - T_\infty \right)$$

(4.3)

where h is the heat transfer coefficient; in W/m^2K, A_s is the surface area in m^2; and T_w and T_∞ are the wall and freestream temperatures respectively in K.

4.3.1 Newton's law vs. the fundamental definition

We admit at this point that Newton's law, in effect, is a definition of h rather than an independent definition of the convective heat transfer itself. However, it is of tremendous practical significance. Despite this, Newton's law has some shortcomings, which we will see first.

1. There is no fundamental derivation or reasoning for the truth of Newton's law. While this is not a serious objection to its validity, it is sufficient reason to not elevate it conceptually to the same level as Eq. (4.2).
2. The law can be meaningful as a definition only if h were, at least approximately, a constant. This is unfortunately true only for small temperature differences $(T_w - T_\infty)$.
3. It turns out that h, unlike the thermal conductivity k, is not a property of the fluid but a property of the flow. To see this, imagine that in a forced convection case we increase the speed of blowing air. We can see intuitively that this should increase

the heat transfer. However, in $hA_s(T_w - T_\infty)$ the only term that can exhibit this dependence on this free stream velocity u_∞ is h. So h depends on the flow and not just the fluid. Consequently, we cannot make a few measurements for a given fluid and tabulate the results as we can for k; h is a complex function of flow parameters. That is, $h = f\left(u_\infty, \rho, \mu, k_f, c_p, T_w, T_\infty, \ldots\right)$.

4. The only way to calculate h theoretically or computationally is via Eq. (4.2). That is, given a flow field, we can calculate h only via

$$h = -\frac{k_f \dfrac{\partial T}{\partial y}\bigg|_{wall}}{\left(T_w - T_\infty\right)} \tag{4.4}$$

So, h is dependent on Eq. (4.2) and, unless h is given beforehand, Newton's law cannot function independently.

Despite these seeming shortcomings, there are excellent and overwhelming reasons for why Newton's law is the basis of engineering practice in heat transfer. The reasons are as follows.

1. **Historical reasons**—For long, engineering practice has equated convective heat transfer with Newton's law. The first estimates of convective heat transfer rates were made by correlating Newton's law of cooling with experimental results. There is, consequently, a lot of accumulated know-how in the form of empirical formulae, charts, and tables about how h behaves in various situations. This knowledge base is tremendously useful in practical situations. Such an approach is frequently referred to as the "empirical approach" in science.
2. **Decoupling of physical effects**—We understand intuitively that higher temperature differences would lead to higher heat transfer. Similarly, faster bulk motion leads to higher heat transfer. The fundamental definition in Eq. (4.2) aggregates both these distinct mechanisms into a single term, $\dfrac{\partial T}{\partial y}$. In contrast, Newton's law decouples these two effects into distinct multiplicative terms; u_∞ affects h while $(T_w - T_\infty)$ accounts for temperature differences. Apart from conceptual clarity, this decoupling also allows us to make design and other engineering judgements in practice.
3. **Useful in making comparisons**—Relatedly, Newton's law is useful in order to make comparisons between situations that share some commonality. For instance, in case we need to compare the relative efficacy of different mechanisms—such as natural versus forced convection, or laminar versus turbulent heat transfer—we could look at situations where we have the same temperature difference and it would be sufficient to compare the heat transfer coefficients. In fact, it is sometimes useful to even derive pseudo-quantities such as the $h_{radiation}$ in order to compare, say, radiative effects with convective effects.
4. **For making back-of-the-envelope calculations**—Finally, for the practicing engineer, it is often possible to make quick, initial estimates on heat transfer or

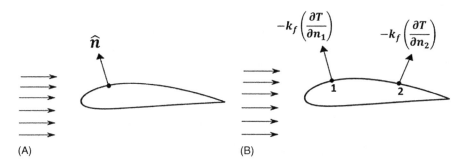

FIGURE 4.3

Convection heat transfer in flow past a curved surface.
(A) Flow past a curved surface, (B) convection heat transfer at different points on a surface.

sizing, etc., by knowing (from experience) the range of h for a particular situation. For example, in many practical situations the convective heat transfer with air as the moving fluid lies around 10 –100 W/m^{2}K. This knowledge, along with Newton's law, allows an engineer to estimate the heat transfer rapidly in many situations.

For the above reasons, Newton's law is the "go-to formula" for convective heat transfer, despite our lack of prior knowledge of h. Due to its centrality to convective heat transfer, the calculation of h is often called the **problem of convection**. Almost all of our efforts in the coming discussions on convection will be centered on the estimation of h exactly or approximately.

4.3.2 Average heat transfer coefficient

Fig. 4.3 shows a situation where we have a flow past a body. We know from our knowledge of fluid flow that both the temperature as well as its gradient at the wall will vary according to the position. So we can infer from Eq. (4.2) that the heat transfer coefficient will vary along the wall. How then did we calculate the heat transfer through Newton's law in earlier chapters for extended surfaces such as fins where h changes along the surface? We did so implicitly through the idea of the *average* heat transfer coefficient, where the average is defined over the surface.

In order to see this, note that the total heat transfer on the surface would be given by the sum of heat transfer rates on small elemental areas through the surface. That is,

$$Q = \int_{Surface} h\left(T_w - T_\infty\right)dA_s \tag{4.5}$$

Since $(T_w - T_\infty)$ is a constant, we have

$$Q = \left(T_w - T_\infty\right)\int_{Surface} h dA_s \tag{4.6}$$

which can be written as

$$Q = \bar{h} A_s \left(T_w - T_\infty \right) \tag{4.7}$$

where $\bar{h}$ is the average heat transfer coefficient and is defined as the area average of the *local* heat transfer coefficient h as

$$\bar{h} = \frac{1}{A_s} \int_{Surface} h \, dA_s \tag{4.8}$$

So, for extended bodies such as fins, cylinders, etc., the heat transfer coefficient that is reported is typically the average heat transfer coefficient $\bar{h}$, which can be used in Newton's law.

4.3.3 Methods of estimating the heat transfer coefficient

Whether local or average, estimating the heat transfer coefficient is the central task of convection. For a given situation, this may be done using one of the three following approaches.

1. **Experimentally**—As briefly mentioned earlier, historically, h was determined experimentally. One method an experimentalist could use would be to supply some known power at the wall to maintain its temperature. This would effectively measure Q, the heat transfer due to convection that the power source needs to supply. The key point here is to quantify the losses, as the heat transfer to the fluid will always be smaller than the input Q. Temperature measuring devices like thermocouples or thermometers at various locations would determine the temperatures. Hence, since we know Q and $\Delta T = T_w - T_\infty$ it is possible to determine h for a wide variety of situations. The accompanying uncertainties in h also need to be quantified. This approach of measuring heat transfer directly is known as experimental heat transfer.

2. **Theoretically**—We need $\dfrac{\partial T}{\partial y}$ at the wall to calculate h through Eq. (4.2). This could be done theoretically if we knew the temperature gradient at the wall. It turns out that, in general, this can be computed only if the whole temperature and velocity field can be computed analytically. Carrying out the analytical solution is possible only in simple cases. Theoretical approaches also exist for approximating h. However, these work only for fairly simple flows.

3. **Computationally**—The approach is in some ways very similar to the theoretical one in that we estimate h directly from $\dfrac{\partial T}{\partial y}$ at the wall. However, the whole field is now estimated via a computational solution of the governing equations (We will see an outline of how to do this in the chapter on numerical heat transfer). The approach is fairly general and is now a very commonly used method for practical engineering situations. However, it is not good to completely rely on computations, and some experimental validation is always desirable in practice.

If we wish to determine h via the theoretical or computational route, we will need to solve for the whole flow. For this, we need the equations for the full, convective flow.

4.4 Governing equations

Unlike the conduction case, in convection we have relative motion of the flow. So, while in conduction, we could look at the energy equation in isolation, as velocity did not play a role in it, in convection, we can no longer decouple the energy equation from the flow equations. As you will see, velocity makes an appearance in the full energy equation with convective terms. Therefore, all conservation equations need to be dealt with simultaneously, and hence we need the equations of mass, momentum, and energy.

4.4.1 General approach to conservation laws

Our governing equations (often called conservation equations) are balance laws. We will be deriving all conservation laws from a control volume perspective, where we look at a fixed region in space and account for the influx and efflux of our quantity of interest. Heuristically, our balance equations look as follows:

$$\text{Rate of change of quantity} = \text{Influx} - \text{Efflux} + \text{Source} \qquad (4.9)$$

We now apply this balance equation to mass, momentum, and energy. While this is a mathematical equation, notice in the derivations below how our knowledge of physics and domain-specific knowledge (such as Fourier's Law, Newton's law of viscosity, etc.) comes in how we express the influx, efflux, and source terms.

NOTE: Some readers may find the derivations terse as well as tedious. They may skip directly to the summary of the equations at the end of the section. However, our opinion is that practitioners will gain a better understanding of the physical significance of the terms and also the limitations of the equations by going through the derivations carefully and noting the assumptions made.

4.4.2 Law of conservation of mass

Consider a two-dimensional rectangular control volume of fluid of dimensions Δx and Δy with unit dimension in the direction perpendicular to the plane of the paper, as shown in the inset of Fig. 4.4A.

Let us apply our balance equation to the conservation of mass for this control volume. Assuming no sources of mass within the volume, we obtain

$$\frac{\partial m}{\partial t} = \sum_{in} \dot{m} - \sum_{out} \dot{m} \qquad (4.10)$$

From Fig. 4.4B, the terms in Eq. (4.10) can be written out as,

$$\frac{\partial \rho}{\partial t} \Delta x \Delta y = -\left(\left(\rho u + \frac{\partial(\rho u)}{\partial x} \Delta x \right) \Delta y + \left(\rho v + \frac{\partial(\rho v)}{\partial y} \Delta y \right) \Delta x \right) + (\rho u \Delta y + \rho v \Delta x) \qquad (4.11)$$

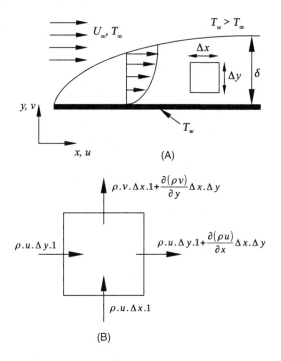

FIGURE 4.4

(A) Typical control volume employed in the derivation of the continuity equation for a two-dimensional flow, and (B) enlarged view of the control volume.

Rearranging and canceling terms, we obtain

$$\frac{\partial \rho}{\partial t} + \frac{\partial(\rho u)}{\partial x} + \frac{\partial(\rho v)}{\partial y} = 0 \tag{4.12}$$

This is the conservation of mass equation (also known as the *continuity* equation) for a general, compressible fluid. For an incompressible flow, we can approximate the density to be a constant. So the above equation is simplified as follows:

$$\frac{\partial(u)}{\partial x} + \frac{\partial(v)}{\partial y} = 0 \tag{4.13}$$

This is the continuity equation for a two-dimensional, incompressible flow in Cartesian coordinates. Similarly, for a three-dimensional flow in Cartesian coordinates, the equations for a compressible flow are

$$\frac{\partial \rho}{\partial t} + \frac{\partial(\rho u)}{\partial x} + \frac{\partial(\rho v)}{\partial y} + \frac{\partial(\rho w)}{\partial z} = 0 \tag{4.14}$$

For a three-dimensional, incompressible case

$$\frac{\partial(u)}{\partial x} + \frac{\partial(v)}{\partial y} + \frac{\partial(w)}{\partial z} = 0 \tag{4.15}$$

If we change coordinate systems to cylindrical coordinates, we will obtain, for incompressible flow

$$\frac{\partial(v_r)}{\partial r} + \frac{v_r}{r} + \frac{1}{r}\frac{\partial(v_\theta)}{\partial \theta} + \frac{\partial(v_z)}{\partial z} = 0 \tag{4.16}$$

The continuity equation in spherical coordinates (r, θ, ϕ) can be similarly derived. This is left as an exercise to the reader.

4.4.3 Momentum equations

The law governing momentum balance is Newton's second law; that is, the rate of change of momentum is equal to the net external forces. For the momentum equation, the source term in Eq. (4.9) will be the (vectorial) sum of all the external forces $\sum \vec{F}_{ext}$.

So the balance Eq. (4.9) applied to momentum balance becomes

$$\frac{\partial(m\vec{V})}{\partial t} = \sum_{in} \dot{m}\vec{V} - \sum_{out} \dot{m}\vec{V} + \sum \vec{F}_{ext} \tag{4.17}$$

Note that this is a vector equation and has two components—the x and y momentum equations. We use the terminology u, v for the x, y components of the velocity respectively. Let us now look at each term individually in the x momentum equation.

Consider Fig. 4.5. The net rate of change of x momentum in the control volume is given by

$$\frac{\partial(mu)}{\partial t} = \frac{\partial(\rho u)}{\partial t} \Delta x.\Delta y.1 \tag{4.18}$$

Net flux of momentum in the control volume is given by

$$\sum_{in} \dot{m}u - \sum_{out} \dot{m}u = -\frac{\partial(\rho u.u)}{\partial x} \Delta x.\Delta y.1 - \frac{\partial(\rho u.v)}{\partial y} \Delta x.\Delta y.1 \tag{4.19}$$

A key term is the sum of external forces acting on the control volume. In a *moving* fluid, there are two possible sources of external force in a control volume:

1. **Surface forces**—These are forces that act by contact and consist of the normal and shear stresses. These are denoted, respectively, by σ and τ. Physically, the origins of the normal stresses are from pressure and the viscous forces, whereas the shear stresses originate purely from the viscous force.

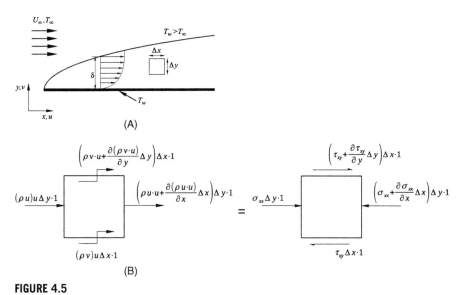

FIGURE 4.5

(A) Typical control volume employed in the derivation of the momentum equation for a two-dimensional flow; (B) the conservation of momentum in x-direction for an infinitesimal two-dimensional control volume.

2. **Body forces**—These are forces that act from a distance. This is denoted by $\vec{X}$ here. The most common force of this sort is the gravitational force. In flows involving electromagnetic fields, this may be due to electromagnetic force as well.

Fig. 4.6 shows the surface forces acting in the x-direction. Using this and Fig. 4.5, we may write the net external force in the x-direction as,

$$\sum (F_x) = \frac{\partial(\sigma_{xx})}{\partial x}\Delta x.\Delta y.1 + \frac{\partial(\tau_{xy})}{\partial x}\Delta x.\Delta y.1 + X.\Delta x.\Delta y.1 \tag{4.20}$$

Substituting terms and rearranging, we obtain

$$\frac{\partial \rho u}{\partial t} + \frac{\partial \rho u.u}{\partial x} + \frac{\partial \rho v.u}{\partial y} = \frac{\partial \sigma_{xx}}{\partial x} + \frac{\partial \tau_{xy}}{\partial y} + X \tag{4.21}$$

It is important to note here what this equation denotes.

1. $\dfrac{\partial \rho u}{\partial t} + \dfrac{\partial \rho u.u}{\partial x} + \dfrac{\partial \rho v.u}{\partial y}$ is the inertia of the fluid element in the x-direction. The first term represents the local rate of change of momentum, whereas the other two terms are due to advection of momentum.

2. $\dfrac{\partial \sigma_{xx}}{\partial x} + \dfrac{\partial \tau_{xy}}{\partial y}$ is the stress on the fluid element or control volume due to normal and shear forces in the x-direction.

3. X is the body force per unit volume in the x-direction.

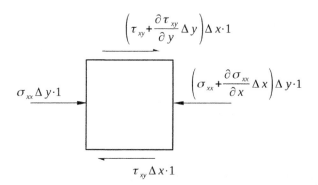

$$\left(\tau_{xy}+\frac{\partial \tau_{xy}}{\partial y}\Delta y\right)\Delta x\cdot 1$$

$$\left(\sigma_{xx}+\frac{\partial \sigma_{xx}}{\partial x}\Delta x\right)\Delta y\cdot 1$$

$$\sigma_{xx}\Delta y\cdot 1$$

$$\tau_{xy}\Delta x\cdot 1$$

FIGURE 4.6

The conservation of momentum in x-direction for an infinitesimal two-dimensional control volume.

In summary, the equation physically says that the inertial forces on the fluid are balanced by the stresses and body forces acting upon it.

This equation, unfortunately, is still unusable in correct form because of some unknown terms—X, σ, and τ, which are not in terms of the variables we already know. X is the body force per unit volume and usually does not pose a problem because it is a property we specify based on the nature of the body force. The real challenge before us is to relate σ_{xx} and τ_{xy} to u and v or their derivatives so that u,v remain the only unknowns of the equation. This relation depends on the nature of the fluid's molecular arrangement and is hence called a constitutive relationship.

The most common type of fluid is called a Newtonian fluid (water, oil, etc., are examples). This is a fluid in which the stress is proportional to rate of deformation or strain. For the normal stress, this is mathematically given by

$$\sigma_{xx}+P=2\mu\frac{\partial u}{\partial x}-\frac{2}{3}\mu\left(\frac{\partial u}{\partial x}+\frac{\partial v}{\partial y}\right) \tag{4.22}$$

NOTE: The appearance of 2/3 in the above equation is referred to as the Stokes' hypothesis and is not always accurate for compressible flow. For incompressible flows, however, conservation of mass gives $\frac{\partial u}{\partial x}+\frac{\partial v}{\partial y}=0$ and we obtain $\sigma_{xx}+P=2\mu\frac{\partial u}{\partial x}$, making Stokes' hypothesis irrelevant.

The shear stress relationship for Newtonian fluids is

$$\tau_{xy}=\mu\left(\frac{\partial u}{\partial y}+\frac{\partial v}{\partial x}\right) \tag{4.23}$$

Substituting for the normal and shear stresses in Eq. (4.21) and rearranging, the X momentum equation becomes

$$\rho\left(u\frac{\partial u}{\partial x}+v\frac{\partial u}{\partial y}\right)=\mu\left(\frac{\partial^2 u}{\partial x^2}+\frac{\partial^2 u}{\partial y^2}\right)-\frac{\partial P}{\partial X}+X \tag{4.24}$$

By an exactly similar procedure, the Y-momentum equation can be derived as follows,

$$\rho\left(\frac{\partial v}{\partial t}+u\frac{\partial v}{\partial x}+v\frac{\partial v}{\partial y}\right)=\mu\left(\frac{\partial^2 v}{\partial x^2}+\frac{\partial^2 v}{\partial y^2}\right)-\frac{\partial P}{\partial Y}+Y \tag{4.25}$$

Eqs. (4.24) and (4.25) together are frequently referred to as the Navier-Stokes equations. Once again, it is worthwhile to understand the physical significance of the terms. The left hand signifies the inertial forces. The right-hand side has a sum of viscous, pressure, and body forces respectively. Therefore, the momentum equation is, in essence, a balance equation for the inertial, viscous, pressure, and body forces.

4.4.4 Energy equation

Applying the balance equation to the energy equation, we obtain

$$\Delta x \Delta y \frac{\partial \rho e}{\partial t}=E_{in}-E_{out}+E_{source} \tag{4.26}$$

Let us take this term-by-term.

The LHS $\frac{\partial \rho e}{\partial t}$ represents the rate of change of energy in the control volume. Per unit mass, E represents the total energy that is given by $e+\frac{V^2}{2}$, where e is the internal energy and $\frac{V^2}{2}=\frac{u^2+v^2}{2}$ is

$$\frac{\partial(\rho e)}{\partial t}=\frac{\partial \rho\,(e+V^2/2)}{\partial t} \tag{4.27}$$

As represented in Fig. 4.7, physically the energy exchange on the boundaries of the volume happens due to conduction and advection.

Therefore,

$$E_{in}=E_{cond,x}+E_{adv,x}+E_{cond,y}+E_{adv,y} \tag{4.28}$$

$$E_{out}=E_{cond,x+\Delta x}+E_{adv,x+\Delta x}+E_{cond,y+\Delta y}+E_{adv,y+\Delta y} \tag{4.29}$$

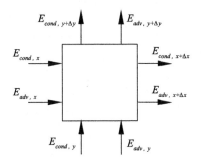

FIGURE 4.7

The conservation of energy in an infinitesimal two-dimensional control volume.

So,

$$E_{in} - E_{out} = \left(E_{cond,x} - E_{cond,x+\Delta x}\right) + \left(E_{adv,x} - E_{adv,x+\Delta x}\right)$$
$$+ \left(E_{cond,y} - E_{cond,y+\Delta y}\right) + \left(E_{adv,y} - E_{adv,y+\Delta y}\right) \tag{4.30}$$

The conduction terms can now be further expanded as

$$E_{cond,x} = -k\frac{\partial T}{\partial x}\bigg|_x \Delta y \tag{4.31}$$

$$E_{cond,x+\Delta x} = -k\frac{\partial T}{\partial x}\bigg|_{x+\Delta x} \Delta y$$
$$= -k\frac{\partial T}{\partial x}\bigg|_x \Delta y + \frac{\partial}{\partial x}\left(-k\frac{\partial T}{\partial x}\bigg|_x\right)\Delta x\ \Delta y \tag{4.32}$$

So considering conduction terms alone

$$E_{in} - E_{out}(conduction) = \frac{\partial}{\partial x}\left(k\frac{\partial T}{\partial x}\right)\Delta x \Delta y + \frac{\partial}{\partial y}\left(k\frac{\partial T}{\partial y}\right)\Delta x \Delta y \tag{4.33}$$

Now let us consider the advection terms.

$$E_{adv,x} = (\rho u \Delta y)\left(e + \frac{V^2}{2}\right) \tag{4.34}$$

Physically, the first multiplicative term in the **RHS** of the above equation represents the mass and the second term represents the specific energy. Now, by following an exercise similar to conduction, we obtain for advection

$$E_{in} - E_{out}(advection) = -\frac{\partial}{\partial x}\left(\rho u\left(e + \frac{u^2}{2}\right)\right)\Delta y\ \Delta x - \frac{\partial}{\partial y}\left(\rho v\left(e + \frac{v^2}{2}\right)\right)\Delta y\ \Delta x \tag{4.35}$$

We now move on to the source terms E_{source}. These sources are internal sources of heat generation; work done by the body forces; and normal and shear stresses.

As earlier, the internal sources of heat generation per unit volume are denoted by q_v. The work done by the body forces is given by $\vec{X} \cdot \vec{V} = Xu + Yv$
The work done by the normal and shear stresses is given by $W_{net,x} + W_{net,y}$ where

$$W_{net,x} = \frac{\partial}{\partial x}(\sigma_{xx}u)\Delta x \Delta y + \frac{\partial}{\partial y}(\tau_{xy}u)\Delta x \Delta y \tag{4.36}$$

$$W_{net,y} = \frac{\partial}{\partial y}(\sigma_{yy}v)\Delta x \Delta y + \frac{\partial}{\partial x}(\tau_{xy}v)\Delta x \Delta y \tag{4.37}$$

Substituting all the individual terms in the balance equation, we obtain

$$\frac{\partial}{\partial t}\rho(e+V^2/2)+\frac{\partial}{\partial x}\left(\rho u\left(e+\frac{u^2+v^2}{2}\right)\right)+\frac{\partial}{\partial y}\left(\rho v\left(e+\frac{u^2+v^2}{2}\right)\right)$$

$$=\frac{\partial}{\partial x}\left(k\frac{\partial T}{\partial x}\right)+\frac{\partial}{\partial y}\left(k\frac{\partial T}{\partial y}\right)+Xu+Yv+q_v \tag{4.38}$$

$$+\frac{\partial}{\partial x}(\sigma_{xx}u)+\frac{\partial}{\partial y}(\tau_{xy}u)+\frac{\partial}{\partial y}(\sigma_{yy}v)+\frac{\partial}{\partial x}(\tau_{xy}v)$$

The physical meaning of the terms is as follows:

- $\frac{\partial}{\partial t}\rho(e+V^2/2)$ is the local time rate of change of the total energy stored in the control volume.

- $\frac{\partial}{\partial x}\left(\rho u\left(e+\frac{u^2+v^2}{2}\right)\right)+\frac{\partial}{\partial y}\left(\rho v\left(e+\frac{u^2+v^2}{2}\right)\right)$ is the energy flux due to advection.

- $\frac{\partial}{\partial x}\left(k\frac{\partial T}{\partial x}\right)+\frac{\partial}{\partial y}\left(k\frac{\partial T}{\partial y}\right)$ is the conductive flux, which you would recognize from the energy equation in the conduction chapter as well.

- $Xu+Yv$ is the work done by the body forces.

- $\frac{\partial}{\partial x}(\sigma_{xx}u)+\frac{\partial}{\partial y}(\sigma_{yy}v)$ is the work done by the normal forces.

- $\frac{\partial}{\partial y}(\tau_{xx}u)+\frac{\partial}{\partial x}(\tau_{xy}v)$ is the work done by the shear forces.

Eq. (4.38) is not yet usable. We need to

- Apply a constitutive law for σ and τ. We will assume the Newtonian fluid with Stokes' hypothesis.
- Make any required assumptions about the thermal conductivity. We will assume that k is a constant.
- Make assumptions about compressibility. We will assume that the flow is incompressible.
- For an incompressible flow, we can also apply $c_p = c_v = c$
- Incorporate the fact that the flow field also satisfies the continuity and momentum equations and make any further simplifications of terms.

On making all the above assumptions, and after a couple of pages of tedious algebra, it is possible to reduce the energy equation to the following from[a]

$$\rho C_p\left(\frac{\partial T}{\partial t}+u\frac{\partial T}{\partial x}+v\frac{\partial T}{\partial y}\right)=k\nabla^2 T+q_v+\mu\phi \tag{4.39}$$

where

$$\phi=\left(\frac{\partial u}{\partial y}+\frac{\partial v}{\partial x}\right)^2+2\left(\left(\frac{\partial u}{\partial x}\right)^2+\left(\frac{\partial v}{\partial y}\right)^2\right) \tag{4.40}$$

---

[a]The tedious derivation is available in many heat transfer textbooks and has been left here as an exercise to the motivated reader.

The terms in the above equation have the following physical meanings:

- The left-hand side represents the total rate of change of internal energy of a fluid element following the fluid. It is a combination of the local rate of change and the advective flux of the internal energy.
- $k\nabla^2 T$ is the energy flux due to conduction.
- q_v is the volumetric heat generation term.
- $\mu\varphi$ is the heat dissipation due to viscous forces.

In summary, the energy equation shows that the *internal energy of a fluid element changes due to a combination of conduction, heat generation, or viscous dissipation.* This is not a trivial statement. Note, for instance, that pressure or gravitational work does not play a direct role in internal energy change.

4.4.5 Summary of equations

For a two-dimensional, incompressible flow of constant property fluid the governing equations are,

$$\frac{\partial u}{\partial x} + \frac{\partial v}{\partial y} = 0 \tag{4.41}$$

$$\rho\left(\frac{\partial u}{\partial t} + u\frac{\partial u}{\partial x} + v\frac{\partial u}{\partial y}\right) = \mu\left(\frac{\partial^2 u}{\partial x^2} + \frac{\partial^2 u}{\partial y^2}\right) - \frac{\partial P}{\partial x} + X \tag{4.42}$$

$$\rho\left(\frac{\partial v}{\partial t} + u\frac{\partial v}{\partial x} + v\frac{\partial v}{\partial y}\right) = \mu\left(\frac{\partial^2 v}{\partial x^2} + \frac{\partial^2 v}{\partial y^2}\right) - \frac{\partial P}{\partial y} + Y \tag{4.43}$$

$$\rho C_p\left(\frac{\partial T}{\partial t} + u\frac{\partial T}{\partial x} + v\frac{\partial T}{\partial y}\right) = k\nabla^2 T + q_v + \mu\phi \tag{4.44}$$

These are four equations in four unknowns—namely u, v, T, and P; q_v is known in the problem a priori. Hence, the problem satisfies closure in the mathematical sense. Given the appropriate initial and boundary conditions, this is self-sufficient to determine the solutions to any convection problem.

4.5 Summary

This chapter covers the concepts of the heat transfer coefficient, the full equations of fluid motion, and the physical significance of the terms. Despite having the full equations, a complete picture of convection is impossible without an understanding of boundary layer theory. We will be looking at boundary layer theory along with its application to forced convection in the next chapter.

References

Pritchard, P.J., John, W., 2016. Fox and McDonald's Introduction to Fluid Mechanics. Wiley, New York.

Incropera, F.P., Lavine, A.S., Bergman, T.L., DeWitt, D.P., 2007. Fundamentals of Heat and Mass Transfer. Wiley, New York.

Forced convection

5

5.1 Introduction

In the previous chapter, the basic ideas of convection heat transfer and heat transfer coefficient were introduced, along with the full set of governing equations of fluid flow and heat transfer. The physical significance of each of the terms in these equations was elucidated. In this chapter, we first introduce the boundary layer theory through the order of magnitude analysis, followed by nondimensionalization of the governing equations. Next, the integral method is developed for flow over a flat plate. Further, correlations for cylinder and sphere are presented. These are followed by a presentation of forced convection in tubes and ducts. The analytical solution is presented for a simple laminar case, and the chapter ends with a quick introduction to turbulent flow and correlations for these.

5.2 Approximation using order of magnitude analysis

In this section, we look at Prandtl's famous order of magnitude analysis leading to the boundary layer theory, which is one of the cornerstones of fluid mechanics and convective heat transfer. Even before this analysis, Prandtl argued that near the wall, the product of $\mu(\partial u/\partial y)$ cannot be neglected even though μ is very small, as $\partial u/\partial y$ will be very high near the wall so that the product of the two is not a quantity that can be neglected. This announced the birth of fluid mechanics.

The governing equations for two-dimensional, laminar, incompressible flow were derived in the previous chapter. For the case of steady flow, the governing equations are

Continuity equation

$$\frac{\partial u}{\partial x} + \frac{\partial v}{\partial y} = 0 \tag{5.1}$$

x-momentum equation

$$\rho\left(u\frac{\partial u}{\partial x} + v\frac{\partial u}{\partial y}\right) = \mu\left(\frac{\partial^2 u}{\partial x^2} + \frac{\partial^2 u}{\partial y^2}\right) - \frac{\partial P}{\partial x} + X \tag{5.2}$$

y-momentum equation

$$\rho\left(u\frac{\partial v}{\partial x} + v\frac{\partial v}{\partial y}\right) = \mu\left(\frac{\partial^2 v}{\partial x^2} + \frac{\partial^2 v}{\partial y^2}\right) - \frac{\partial P}{\partial y} + Y \tag{5.3}$$

Heat Transfer Engineering. http://dx.doi.org/10.1016/B978-0-12-818503-2.00005-8

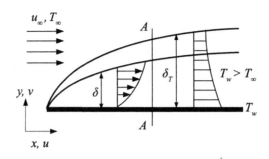

FIGURE 5.1

Boundary layer over a flat plate.

We now perform an order of magnitude analysis of these equations by introducing the concept of the boundary layer. We neglect body forces X and Y in the analysis. Before we begin, we need to understand what exactly a boundary layer is. Let us take an example of a flow parallel over a flat plate, as shown in Fig. 5.1.

Over the surface of the plate, at any section, say AA, starting from the leading edge (i.e., at $x = 0$), we can intuit that the velocity increases from $u = 0$ at $y = 0$ to $u \rightarrow u_\infty$, as y $\rightarrow \infty$. If we take a cutoff of $0.99\ u_\infty$, and see the value of y at that point, for a particular x, that value of y is termed as δ, the boundary layer thickness, which should be a function of x, so that without any loss of generality, we denote it as $\delta(x)$. The locus of all these δ's at various values of x is known as the boundary layer. The region outside the boundary layer is free from the effects of the viscous shear that is present at the wall. The region within the boundary layer is thus the region where a free stream moving at u_∞ is "braked" by the presence of the wall. The key quantity that determines the amount of braking and the pressure loss and attendant required pumping power thereof is the dynamic viscosity of the fluid, μ. From the preceding discussion, it is also clear that some sort of simplification to the governing equations should be possible if we can conjecture that $\delta/x \ll 1$. Stated explicitly, if the plate is, say, 1 m long, δ at $x = 1$ m is of the order of a few mm, under this assumption. This is a game-changing assumption, and we will see how this results in dramatic simplification of the governing equations. Let us perform an order of magnitude analysis. The scales for length x, y, and velocity u are L, δ, and u_∞, respectively. From the scales of x and y, we have

$$\delta \ll L \qquad (5.4)$$

All these scales ensure that x/L, y/δ, u/u_∞ vary from 0 to 1. Substituting for the above scales in the continuity equation (Eq. (5.1)) leads to the condition that u_∞/L and v/δ have to be comparable.

From the above, we can infer that the scale for the velocity in the y direction is

$$v \approx \frac{u_\infty \delta}{L} \qquad (5.5)$$

Substituting for slenderness assumption $\delta \ll L$ (Eq. (5.4)) in Eq. (5.5), we have

$$v \ll u_\infty \qquad (5.6)$$

Hence, the v velocity is much less compared to the u velocity. This is the first key result of the scaling or order of magnitude analysis. Furthermore, as the length scales of x and y are L and δ respectively, one can infer that

$$\frac{\partial u}{\partial y} \gg \frac{\partial u}{\partial x} \qquad (5.7)$$

$$\frac{\partial v}{\partial y} \gg \frac{\partial v}{\partial x} \qquad (5.8)$$

Moreover, as $u \gg v$, this leads to

$$\frac{\partial u}{\partial y} \gg \frac{\partial v}{\partial y} \qquad (5.9)$$

Looking at the scales for the x-momentum equation term by term, we have

$$\rho\left(u_\infty \frac{u_\infty}{L},\ u_\infty \frac{u_\infty}{L}\right) = \mu\left(\frac{u_\infty}{L^2},\ \frac{u_\infty}{\delta^2}\right) - \frac{\partial P}{\partial x} \qquad (5.10)$$

Let us look at the right-hand side. Since scales for pressure are as yet not defined by us, we retain the terms as they are.

We know that $\delta \ll L$, and this leads to

$$\frac{\partial^2 u}{\partial y^2} \gg \frac{\partial^2 u}{\partial x^2} \qquad (5.11)$$

Hence, the x-momentum equation can be simplified as

$$\rho\left(u\frac{\partial u}{\partial x} + v\frac{\partial u}{\partial y}\right) = \mu \frac{\partial^2 u}{\partial y^2} - \frac{\partial P}{\partial x} \qquad (5.12)$$

Now, consider the y-momentum equation (Eq. 5.3). The order of the equation is $u_\infty . u_\infty \delta / L^2$. The order of the x-momentum equation is $u_\infty . u_\infty / L$. Therefore, the y-momentum equation has an order of δ / L times the x-momentum equation. However, $\delta / L \ll 1$ and, in view of this, the y–momentum equation can be neglected in relation to the x-momentum equation. We now have a key piece of information about the pressure, which is as follows.

$$\frac{\partial p}{\partial y} = 0 \qquad (5.13)$$

Let us now apply the x-momentum equation to the free stream, where

$$u = u_\infty \qquad (5.14)$$

$$\frac{\partial u}{\partial x} = 0 \qquad (5.15)$$

$$v = 0 \tag{5.16}$$

$$\frac{\partial^2 u}{\partial y^2} = 0 \tag{5.17}$$

In view of all this,

$$\frac{\partial p}{\partial x} = \frac{dp}{dx} = 0. \tag{5.18}$$

Now, at any vertical section, say, AA along x (see Fig. 5.1), if we traverse an instrument to measure the variation of pressure, since $\partial p / \partial y = 0$ at all points on section AA, as we traverse from the free stream to the inside of the boundary layer, the pressure must be the same. Because $dp/dx = 0$ at the outside boundary layer with $p = p_\infty$, it stands to reason that at the outer edge of the boundary layer too, $p = p_\infty$, and by extending the above arguments, $\partial p / \partial x, dp/dx = 0$ within the boundary layer too. So, by a clever use of the information that $\partial p / \partial y$ is zero within the boundary layer and that p_∞ is a constant outside the boundary layer, we have achieved one more major simplification. In fact, this is equally profound as now pressure is no longer a variable in the boundary layer equations. Hence $dp/dx = 0$ within the boundary layer too. Please note this in no way means that $\Delta p = 0$. This simplification simply allows us to ignore the pressure gradient term in an analysis of boundary flow over a flat plate. If, on the other hand, the free stream velocity u_∞ is not a constant, but a function of x, as in the case of flow over variable area geometries like a wedge, then dp/dx within the boundary can still be found from the dp/dx at the free stream, which itself can be determined using the Bernouli's equation in the free stream. In short, the pressure distribution in the free stream is imposed within the boundary layer. This fact is used for rapid calculations of net pressure forces in aerodynamic flows.

The energy equation was given in Chapter 4 as eqn. 4.39. For the case of steady flow with no heat generation, it reduces to

$$\rho c_p \left[\frac{\partial T}{\partial x} + \frac{\partial T}{\partial y} \right] = k\nabla^2 T + \mu \phi \tag{5.19}$$

With the understanding that δ_T is the thermal boundary layer thickness (see Fig. 5.1) is much less than L, we have

$$\frac{\partial T}{\partial y} \gg \frac{\partial T}{\partial x} \tag{5.20}$$

The thermal boundary layer thickness, δ_T, is defined as the value of y, where

$$T = T_\infty + 0.01(T_w - T_\infty)$$

With regard to the viscous dissipation term,

$$\frac{\partial u}{\partial y} \gg \frac{\partial v}{\partial x} \tag{5.21}$$

$$\frac{\partial u}{\partial y} \gg \frac{\partial u}{\partial x} \tag{5.22}$$

$$\frac{\partial u}{\partial y} \gg \frac{\partial v}{\partial y} \tag{5.23}$$

In view of this, the viscous dissipation term becomes $\mu\left(\partial u/\partial y\right)^2$.

Following the order of magnitude analysis, the governing equations are simplified as follows for the case of negligible viscous dissipation:

Continuity equation

$$\frac{\partial u}{\partial x}+\frac{\partial v}{\partial y}=0 \qquad (5.24)$$

Momentum equation

$$u\frac{\partial u}{\partial x}+v\frac{\partial u}{\partial y}=v\frac{\partial^2 u}{\partial y^2} \qquad (5.25)$$

Energy equation

$$\rho C_p\left(u\frac{\partial T}{\partial x}+v\frac{\partial T}{\partial y}\right)=k\frac{\partial^2 T}{\partial y^2}+\mu\left(\frac{\partial u}{\partial y}\right)^2 \qquad (5.26)$$

The above equations are referred to as the boundary layer equations. Eqs. (5.24–5.26) are three equations in three unknowns, u, v, and T, and hence they satisfy closure. Mathematically, Eqs. (5.25) and (5.26) are parabolic in nature and are easier to solve compared to the original N-S equations, and the equation of energy (without simplification).

5.3 Nondimensionalization of the governing equations

We now move on to nondimensionalizing or normalizing the boundary layer equations. To make matters simple, we drop the viscous dissipation term in Eq. 5.26, which becomes significant only for high speed flows. The motivation for this nondimensionalization is to extract the pertinent dimensionless parameters that govern the fluid flow and heat transfer. Pertinent dimensionless parameters can also be obtained using the Buckingham's Pi theorem.

We now normalize (or nondimensionalize) the governing equations as follows.

Let

$$u^+=\frac{u}{u_\infty} \qquad (5.27)$$

$$v^+=\frac{v}{u_\infty} \qquad (5.28)$$

$$x^+=\frac{x}{L} \qquad (5.29)$$

$$y^+=\frac{y}{L} \qquad (5.30)$$

$$\phi=\frac{\left(T-T_\infty\right)}{\left(T_w-T_\infty\right)} \qquad (5.31)$$

and

$$p^+ = \frac{p}{\rho u_\infty^2}$$

(5.32)

Using the above, the continuity equation becomes

$$\frac{\partial u^+}{\partial x^+} + \frac{\partial v^+}{\partial y^+} = 0$$

(5.33)

The momentum equation becomes

$$\frac{u_\infty^2}{L} \cdot u^+ \frac{\partial u^+}{\partial x^+} + \frac{u_\infty^2}{L} \cdot v^+ \frac{\partial u^+}{\partial y^+} = -\frac{1}{\rho} \cdot \frac{\rho u_\infty^2}{L} \cdot \frac{dp^+}{dx^+} + \frac{\mu}{\rho} \cdot \frac{u_\infty}{L^2} \cdot \frac{\partial^2 u^+}{\partial y^{+2}}$$

(5.34)

Upon simplification, we get

$$u^+ \frac{\partial u^+}{\partial x^+} + v^+ \frac{\partial u^+}{\partial y^+} = -\frac{dp^+}{dx^+} + \frac{v}{u_\infty L} \cdot \frac{\partial^2 u^+}{\partial y^{+2}}$$

(5.35)

where

$$\frac{v}{u_\infty L} = \frac{1}{Re_L}$$

(5.36)

$\therefore$ The x-momentum equation turns out to be

$$u^+ \frac{\partial u^+}{\partial x^+} + v^+ \frac{\partial u^+}{\partial y^+} = -\frac{dp^+}{dx^+} + \frac{1}{Re_L} \cdot \frac{\partial^2 u^+}{\partial y^{+2}}$$

(5.37)

Similarly, the boundary layer energy equation becomes

$$u^+ \frac{\partial \phi}{\partial x^+} + v^+ \frac{\partial \phi}{\partial y^+} = \frac{k}{\rho c_p u_\infty L} \cdot \frac{\partial^2 \phi}{\partial y^{+2}}$$

(5.38)

$$\frac{k}{\rho c_p u_\infty L} = \frac{\alpha v}{v u_\infty L} = \frac{1}{\frac{v}{\alpha} \cdot \frac{u_\infty L}{v}} = \frac{1}{Re_L Pr}$$

(5.39)

$\therefore$ The energy equation becomes

$$u^+ \frac{\partial \phi}{\partial x^+} + v^+ \frac{\partial \phi}{\partial y^+} = \frac{1}{Re_L Pr} \cdot \frac{\partial^2 \phi}{\partial y^{+2}}$$

(5.40)

In the light of the above results, the following becomes evident:

$$u^+ = f_1\left(x^+, y^+, Re_L, \frac{dp^+}{dx^+}\right)$$

(5.41)

The shear stress at the wall (τ_w) is given by

$$\tau_w = \mu\left(\frac{\partial u}{\partial y}\right)_{y=0}$$

(5.42)

$$\mu\left(\frac{\partial u}{\partial y}\right)_{y=0} = \frac{\mu u_\infty}{L}\left(\frac{\partial u^+}{\partial y^+}\right)_{y^+=0} \tag{5.43}$$

The local skin friction coefficient is given by

$$c_{f,x} = \frac{\tau_w}{\frac{1}{2}\rho u_\infty^2} \tag{5.44}$$

$$c_{f,x} = \frac{\frac{\mu u_\infty}{L}\left(\frac{\partial u^+}{\partial y^+}\right)_{y^+=0}}{\frac{1}{2}\rho u_\infty^2} \tag{5.45}$$

$$c_{f,x} = \frac{2}{Re_L}\left(\frac{\partial u^+}{\partial y^+}\right)_{y^+=0} \tag{5.46}$$

From the functional form of u^+ given in Eq. (5.41)

$$c_{f,x} = f_2\left(x^+, Re_L, \frac{dp^+}{dx^+}\right) \tag{5.47}$$

For a flat plate $\dfrac{dp^+}{dx^+} = 0$

$$\therefore c_{f,x} = f_2(x^+, Re_L) \tag{5.48}$$

For any other given geometry $\dfrac{dp^+}{dx^+}$ can be obtained from the Bernouli's equation.

$$\therefore c_{f,x} = \frac{2}{Re_L} f_2\left(x^+, Re_L\right) \tag{5.49}$$

When we take an average from $x = 0$ to $x = L$ to obtain the average shear stress and the average skin friction coefficient, the x^+ dependence also vanishes!

$$\tau_w = f_3\left(Re_L\right) \text{ alone} \tag{5.50}$$

This is indeed a revelation after so much toil.
Similarly, if we consider the dimensionless form of the energy equation

$$\phi = f_4\left(x^+, y^+, Re_L, Pr, \frac{dp^+}{dx^+}\right) \tag{5.51}$$

For a flat plate, $\dfrac{dp^+}{dx^+} = 0$

$$h(T_w - T_\infty) = -k_f\left(\frac{\partial T}{\partial y}\right)_{y=0} \quad (\text{At the wall, } q_{conv} = q_{cond}) \tag{5.52}$$

$$h(T_w - T_\infty) = -\frac{k_f(T_w - T_\infty)}{L}\left(\frac{\partial \phi}{\partial y^+}\right)_{y^+=0} \tag{5.53}$$

$$\therefore h = -\frac{k_f}{L}\left(\frac{\partial \phi}{\partial y^+}\right)_{y^+=0}$$

(5.54)

$$\frac{Nu_x k_f}{L} = -\frac{k_f}{L}\left(\frac{\partial \phi}{\partial y^+}\right)_{y^+=0}$$

(5.55)

where Nu_x is the local Nusselt number (the dimensionless heat transfer coefficient) given by hL/k_f. The local Nusselt number can also be defined based on x as the length scale. Here, it is based on L for convenience.

$$\therefore Nu_x = f_5\left(x^+, R_{eL}, Pr\right)$$

(5.56)

The Nusselt number averaged over the length L, known as the mean or average Nusselt number is given by

$$\overline{Nu_L} = f_6\left(Re_L, Pr\right)$$

(5.57)

The above development is profound as it has helped us obtain key dimensionless parameters in convective fluid flow and heat transfer. Even if one is an experimentalist and wants to stay away from computing to solve the Navier-Stokes equations and the energy equation, the nondimensionalization procedure reveals to us the contours of how the skin friction coefficient and Nusselt number correlations will turn out to be.

5.4 Approximate solution to the boundary layer equations

Consider the momentum equation in the x-direction inside the boundary layer (Eq. 5.25)

$$u\frac{\partial u}{\partial x} + v\frac{\partial u}{\partial y} = \frac{\mu}{\rho}\frac{\partial^2 u}{\partial y^2}$$

Integrating the above equation from 0 to δ in y, we have

$$\int_0^\delta u\frac{\partial u}{\partial x}dy + \int_0^\delta v\frac{\partial u}{\partial y}dy = \int_0^\delta \frac{\mu}{\rho}\frac{\partial^2 u}{\partial y^2}dy$$

(5.58)

$$\int_0^\delta u\frac{\partial u}{\partial x}dy + \int_0^\delta v\frac{\partial u}{\partial y}dy = -\frac{\mu}{\rho}\frac{\partial u}{\partial y}\bigg|_{y=0}$$

(5.59)

Eq. (5.59) can be written as

$$2\int_0^\delta u\frac{\partial u}{\partial x}dy + \int_0^\delta \frac{\partial(uv)}{\partial y}dy - \int_0^\delta u\frac{\partial v}{\partial y}dy - \int_0^\delta u\frac{\partial u}{\partial x}dy = -\frac{\tau_w}{\rho}$$

(5.60)

where τ_w is the shear stress at the wall. Eq. (5.60) can be rearranged as given below.

$$2\int_0^\delta u\frac{\partial u}{\partial x}dy + uv\big|_{y=\delta} - \int_0^\delta u\left(\frac{\partial u}{\partial x} + \frac{\partial v}{\partial y}\right)dy = -\frac{\tau_w}{\rho}$$

(5.61)

The term within the brackets in Eq. (5.61) is nothing but a statement of the continuity equation and as a consequence is 0. In view of this, Eq. (5.61) becomes

$$2\int_0^\delta u\frac{\partial u}{\partial x}dy + uv\big|_{y=\delta} = -\frac{\tau_w}{\rho} \tag{5.62}$$

From the definition of the boundary layer, we know that $u = u_\infty$ at $y = \delta$.

Now we need to determine the value of v at $y = \delta$.

Let us take recourse to the continuity equation Eq. (5.24)

$$\frac{\partial u}{\partial x} + \frac{\partial v}{\partial y} = 0$$

Integrating the above continuity equation from $y = 0$ to $y = \delta$, we have

$$\int_0^\delta \frac{\partial u}{\partial x}dy = -\int_0^\delta \frac{\partial v}{\partial y}dy = -v\big|_{y=\delta} - v\big|_{y=0} \tag{5.63}$$

$$\therefore v_\delta = -\int_0^\delta \frac{\partial u}{\partial x}dy \tag{5.64}$$

Substituting Eq. (5.64) in Eq. (5.62) we get

$$2\int_0^\delta u\frac{\partial u}{\partial x}dy - u_\infty \int_0^\delta \frac{\partial u}{\partial x}dy = -\frac{\tau_w}{\rho} \tag{5.65}$$

$$\int_0^\delta \frac{\partial}{\partial x}\left(u^2 - u_\infty u\right)dy = -\frac{\tau_w}{\rho} \tag{5.66}$$

From Leibniz rule, the differential and integral signs can be interchanged. Additionally, the equation for boundary layer thickness becomes only a function in just x, consequent upon integration across y. In view of the above,

$$\frac{d}{dx}\int_0^\delta u(u - u_\infty)dy = -\frac{\tau_w}{\rho} \tag{5.67}$$

The above equation is known as the integral momentum equation. Consider the energy equation for negligible viscous dissipation

$$u\frac{\partial T}{\partial x} + v\frac{\partial T}{\partial y} = \alpha\frac{\partial^2 T}{\partial y^2} \tag{5.68}$$

Integrating the above equation from 0 to δ_T in y, we have

$$\int_0^{\delta_T} u\frac{\partial T}{\partial x}dx + \int_0^{\delta_T} v\frac{\partial T}{\partial y}dy = \int_0^{\delta_T} \alpha\frac{\partial^2 T}{\partial y^2}dy \tag{5.69}$$

$$\int_0^{\delta_T} u\frac{\partial T}{\partial x}dy + \int_0^{\delta_T} v\frac{\partial T}{\partial y}dy = -\alpha\frac{\partial T}{\partial y}\bigg|_{y=0} \tag{5.70}$$

$$\int_0^{\delta_T} \frac{\partial uT}{\partial x}dy + vT\big|_{y=\delta_T} - vT\big|_{y=0} = -\alpha\frac{\partial T}{\partial y}\bigg|_{y=0} \tag{5.71}$$

$T_{y=\delta_T} = T_\infty$ and v at δ_T is given by Eq. (5.64) with changed upper limit of integration, as given below.

$$\int_0^{\delta_T} \frac{\partial uT}{\partial x}dy - T_\infty \int_0^{\delta_T} \frac{\partial u}{\partial x}dy = -\alpha \frac{\partial T}{\partial y}\Big|_{y=0} \tag{5.72}$$

$$\int_0^{\delta_T} \frac{\partial(uT - uT_\infty)}{\partial x}dy = -\alpha \frac{\partial T}{\partial y}\Big|_{y=0} \tag{5.73}$$

Using the Leibniz rule and changing $\partial/\partial x$ to be d/dx, we get the following equation.

$$\frac{d}{dx}\int_0^{\delta_T} u(T - T_\infty)dy = -\alpha \frac{\partial T}{\partial y}\Big|_{y=0} \tag{5.74}$$

Eq. (5.74) is known as the integral energy equation. The momentum integral equation (Eq. (5.67)) can also be written as

$$\frac{d}{dx}\int_0^{\delta} u(u - u_\infty)dy = -v \frac{\partial u}{\partial y}\Big|_{y=0} \tag{5.75}$$

Similarly, the integral form of the species transport equation can be derived as

$$\frac{d}{dx}\int_0^{\delta} u(\rho_A - \rho_{A,\infty})dy = -D_{AB} \frac{\partial \rho_A}{\partial y}\Big|_{y=0} \tag{5.76}$$

Eq. (5.76) is useful in the study of convective mass transfer, where D_{AB} is the binary diffusion coefficient and ρ_A is the density of the medium that is flowing and is under consideration.

Solution to integral momentum and energy equations with trial velocity and temperature profiles

Consider a linear profile for velocity

$$\frac{u}{u_\infty} = a + b\left(\frac{y}{\delta}\right) \tag{5.77}$$

The next step is to apply the boundary conditions on the assumed velocity profiles to obtain the constants in the assumed profile. The boundary conditions applicable for the problem under consideration are:

$$\text{At } y = 0, u = 0 \tag{5.78}$$

$$\text{At } y = \delta, u = u_\infty \tag{5.79}$$

Substituting these boundary conditions into Eq. (5.77), we get $a = 0$ and $b = 1$. Hence, the profile assumed in Eq. (5.77) becomes

$$\frac{u}{u_\infty} = \frac{y}{\delta} \tag{5.80}$$

Rearranging Eq. (5.75) and substituting Eq. (5.80) into it, we get

$$u_\infty \frac{d}{dx} \int_0^\delta \frac{u}{u_\infty}\left(1 - \frac{u}{u_\infty}\right) dy = \frac{v}{\delta} \tag{5.81}$$

$$u_\infty \frac{d}{dx} \int_0^\delta \frac{y}{\delta}\left(1 - \frac{y}{\delta}\right) dy = \frac{v}{\delta} \tag{5.82}$$

$$u_\infty \frac{d}{dx}\left(\frac{y^2}{2\delta}\Big|_{y=0}^{y=\delta} - \frac{y^3}{3\delta^2}\Big|_{y=0}^{y=\delta}\right) = \frac{v}{\delta} \tag{5.83}$$

$$u_\infty \frac{d}{dx}\left(\frac{\delta}{6}\right) = \frac{v}{\delta} \tag{5.84}$$

$$\frac{d}{dx}\left(\frac{\delta^2}{12}\right) = \frac{v}{u_\infty} \tag{5.85}$$

Integrating Eq. (5.85), we get the following equation.

$$\frac{\delta^2}{12} = \frac{vx}{u_\infty} + A \tag{5.86}$$

At $x = 0$, $\delta = 0$ and so $A = 0$. Therefore, the solution becomes

$$\frac{\delta^2}{12} = \frac{1}{Re_x} \tag{5.87}$$

or

$$\frac{\delta}{x} = \sqrt{\frac{12}{Re_x}} = \frac{3.464}{\sqrt{Re_x}} \tag{5.88}$$

A cubic profile for velocity yields

$$\frac{\delta}{x} = \frac{4.64}{\sqrt{Re_x}} \tag{5.89}$$

The exact solution to this by Blasius (1921), using the technique of similarity transformation (similar to the one we saw in unsteady conduction)

$$\frac{\delta}{x} = \frac{4.92}{\sqrt{Re_x}} \tag{5.90}$$

As expected, the assumption of a cubic profile for velocity gives a solution close to the exact solution of Blasius.

We now assume a cubic profile for temperature in order to get an estimate of the thermal boundary layer thickness, δ_T.

$$\frac{T - T_\infty}{T_w - T_\infty} = a + b\left(\frac{y}{\delta_T}\right) + c\left(\frac{y}{\delta_T}\right)^2 + d\left(\frac{y}{\delta_T}\right)^3 \tag{5.91}$$

There are four constants in Eq. (5.91), so we need four boundary conditions to determine the constants.

$$\text{At } y = 0, \, T = T_w \tag{5.92}$$

$$\text{At } y = \delta_T, T = T_\infty \tag{5.93}$$

$$\text{At } y = \delta_T, \frac{\partial T}{\partial y} = 0 \tag{5.94}$$

At $y = 0$, the energy equation is satisfied at the wall. On the wall, $u = 0$ and $v = 0$, so the energy equation reduces to

$$\frac{\partial^2 T}{\partial y^2} = 0 \tag{5.95}$$

Plugging these boundary conditions into Eq. (5.91), we get the values of a, b, c, d equal to be 1, $-3/2$, 0, 1/2, respectively.

Hence, the cubic temperature profile turns out to be

$$\theta = \frac{(T - T_\infty)}{(T_w - T_\infty)} = 1 - \frac{3}{2}\left(\frac{y}{\delta_T}\right) + \frac{1}{2}\left(\frac{y}{\delta_T}\right)^3 \tag{5.96}$$

Substituting Eq. (5.96) into the integral energy equation Eq. (5.74), we get

$$\frac{d}{dx} \int_0^{\delta_T} u \left(1 - \frac{3}{2}\frac{y}{\delta_T} + \frac{1}{2}\left(\frac{y}{\delta_T}\right)^3\right)(T_w - T_\infty)dy = \alpha \frac{3}{2\delta_T}(T_w - T_\infty) \tag{5.97}$$

At this juncture, in order to get an estimate of δ_T closer to experiments, we would like to use a cubic profile for velocity, and we also know that $\delta/x = 4.64/\sqrt{Re_x}$ for this profile. (This is given as an exercise problem at the end of this chapter.) Now, if we assume a cubic profile for velocity we get

$$\frac{u}{u_\infty} = \frac{3}{2}\left(\frac{y}{\delta}\right) - \frac{1}{2}\left(\frac{y}{\delta}\right)^3 \tag{5.98}$$

Substituting Eq. (5.98) in Eq. (5.97) Upon integrating Eq. (5.99),

$$\frac{d}{dx} \int_0^{\delta_T} u_\infty \left(\frac{3}{2}\frac{y}{\delta} - \frac{9}{4}\frac{y^2}{\delta\delta_T} + \frac{3}{4}\frac{y^4}{\delta\delta_T^3} - \frac{1}{2}\frac{y^3}{\delta^3} + \frac{3}{4}\frac{y^4}{\delta^3\delta_T} - \frac{1}{4}\frac{y^6}{(\delta\delta_T)^3}\right)dy = \frac{3\alpha}{2\delta_T} \tag{5.99}$$

$$\frac{d}{dx}\left(\frac{3\delta_T^2}{4\delta} - \frac{9\delta_T^3}{12\delta\delta_T} + \frac{3\delta_T^5}{20\delta\delta_T^3} - \frac{\delta_T^4}{8\delta^3} + \frac{3\delta_T^5}{20\delta^3\delta_T} - \frac{\delta_T^7}{28(\delta\delta_T)^3}\right) = \frac{3\alpha}{2u_\infty\delta_T} \tag{5.100}$$

$$\frac{d}{dx}\left(\frac{3}{20}\frac{\delta_T^2}{\delta} - \frac{3}{280}\frac{\delta_T^4}{\delta^3}\right) = \frac{3\alpha}{2u_\infty\delta_T} \tag{5.101}$$

Now, we consider a situation where the following holds.

$$\frac{\delta_T}{\delta} < 1 \tag{5.102}$$

In view of this, the higher order term $\dfrac{\delta_T^4}{\delta^3}$ can be neglected in Eq. (5.101) which then reduces to

$$\frac{d}{dx}\left(\frac{\delta_T^2}{\delta}\right) = \frac{10\alpha}{u_\infty \delta_T} \tag{5.103}$$

Let $\delta_T/\delta = \varepsilon$, Eq. (5.103) then can be written as

$$\frac{d}{dx}\left(\varepsilon^2 \delta\right) = \frac{10\alpha}{u_\infty \delta \varepsilon} \tag{5.104}$$

$$\delta \frac{d\varepsilon^2}{dx} + \varepsilon^2 \frac{d\delta}{dx} = \frac{10\alpha}{u_\infty \delta \varepsilon} \tag{5.105}$$

Substituting Eq. (5.89) in Eq. (5.105),

$$4.64\sqrt{\frac{vx}{u_\infty}}\frac{d\varepsilon^2}{dx} + \varepsilon^2\left[4.64\sqrt{\frac{v}{u_\infty}}\frac{d\sqrt{x}}{dx}\right] = \frac{10\alpha}{4.64\sqrt{u_\infty x v} \times \varepsilon} \tag{5.106}$$

$$\varepsilon x\left[\frac{d\varepsilon^2}{dx} + \frac{\varepsilon^2}{2x}\right] = \frac{0.464}{Pr} \tag{5.107}$$

$$4\varepsilon^2 x\frac{d\varepsilon}{dx} + \varepsilon^3 = \frac{0.928}{Pr} \tag{5.108}$$

$$\frac{4}{3}x\frac{d\varepsilon^3}{dx} + \varepsilon^3 = \frac{0.928}{Pr} \tag{5.109}$$

The solution to the above differential equation is

$$\varepsilon^3 = Ax^{-3/4} + \frac{0.928}{Pr} \tag{5.110}$$

To get A, we need a boundary condition. At $x = 0$, ε^3 is finite. Therefore $A = 0$

$$\therefore \varepsilon^3 = \frac{0.928}{Pr} \tag{5.111}$$

$$\varepsilon = \frac{\delta_T}{\delta} = 0.975\,Pr^{-1/3} \tag{5.112}$$

Substituting for $\delta/x = 4.64/\sqrt{Re_x}$ we have

$$\frac{\delta_T}{x} = 4.64 \times 0.975\ Re^{-1/2} Pr^{-1/3} \tag{5.113}$$

$$\therefore \frac{\delta_T}{x} = 4.52\ Re^{-1/2} Pr^{-1/3} \tag{5.114}$$

The local heat transfer coefficient can now be determined as

$$q = h(T_w - T_\infty) = -k \left.\frac{\partial T}{\partial y}\right|_{y=0} \tag{5.115}$$

We can define a local Nusselt number as

$$Nu_x = \frac{hx}{k_f} \tag{5.116}$$

The Nusselt number is actually the dimensionless heat transfer coefficient, as already mentioned.

$$\frac{Nu_x k_f}{x}(T_w - T_\infty) = -k_f (T_w - T_\infty)\left(\frac{\partial \theta}{\partial y}\right)_{y=0} \tag{5.117}$$

$$Nu_x = \frac{3x}{2\delta_T} \tag{5.118}$$

$$\therefore Nu_x = 0.332 \, Re_x^{1/2} \, Pr^{1/3} \tag{5.119}$$

From Nu_x, we get an expression for h_x, integrate this from $x = 0$ to $x = L$, and average it over L to get an expression for $\overline{h_L}\left(i.e., \overline{h_L} = 1/L \int_0^L h_x \, dx\right)$.

From this, we get an expression for the average Nusselt number $\overline{Nu_L}$ as follows:

$$\overline{Nu_L} = 0.664 \, Re_x^{1/2} \, Pr^{1/3} \tag{5.120}$$

This is valid for fluids for which $Pr > 1$. We now see why this is so. We know that $\delta_T/\delta = (0.975/Pr)^{1/3}$. We have assumed earlier that $\delta/\delta_T < 1$. Hence for this assumption to be true $Pr > 1$. The above expression is also reasonable for gases with $Pr \approx 1$.

Integral method for fluids with Pr < 1

For liquid metals like liquid sodium, potassium, and mercury, the Prandtl number $Pr \ll 1$. These are highly conducting liquids with $\alpha \gg v$ and have excellent heat transport properties and are used in nuclear reactors to remove fission heat.

Consider the velocity and thermal boundary layers for flow and heat transfer over a heated flat plate. The boundary layers are shown in Fig. 5.2.

From Fig. 5.2, it is clear that for the most part, within the thermal boundary layer $u = u_\infty (\because \delta_T \gg \delta)$. Using the same cubic profile for temperature, as before, we have

$$u_\infty \frac{d}{dx} \int_0^{\delta_T} \left[1 - \frac{3}{2}\frac{y}{\delta_T} + \frac{1}{2}\left(\frac{y}{\delta_T}\right)^3\right](T_w - T_\infty)\,dy = \left(\frac{3\alpha}{2\delta_T}\right)(T_w - T_\infty) \tag{5.121}$$

$$u_\infty \frac{d}{dx}\left(\delta_T - \frac{3}{4}\delta_T + \frac{\delta_T}{8}\right) = \left(\frac{3\alpha}{2\delta_T}\right) \tag{5.122}$$

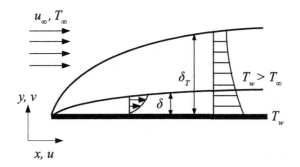

FIGURE 5.2

The velocity and thermal boundary layers in low-Pr fluids.

$$u_\infty \frac{3}{8} \frac{d\delta_T}{dx} = \left(\frac{3\alpha}{2\delta_T} \right) \tag{5.123}$$

$$\delta_T \frac{d\delta_T}{dx} = \left(\frac{4\alpha}{u_\infty} \right) \tag{5.124}$$

$$\frac{d\delta_T^2}{dx} = \left(\frac{8\alpha}{u_\infty} \right) \tag{5.125}$$

Integrating once, we get

$$\delta_T^2 = \frac{8\alpha x}{u_\infty} + A \tag{5.126}$$

At $x = 0$, $\delta_T = 0$, and so $A = 0$

$$\therefore \delta_T = \sqrt{\frac{8\alpha x}{u_\infty}} \tag{5.127}$$

Multiplying and dividing the RHS by $\sqrt{\nu}$,

$$\delta_T = \sqrt{\frac{8\alpha \nu x}{u_\infty \nu}} = \frac{2.82x}{\sqrt{Re_x . Pr}} \tag{5.128}$$

The product of $Re_x . Pr$ is known as the Peclet number, Pe.

$$\frac{\delta_T}{x} = \frac{2.82}{\sqrt{Pe_x}} \tag{5.129}$$

$$Nu_x = -x \left(\frac{\partial \theta}{\partial y} \right)_{y=0} \tag{5.130}$$

$$Nu_x = -x \left(\frac{-3}{2\delta_T} \right) \tag{5.131}$$

Substituting for δ_T in the above expression

$$Nu_x = \frac{3}{2}x\left(\frac{\sqrt{Pe_x}}{2.82x}\right) = 0.532 Pe_x^{1/2} \tag{5.132}$$

By using a procedure similar to what was done for $Pr > 1$ fluids, we get the final form of the expression for the average Nusselt number as

$$\overline{Nu_L} = 1.064\, Pe_L^{1/2} \tag{5.133}$$

From all the above developments, it is clear that the Nusselt number for forced convection, under steady state, takes up the following functional form

$$\overline{Nu_L} = aRe^b\, Pr^c \tag{5.134}$$

For the case of liquid metals, the form of the correlation gets further simplified with $b = c$, thereby introducing a new dimensionless number called the Peclet number. Analytical and semianalytical treatments are possible, as shown above, for flow over a flat plate. However, for other geometries, empirical correlations based on carefully done experiments are available in literature. Forms of these correlations are guided by theory, as mentioned above.

Example 5.1: *Consider a steady, laminar, incompressible, two-dimensional external flow of air (at 30 °C) over a flat plate. The velocity of air is 5 m/s. The length of the plate is 1 m. The plate is maintained at temperature of 90 °C. Under steady state conditions, estimate the average heat transfer coefficient from the plate. Also, calculate the heat transfer from the plate for unit width of the plate.*

Solution:
First, we determine the Reynolds number,

$$Re = \frac{u_\infty L}{\nu}$$

The properties of the fluid are evaluated at the mean film temperature

$$T_f = \frac{T_w + T_\infty}{2}$$
$$T_f = \frac{90 + 30}{2}$$
$$T_f = 60\,°C$$

Substituting the values of properties at 60 °C (from Table 5.2),

$$Re = \frac{5 \times 1}{18.82 \times 10^{-6}}$$

$Re = 2.66 \times 10^5 < 5 \times 10^5$ (More about this criterion is in Section 5.5.)
∴ The flow is laminar.
Now using the correlation for the external flow over a flat plate

$$\overline{Nu_L} = 0.662\, Re^{1/2}\, Pr^{1/3}$$
$$\overline{Nu_L} = 0.662 \times \left(2.66 \times 10^5\right)^{1/2} (0.711)^{1/3}$$
$$\overline{Nu_L} = 304.77$$
$$\overline{Nu_L} = \frac{hL}{k_f} = 304.77$$

$$h = \frac{304.77 \times 0.027}{1}$$

$$h = 8.23 \ \text{W/m}^2\,\text{K}$$

The rate of heat transfer is given by

$$q = hA\Delta T$$
$$q = 494.1 \ \text{W/m}^2$$

Flow over a cylinder

Flow over a cylinder is commonly encountered in several heat transfer applications like evaporator, condenser, recuperator, automative radiator, and so on. The special difficulty associated with flow over a cylinder is the curvature and its effect on the boundary layer development. Consider a flow over a cylinder, as shown in Fig. 5.3.

The key difference between the flat plate and this flow is that U_∞ is no longer a constant and varies with x, as indicated in the figure. For $\theta = 0$ to $90°$, $u_\infty(x)$ increases

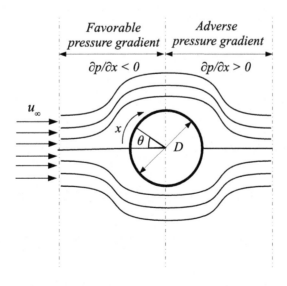

FIGURE 5.3

Development and separation of boundary layer over a cylinder.

with x and this results in a favorable pressure gradient, whereas for $90° \leq \theta \leq 180°$, the fluid decelerates consequent upon which there is an adverse pressure gradient, and separation of the boundary layer occurs from the wall the condition for which is mathematically given by $\partial u/\partial y = 0$. When the boundary layer detaches at the surface, a wake is formed in the downstream. Please refer to Incropera et al. (2013) for a more comprehensive discussion on this special flow. If $Re_D \leq 2 \times 10^5$, the flow stays laminar and separation occurs at $\theta = 80°$. However, beyond a Re_D of 2×10^5, transition to turbulence (more on this in Section 5.5) occurs and separation takes place at $\theta = 140°$.

For flow over a cylinder, the correlation for Nusselt number (Nu_D) proposed by Zukauskas (1972) is of the form

$$\overline{Nu_D} = C Re_D^m Pr^n \left(\frac{Pr}{Pr_s} \right)^{0.25} \tag{5.135}$$

The constants C, m are reported in Table 5.1.

This correlation is valid for $0.7 \leq Pr \leq 500$ and $1 \leq Re_D \leq 10^6$. If $Pr \leq 10$ then $n = 0.37$ and if $Pr > 10$, then $n = 0.36$. Churchill and Bernstein (1977) have proposed a single comprehensive equation as follows

$$\overline{Nu_D} = 0.3 + \frac{0.62 Re_D^{1/2} Pr^{1/3}}{0.25 \left[1 + \left(\frac{0.4}{Pr} \right)^{2/3} \right]} \left[1 + \left(\frac{Re_D}{28200} \right)^{5/8} \right]^{4/5} \tag{5.136}$$

The above equation is recommended for all the values of $Re_D Pr \geq 0.2$

Flow over a sphere

Whitaker (1972) proposed a correlation for the average Nusselt number for flow over a sphere as follows.

$$\overline{Nu_D} = 2 + \left(0.4 Re_D^{1/2} + 0.06 Re_D^{2/3} \right) Pr^{2/5} \left(\frac{\mu}{\mu_s} \right)^{1/2} \tag{5.137}$$

This correlation is valid for $0.71 \leq Pr \leq 380, 3.5 \leq Re_D \leq 7.6 \times 10^6$, and $1 \leq \mu/\mu_s \leq 3.2$. All properties are to be evaluated at T_∞ except for μ_s, which alone is to be evaluated at the surface (or wall temperature) $T_s (\text{or } T_w)$.

Table 5.1 Constants in Eq. (5.135).

Re	C	m
0.4–4	0.989	0.330
4–40	0.911	0.385
40–4000	0.683	0.466
4000–40000	0.193	0.618
40000–400000	0.027	0.805

Heat transfer in flows across a bank of tubes

Flow and heat transfer over a bank of tubes or rod bundles frequently occur in industrial practice, as, for example, in heat exchangers, regenerators (for example, in an application where exhaust gases preheat the incoming fluid, say air or water), and in nuclear fuel rods. A variety of configurations are possible. Two frequently encountered ones are shown below in Figs. 5.4A and 5.4B.

In the above two figures, there are three rows of tubes. The difference between the two is that in (a) the arrangement is aligned and is frequently referred to as "in-line arrangement" while in (b) the arrangement of the tubes is staggered. There are two key geometrical parameters that enter the problem. These are the transverse pitch S_T and the longitudinal pitch S_L. For such flows, it is intuitive to expect that the average Nusselt number is a function of the pertinent parameters as follows.

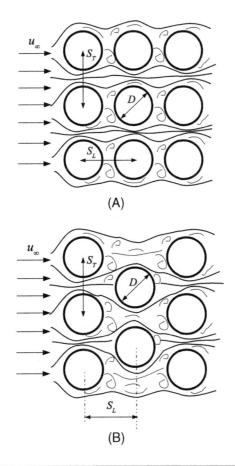

(A)

(B)

FIGURE 5.4

Arrangement of (A) aligned and (B) staggered rows of tubes in tube/rod bundle.

$$\overline{Nu_D} = f\left(Re, Pr, S_T, S_L\right) \tag{5.138}$$

Following the development presented in the earlier portions of this chapter, the Nusselt number should vary as

$$\overline{Nu_D} = aRe^b Pr^{1/3} S_T^c S_L^d, \text{or} \tag{5.139}$$

$$\overline{Nu_D} = aRe^b Pr^{1/3} \left(S_T / S_L\right)^e \tag{5.140}$$

Professor Zukauskas from Lithuania did extensive work and has presented reviews of heat transfer in these geometries. In fact, he synthesized the work of all the investigators before him and proposed a generic expression for Nu_D as

$$\overline{Nu_D} = Pr^{0.36} \left(Pr / Pr_w\right)^n Z\left(Re_D\right) \tag{5.141}$$

where $n = 0$ for gases and 0.25 for liquids. All properties need to be evaluated at the bulk fluid temperature, while Pr_w is evaluated at the wall temperature T_w. For $10^2 \leq Re_D \leq 10^3$

$$\text{Aligned rows} : Z = 0.52 Re_D^{0.5}$$

$$\text{Staggered rows} : Z = 0.71 Re_D^{0.5}$$

For $10^3 \leq Re_D \leq 2 \times 10^5$

$$\text{Aligned rows} : Z = 0.27 Re_D^{0.63}, S_T / S_L \geq 0.7$$

$$\text{Staggered rows} : Z = 0.35\left(S_T / S_L\right)^{0.2} Re_D^{0.6}, S_T / S_L \leq 2$$

For even higher Reynolds numbers and for liquid metals, please refer to Lienhard and Lienhard (2020), who have presented a comprehensive account of heat transfer across tube bundles.

Example 5.2: *Consider a long pin fin of diameter 8 mm. The fin is made of aluminum* $(k = 205 \text{ W}/mK)$ *and is maintained at a base temperature of 100 °C. Air at 30 °C flows over the fin with a velocity of 8 m/s.*
a. *Determine the heat transfer coefficient from the fin.*
b. *Determine the net heat transfer from the fin.*
c. *If the diameter of the fin is changed to 14 mm, with all other conditions being the same, determine the heat transfer coefficient and the heat transfer from the fin.*

Solution:
The Reynolds number, $Re_D = \dfrac{u_\infty D}{v}$

Air properties at $T_m = \dfrac{T_w + T_\infty}{2} = \dfrac{100 + 30}{2} = 65 \,°C = 338 \text{ K}.$

Now, we obtain the air properties at 338 K from Table 5.2 given on the next page.

Table 5.2 Thermophysical properties of air at atmospheric pressure (Kadoya et al., 1985; Jacobsen, R. T., et al., 1992).

T(K)	ρ (kg/m³)	c_p (J/kg.K)	k (W/m.K)	μ (kg/m.s) ×10⁶	α (m²/s) ×10⁵	ν (m²/s) ×10⁶	Pr
100	3.605	1039	0.00941	7.11	0.251	1.97	0.784
135	2.739	1020	0.01267	9.38	0.486	3.65	0.757
150	2.368	1012	0.01406	10.35	0.587	4.37	0.745
170	2.128	1010	0.01578	11.54	0.764	5.63	0.739
200	1.769	1007	0.01836	13.33	1.031	7.54	0.731
225	1.591	1007	0.02038	14.69	1.305	9.46	0.726
250	1.412	1006	0.02241	16.06	1.578	11.37	0.721
260	1.358	1006	0.02329	16.49	1.705	12.14	0.712
270	1.308	1006	0.02400	16.99	1.824	12.99	0.712
273	1.294	1006	0.02400	17.13	1.841	13.25	0.712
280	1.261	1006	0.02473	17.47	1.879	13.85	0.711
290	1.217	1006	0.02544	17.95	2.078	14.75	0.710
300	1.177	1007	0.02623	18.57	2.213	15.78	0.713
310	1.139	1007	0.02684	18.89	2.340	16.59	0.709
320	1.103	1008	0.02753	19.35	2.476	17.54	0.708
330	1.070	1008	0.02821	19.81	2.616	18.51	0.708
340	1.038	1009	0.02888	20.25	2.821	19.51	0.707
350	1.008	1009	0.02984	20.90	2.931	20.73	0.707
360	0.983	1010	0.03052	21.34	3.089	21.82	0.706
380	0.932	1012	0.03191	22.22	3.405	24.01	0.705
400	0.882	1014	0.03328	23.10	3.721	26.19	0.704
420	0.843	1017	0.03459	23.93	4.059	28.55	0.704
440	0.803	1020	0.03590	24.76	4.397	30.92	0.703
450	0.784	1021	0.03656	25.17	4.567	32.10	0.703
460	0.768	1023	0.03719	25.56	4.746	33.37	0.703
480	0.737	1027	0.03845	26.35	5.105	35.91	0.704
500	0.706	1030	0.03971	27.13	5.464	38.45	0.704
550	0.642	1040	0.04277	29.02	6.412	45.24	0.706
600	0.588	1051	0.04573	30.82	7.400	52.42	0.708
650	0.543	1063	0.04863	32.57	8.430	60.01	0.712
700	0.504	1075	0.05146	34.25	9.498	67.96	0.715
750	0.471	1087	0.05425	35.88	10.612	76.23	0.719
800	0.441	1099	0.05699	37.47	11.761	84.97	0.723
900	0.392	1121	0.06237	40.52	14.193	103.42	0.728
1000	0.353	1142	0.06763	43.43	16.792	123.13	0.733

$$\rho = 1.038 \text{ kg/m}^3$$
$$c_p = 1009 \text{ J/kg.K}$$
$$v = 1.951 \times 10^{-5}$$
$$k = 0.0289 \text{ W/mK}$$
$$Pr = 0.707$$

(For convenience, all properties were taken at 340 K.)

$$\therefore Re_D = \frac{u_\infty D}{v} = \frac{8 \times 8 \times 10^{-3}}{19.51 \times 10^{-6}} = 3280$$

$\therefore$ The constants C and m from Table 5.1 corresponding to this Reynolds number are $C = 0.683$ and $m = 0.466$.

$$\overline{Nu_D} = 0.683 Re^{0.466} Pr^{0.37} \left(\frac{Pr}{Pr_s} \right)^{0.25}$$
$$Pr_s = 0.706$$

(It is to be noted that for air, unless the temperature difference between the surface and the surrounding is very high (100 to 150 °C), the Prandtl number correction parameter will have a negligible role in the determination of the Nusselt number.)

a. For the problem under consideration,

$$\overline{Nu_D} = 0.683 \times (3280)^{0.466} \times (0.707)^{0.37} \times \left(\frac{0.707}{0.706} \right)^{0.25}$$
$$\overline{Nu_D} = 26.13$$
$$h = \frac{\overline{Nu}.k}{D} = \frac{25.87 \times 0.0289}{8 \times 10^{-3}}$$
$$h = 94.43 \text{ W/m}^2\text{K}$$

b.

$$Q = \sqrt{hpAk\theta_b}$$
$$Q = \sqrt{94.43 \times \pi \times 8 \times 10^{-3} \times \pi \times \frac{(8 \times 10^{-3})^2}{4} \times 205} \times 70$$
$$Q = 10.94 \text{ W}$$

c. When the fin diameter is increased to 14 mm

$$Re_D = 3280 \times \frac{14}{8} = 5740$$

Now, the constants C and m change to 0.193 and 0.618, respectively.

$$\therefore \overline{Nu_D} = 0.193 \times (5740)^{0.618} \times (0.707)^{0.37} \times \left(\frac{0.707}{0.706} \right)^{0.25}$$
$$\therefore \overline{Nu_D} = 35.72$$

$$h = \frac{\overline{Nu_D}.k}{D} = \frac{35.72 \times 0.0289}{14 \times 10^{-3}} = 73.75 \text{ W/m}^2\text{K}$$

$$Q = \sqrt{h.p.A.k\theta_b}$$

$$Q = \sqrt{73.75 \times \pi \times 14 \times 10^{-3} \times \pi \times \frac{(14 \times 10^{-3})^2}{4} \times 205 \times 70}$$

$$Q = 22.37 \text{ W}$$

The key takeaway from this problem is that if we had tried to solve this problem with knowledge of only fin heat transfer but without any idea of how the heat transfer coefficient will change for a change in diameter, our solution would have been

$$\frac{Q_1}{Q_2} = \frac{\sqrt{h.p_2.A_2.k \times \theta_b}}{\sqrt{h.p_1.A_1.k \times \theta_b}} = \frac{\sqrt{p_2 A_2}}{\sqrt{p_1 A_1}} = \frac{\sqrt{D_2^3}}{\sqrt{D_1^3}} = \left(\frac{D_2}{D_1}\right)^{1.5}$$

$$\therefore \frac{Q_2}{Q_1} = 2.311$$

The actual ratio of Q_2/Q_1 We obtained is 2.04 $(i.e., 22.37 \text{ W}/10.94 \text{ W})$, which leads to a significantly different result by about 15%.

Example 5.3: *A heated sphere made of aluminum (k = 205 W/mK) of 8 mm diameter at 100 °C cools in a current of air at 30 °C, with a free stream velocity of 8 m/s. The density and specific heat of aluminum are 2700 kg/m³ and 900 J/kgK respectively.*

a. Determine the heat transfer coefficient from the sphere.
b. Determine the time constant of the sphere.
c. How much time will it take for the sphere to cool to 50 °C.

Neglect radiation from the sphere.

Solution:

a. The Reynolds number Re_D, is given by

$$Re_D = \frac{u_\infty D}{v} = \frac{8 \times 8 \times 10^{-3}}{19.51 \times 10^{-6}} = 3280$$

Re_D, is the same as in part (a) of Example 5.2.

$$\overline{Nu_D} = 2 + \left[0.4(3280)^{0.5} + 0.06(3280)^{0.667}\right](0.707)^{0.4}\left(\frac{1.857 \times 10^{-5}}{2.025 \times 10^{-5}}\right)^{0.25}$$

$$\overline{Nu_D} = 2 + [22.91 + 13.28] \times 0.87 \times 0.979 = 32.83$$

$$h = \frac{\overline{Nu_D}k}{D} = \frac{32.83 \times 0.0289}{8 \times 10^{-3}} = 118.58 \text{ W/m}^2\text{K}$$

b. Time constant

$$\tau = \frac{mc_p}{hA}$$

$$\tau = \frac{\rho V c_p}{hA} = \frac{2700 \times \frac{4}{3}\pi\left(4\times10^{-3}\right)^3 \times 900}{118.58 \times 4\pi\left(4\times10^{-3}\right)^2}$$

$$\tau = 27.33 \ s$$

c. Let us now calculate the Biot number.

$$Bi = \frac{hr}{3k} = \frac{118.58\times4\times10^{-3}}{3\times205}$$
$$Bi = 0.00077$$

Since $Bi < 0.1$, the lumped capacitance method is valid. Recalling the solution for a spatially lumped object undergoing cooling (from Chapter 3),

$$\therefore \frac{\left(T - T_\infty\right)}{\left(T_i - T_\infty\right)} = e^{-t/T}$$

$$\frac{\left(50 - 30\right)}{\left(100 - 30\right)} = e^{-t/27.33}$$

$$0.2857 = e^{-t/27.33}$$

$$\therefore t = 34.24 \ s$$

5.5 **Turbulent flow**

Consider flow and heat transfer over a flat plate. Up to $Re_L = 5\times10^5$, the flow stays laminar and beyond this there is transition to turbulence. The boundary layer grows thicker and eddies, and consequent mixing will be in the transition region and in the fully turbulent region beyond the transition. A depiction of the boundary layer flow over a flat plate is shown in Fig. 5.5.

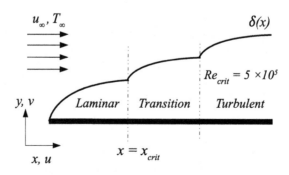

FIGURE 5.5

Forced convection boundary layer development over a horizontal flat plate depicting the laminar, transition, and turbulent regions of flow.

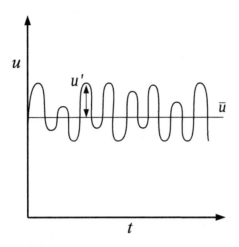

FIGURE 5.6

Typical variation of velocity with time for a turbulent flow.

In the turbulent regions of flow, we encounter time-dependent velocities and temperatures, and these dependencies are characterized by fluctuations.

For example, the horizontal velocity u may be expressed as

$$u = \bar{u} + u' \tag{5.142}$$

In the above equation, u is the mean velocity and u' is the fluctuating component. The velocity u typically varies with time, as shown in Fig. 5.6.

Similarly,

$$v = \bar{v} + v' \tag{5.143}$$

and

$$T = \bar{T} + T' \tag{5.144}$$

The average velocity $\bar{u}$ is given by

$$\bar{u} = \frac{1}{T_0} \int_0^{T_0} u\, dt \tag{5.145}$$

T_0 is the integrating time and must be larger than the slowest fluctuations associated with velocity u or any variable for that matter.

For two-dimensional, steady, incompressible flow with constant properties, using the expressions for u, v, and T, with the inclusion of fluctuating components, the governing equations after applying the boundary layer assumptions become

$$\frac{\partial \bar{u}}{\partial x} + \frac{\partial \bar{u}}{\partial y} = 0 \tag{5.146}$$

$$\rho \left(\bar{u} \frac{\partial \bar{u}}{\partial x} + \bar{v} \frac{\partial \bar{u}}{\partial y} \right) = \frac{\partial}{\partial y} \left(\mu \frac{\partial \bar{u}}{\partial y} - \rho \overline{u'v'} \right) \tag{5.147}$$

and

$$\rho c_p \left(\bar{u} \frac{\partial \bar{T}}{\partial x} + \bar{v} \frac{\partial \bar{T}}{\partial y} \right) = \frac{\partial}{\partial y} \left(k \frac{\partial \bar{T}}{\partial y} - \rho c_p \overline{v'T'} \right) \tag{5.148}$$

The above equations are obtained after we perform time averaging of the governing equations. Let us demonstrate this for term uv.

Consider the term

$$\overline{uv} = \overline{(\bar{u} + u')(\bar{v} + v')} = \overline{\bar{u}\bar{v}} + \overline{u'v'} + \overline{u'\bar{v}} + \overline{\bar{u}v'} \tag{5.149}$$

However, as fluctuations have a zero mean

$$\overline{u'} = 0 \tag{5.150}$$

and

$$\overline{v'} = 0 \tag{5.151}$$

But,

$$\overline{u'v'} \neq 0 \tag{5.152}$$

$$\therefore \overline{uv} = \bar{u}\bar{v} + \overline{u'v'} \tag{5.153}$$

(with the understanding that $\overline{\bar{u}\bar{v}} = \bar{u}\bar{v}$).

A similar procedure can be followed for other terms in the momentum and energy equations.

Now, we can define two new quantities, namely, total shear stress, τ_{total}, and total heat flux, q_{total}, as follows

$$\tau_{total} = \mu \frac{\partial \bar{u}}{\partial y} - \rho \overline{u'v'} \tag{5.154}$$

and

$$q_{total} = -\left(k \frac{\partial \bar{T}}{\partial y} - \rho c_p \overline{v'T'} \right) \tag{5.155}$$

Now we introduce two new quantities, namely, eddy viscosity of momentum, ε_m, and eddy viscosity of heat, ε_H, as follows.

$$\rho \varepsilon_m \frac{\partial \bar{u}}{\partial y} = -\rho \overline{u'v'} \tag{5.156}$$

and

$$\rho c_p \varepsilon_H \frac{\partial \bar{H}}{\partial y} = -\rho c_p \overline{v'T'} \tag{5.157}$$

$$\therefore \tau_{total} = \rho (v + \varepsilon_m) \frac{\partial \bar{u}}{\partial y} \tag{5.158}$$

and

$$q_{total} = \rho c_p \left(\alpha + \varepsilon_H\right) \frac{\partial \overline{T}}{\partial y} \tag{5.159}$$

The challenge though is to get the two quantities ε_m and ε_H. We now have three equations and five unknowns, namely, u, v, T, ε_m, and ε_H. This is the fundamental problem of closure in turbulence modeling. We need to relate ε_m and ε_H to the average quantities $\overline{u}$, $\overline{v}$, or $\overline{T}$ so that we can effectively handle the problem of closure. This activity is actually a separate field by itself and is known as turbulence modeling.

A quick look at the qualitative nature of the velocity profiles in laminar and turbulent boundary layers is in order. The turbulent velocity profile is sharp, with a steep slope at $y = 0$, denoting higher shear stress at the wall (see Fig. 5.7). Discerning readers will quickly realise that the higher shear stress at the wall will lead to more heat transfer. While the former is not desirable, the latter is.

The ratio of ε_m to ε_H is known as turbulent Prandtl number denoted by Pr_T. Similar to α being nearly equal to v in gases, $\varepsilon_m / \varepsilon_H$ is nearly equal to 1.

$$\therefore \frac{\varepsilon_m}{\varepsilon_H} = Pr_T \approx 1 \left(\text{for gases}\right) \tag{5.160}$$

For $Pr \gg 1$ fluids, based on boundary layer measurements, the recommended value of Pr_T is around 0.9. The idea of a turbulent Prandtl number, which is strictly not a physical property and is apparently intangible, is a useful construct in heat transfer engineering, as from five unknowns, we have eliminated one.

Reynolds analogy

Let us now see if we establish a relationship between momentum transfer and heat transfer. If we are successful in doing this, then by performing the easier-to-do velocity measurements together with the use of Newton's law of viscosity, we can

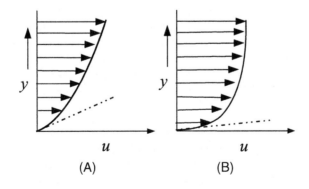

FIGURE 5.7

Figure showing a qualitative variation of horizontal velocity in (A) laminar and (B) turbulent boundary layers over a flat plate.

obtain τ_w, and from this we can get q. This will indeed be a godsend, and inductively we can extend the logic or more precisely the "analogy" to mass transfer itself. The idea of obtaining heat transfer results from cold flow measurements is plain brilliant!

Let us use the results of the integral solution to see if we can establish the analogy quantitatively.

$$\frac{Q}{\tau_w} = \frac{hA(T_w - T_\infty)}{c_f \times \frac{1}{2}\rho u_\infty^2} \tag{5.161}$$

$$Q = -k\left(\frac{\partial T}{\partial y}\right)_{y=0} \tag{5.162}$$

$$\tau_w = \mu\left(\frac{\partial u}{\partial y}\right)_{y=0} \tag{5.163}$$

Let us now use some results from the analysis we performed a little while ago, with the integral method. Using the cubic velocity profile from this analysis, we have

$$\frac{u}{u_\infty} = \frac{3}{2}\frac{y}{\delta} - \frac{1}{2}\left(\frac{y}{\delta}\right)^2 \tag{5.164}$$

$$\left(\frac{\partial u}{\partial y}\right)_{y=0} = \frac{3u_\infty}{2\delta} \tag{5.165}$$

Similarly the temperature profile turned out to be

$$\theta = \frac{(T - T_\infty)}{(T_w - T_\infty)} = 1 - \frac{3}{2}\frac{y}{\delta_T} + \frac{1}{2}\left(\frac{y}{\delta_T}\right)^3 \tag{5.166}$$

$$\left(\frac{\partial T}{\partial y}\right)_{y=0} = \frac{-3}{2\delta_T}(T_w - T_\infty) \tag{5.167}$$

$$\therefore \frac{Q}{\tau_w A} = \left(\frac{hA(T_w - T_\infty)}{c_f \times \frac{1}{2}A\rho u_\infty^2}\right) = \frac{-k \times 2 \times 3 \times \delta \times (T_w - T_\infty)}{2 \times \delta_T \times \mu \times u_\infty \times 3} \tag{5.168}$$

$$\therefore \frac{h(T_w - T_\infty)}{c_f \times \frac{1}{2}\rho u_\infty^2} = \frac{-k \times \delta \times (T_w - T_\infty)}{\delta_T \times \mu \times u_\infty} \tag{5.169}$$

$$\frac{\mu}{k} = \frac{Pr}{c_p} \tag{5.170}$$

$$\therefore \frac{h \times Pr}{\rho u_\infty c_p}\left(\frac{\delta_T}{\delta}\right) = \frac{c_f}{2} \tag{5.171}$$

We know that $\delta_T/\delta \approx Pr^{-1/3}$. Substituting for δ_T/δ in Eq. (5.171)

$$\frac{h \times Pr \times Pr^{-1/3}}{\rho u_\infty c_p} = \frac{c_f}{2} \tag{5.172}$$

$$\therefore \frac{c_f}{2} = \left(\frac{h}{\rho u_\infty c_p}\right) Pr^{2/3} \tag{5.173}$$

The term within the brackets $\left(\dfrac{h}{\rho u_\infty c_p}\right)$ is known as the Stanton number.

$$\therefore \frac{c_f}{2} = St.Pr^{2/3} \tag{5.174}$$

The above is known as the modified Reynolds analogy or the Chilton-Colburn analogy, with the "plain" Reynolds analogy being $\dfrac{c_f}{2} = St$ for $Pr = 1$ fluids.

The term $\left(\dfrac{h}{\rho c_p u_\infty}\right)$ can be simplified as follows.

$$\frac{h}{\rho c_p u_\infty} = \frac{Nu.k}{L.\rho u_\infty c_p} \tag{5.175}$$

We know that

$$\frac{k}{\rho c_p} = \alpha \tag{5.176}$$

$$\frac{h}{\rho u_\infty c_p} = \frac{Nu\ \alpha}{u_\infty}\ \frac{\vartheta}{\vartheta\ L} = \frac{Nu}{Re_L Pr} = St \tag{5.177}$$

$\therefore$ from Eq. (5.173),

$$\frac{c_f}{2} = St.Pr^{2/3} \tag{5.178}$$

In sum, once we have the skin friction coefficient, c_f, using the modified Reynolds analogy, we can determine the Stanton number. From the Stanton number, we can estimate the Nusselt number and hence the heat transfer coefficient and finally the heat flux, q. Getting q or h is the fundamental challenge in convective heat transfer, as we have already emphasized.

The modified Reynolds analogy is valid for $0.6 < Pr < 60$.

The analogy works very well for laminar flows with $dp/dx = 0$, but in turbulent flows, the sensitivity of heat transfer results to dp/dx is not significant, and the analogy holds. The analogy can be applied locally and also for the average coefficients over a geometry.

From boundary layer measurements, for turbulent flow over a flat plate, the following critical information is available.

$$\frac{\delta}{x} = \frac{0.37}{Re_x^{0.2}}$$

$$\text{and } c_{f,x} = \frac{0.0592}{Re_x^{0.2}}, \ 5 \times 10^5 \le Re_x \le 10^8$$

With the above expression, using the modified Reynolds analogy, the local Nusselt number can be determined to be

$$Nu_x = 0.0296 \, Re_x^{0.8} \, Pr^{0.33} \tag{5.179}$$

Again, this is valid for fluids that satisfy $0.6 \le Pr \le 60$.

If the flow is mixed, that is, laminar up to $x = x_{crit}$ and turbulent beyond, the average heat transfer coefficient $\overline{h}_L$ is given by

$$\overline{h}_L = \frac{1}{L} \left[\int_0^{x_{crit}} h_{laminar} \, dx + \int_{x_{crit}}^L h_{turb} \, dx \right] \tag{5.180}$$

Example 5.4: *Consider a heated horizontal plate that is maintained at 100 °C. Air at 30 °C flows over with a velocity of 10 m/s. The plate area is 0.3 m^2, and measurements reveal a drag force of 0.32 N on the plate by the flowing air. Determine the power (in W) that is required to maintain the plate at 100 °C.*

Solution:
We use the modified Reynolds analogy to solve the problem.
Air properties at 65 °C

$$\rho = 1.04 \text{ kg/m}^3, \ \nu = 19.51 \times 10^{-6} \text{ m}^2/\text{s}, \ k = 0.029 \text{ w/mK}, \ Pr = 0.707,$$
$$c_p = 1.009 \text{ kJ/kgK}$$

$$\tau = c_f \cdot \frac{1}{2} \rho u_\infty^2$$

$$\text{Drag force} = \tau \times A_s$$

$$\therefore c_f = \frac{\text{Drag force}}{A_s \cdot \frac{1}{2} \rho u_\infty^2}$$

$$c_f = \frac{0.32}{0.3 \times \frac{1}{2} \times 1.04 \times 10^2} = 0.0205$$

$$\frac{c_f}{2} = \frac{h}{\rho u_\infty c_p}$$

$$\therefore h = \frac{0.0205}{2} \times 1.04 \times 10 \times 1009$$

$$h = 107.6 \text{ W/m}^2\text{K}$$

$$Q = hA(\tau_w - \tau_\infty)$$

$$Q = 107.6 \times 0.3(70) = 2260 \text{ W}$$

Please note that the use of the modified Reynolds number analogy is an absolutely clever way to solve the problem. In fact, the length scale in the problem is not given and so Re_L and $\overline{Nu}_L$ cannot be defined, but then these hardly mattered. We used the original definition of Stanton number, $St = h/\rho u_\infty c_p$, to crack the problem. Additionally, from the friction data, we got the heat transfer. This is vintage engineering heat transfer.

5.6 Internal flows

Forced convection inside tubes and ducts has a very large number of applications, starting from the pipe carrying hot steam from the boiler to the turbine in a thermal power plant, to the tubes carrying the refrigerant in the condenser and evaporator of a refrigerator, and the tubes carrying the coolant in any automobile radiator or in any heat exchanger, for that matter.

The key challenge in the above mentioned applications is the calculation of the heat transfer coefficient h for the internal flow inside the tube in a particular application. Stated explicitly, if a constant heat flux or constant wall temperature on the outside of a tube (due to, say, vapor condensing on the outside), the key challenge is to get "h" for the fluid flowing inside the tube. In the more general case, along with this, the heat transfer coefficient on the outside, the conduction resistance across the tube material and any fouling resistance all have to be considered to obtain the overall heat transfer coefficient, which was briefly introduced in Chapter 1.

The key difference between external flows and internal flows is the concept of "developing" and "fully developed" flow, which can be best understood from a figure.

Consider a circular tube of diameter, d. A fluid enters the tube at $x = 0$ (see Fig. 5.8). Once the fluid enters the duct all over the surface, the presence of wall will be felt by the viscous fluid. Hence, a boundary layer starts developing as indicated in the figure. Once the boundary layer thickness δ becomes $d/2$, the whole of the flow inside the tube becomes a boundary layer flow, as opposed to the flat plate, where beyond $y = \delta$ in the free stream, the effects of fluid viscosity are not felt. The length X at which the whole of the flow becomes a full boundary layer flow is called entry length, and the flow beyond X, is known as "fully developed."

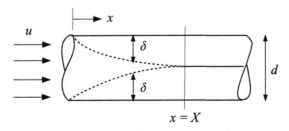

FIGURE 5.8

Concept of developing flow inside a tube.

We can work out the ballpark estimates of this entry length from the boundary layer results applicable for external flow.

We earlier estimated the boundary layer thickness for forced convection over a flat plate to be

$$\frac{\delta}{x} \approx \frac{5}{\sqrt{Re_x}} \tag{5.181}$$

$$\text{or, } \delta \approx 5\sqrt{\frac{vx}{u_\infty}} \tag{5.182}$$

$\delta = d/2$ when $x \approx X$, where X is the entry length.

$$\frac{d}{2} = 5\sqrt{\frac{vX}{u_\infty}} \tag{5.183}$$

$$\frac{d^2}{4} = 25\frac{vX}{u_\infty} \tag{5.184}$$

$$\frac{X}{d} = \frac{1}{100}\frac{u_\infty d}{v}$$
$$\therefore \frac{X}{d} \approx 10^{-2}\frac{u_\infty d}{v} \approx 10^{-2} Re_d \tag{5.185}$$

A more exact estimate of X/d that is frequently used in literature is $0.05 Re_d$ For turbulent flow X/d is approximately 10 and is indeed very short. The idea of fully developed flow can be exploited to our advantage, if we want to get a handle on the analytical treatment of fluid flow and heat transfer in tubes and ducts, as it is intuitively apparent that the governing equation should simplify in the fully developed region. Even so, discerning readers will quickly realize that since the boundary layer thickness is smaller in the developing region of the flow, the local heat transfer coefficient, in whichever way we want to define, will be high. Hence, many short tubes instead of a few longer tubes may do the trick sometimes. Even so, there may be other considerations such as pressure drop, the sheer arrangement of tubes, and so on. So, a decision to work with shorter or longer tubes eventually is an engineering compromise.

5.6.1 Governing equations and the quest for an analytical solution

Consider two-dimensional, steady, laminar, incompressible flow with constant properties inside a pipe. The governing equations derived in Cartesian coordinates will look like what are shown below in the cylindrical coordinates.
Continuity equation

$$\frac{\partial u}{\partial x} + \frac{1}{r}\frac{\partial(vr)}{\partial r} = 0 \tag{5.186}$$

Momentum equations

x−momentum

$$\rho\left(u\frac{\partial u}{\partial x}+v\frac{\partial u}{\partial r}\right)=-\frac{\partial p}{\partial x}+\mu\left(\frac{\partial^2 u}{\partial x^2}+\frac{1}{r}\frac{\partial u}{\partial r}+\frac{\partial^2 u}{\partial r^2}\right)$$

(5.187)

r−momentum

$$\rho\left(u\frac{\partial v}{\partial x}+v\frac{\partial v}{\partial r}\right)=-\frac{\partial p}{\partial r}+\mu\left(\frac{\partial^2 v}{\partial x^2}+\frac{1}{r}\frac{\partial v}{\partial r}-\frac{v}{r^2}+\frac{\partial^2 v}{\partial r^2}\right)$$

(5.188)

The geometry, coordinate axes, and the respective velocities are given in Fig. 5.9. Comparing the scales of the two terms in the continuity equation, we have

$$\frac{v}{D}\approx\frac{U}{x}$$

(5.189)

$$\therefore v\approx\frac{UD}{x}$$

(5.190)

Therefore, as x increases, v decreases. So, as we proceed deeper (i.e., into the fully developed region), $v=0$.

∴ From the continuity equation, in the fully developed flow region, $\dfrac{\partial u}{\partial x}=0.$

$$\therefore u=f(r) \text{ alone}$$

(5.191)

Substituting for $v=0$ in the r−momentum equation, with the understanding that v and all its derivatives are zero in the fully developed region, the r momentum equation reduces to $\partial p/\partial r = 0$. Hence $\partial p/\partial x$ in Eq. (5.187) can be simplified as dp/dx. Eq. (5.187) now becomes

$$\mu\left[\frac{d^2 u}{dr^2}+\frac{1}{r}\frac{du}{dr}\right]=\frac{dp}{dx}$$

(5.192)

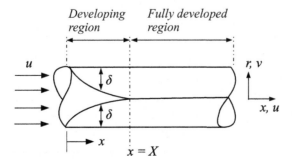

FIGURE 5.9

Problem geometry, coordinates, and velocities for laminar flow inside a pipe.

Please note that Eq. (5.192) is an ordinary differential equation, while Eq. (5.187) is a partial differential equation. The above equation can be solved with the following two boundary conditions.

$$\text{At } r = r_0 \text{ or } d/2, \ u = 0 \ (\text{no slip}) \tag{5.193}$$

$$\text{At } r = 0, \frac{du}{dr} = 0 \ (\text{symmetry}) \tag{5.194}$$

$$\mu \left[\frac{1}{r} \frac{d}{dr} \left(r \frac{du}{dr} \right) \right] = \frac{dp}{dx} \tag{5.195}$$

Integrating twice

$$u = \frac{1}{\mu} \left(\frac{dp}{dx} \right) \frac{r^2}{4} + A \ln r + B \tag{5.196}$$

From the boundary conditions, at $r = r_0$, $u = 0$ and at $r = 0, du/dr = 0$, we have $A = 0$ and $B = -\frac{1}{\mu} \left(\frac{dp}{dx} \right) \left(\frac{r_0^2}{4} \right)$.

$$\therefore u = -\frac{r_0^2}{4\mu} \left(\frac{dp}{dx} \right) \left[1 - \left(\frac{r}{r_0} \right)^2 \right] \tag{5.197}$$

Hence, the velocity profile for fully developed laminar flow in a pipe is parabolic.

The above is also referred to as the Hagen-Poiseulle flow. As we already saw, $p = f(x)$ alone; the x-momentum equation can be rewritten as

$$\frac{dp}{dx} = \mu \left[\frac{d^2u}{dr^2} + \frac{1}{r} \frac{du}{dr} \right] = C \tag{5.198}$$

Since the left-hand side of the above equation is a function of x and the right-hand side of the equation is a function of r alone, both have to be equal to a constant C.

If the entry length x is much smaller than L, where L is the length of the tube under consideration, the relation $dp/dx = C$ can be extended to the "small" entry length too (with an added error, though) so that

$$\frac{dp}{dx} = -\frac{\Delta p}{L} \tag{5.199}$$

Please note that the struggle in fluid mechanics is to get the pressure drop, Δp. The average or mean velocity can now be calculated using the mass flow rate $\dot{m}$

$$\dot{m} = \int_0^{r_0} \rho \cdot u(r) \cdot 2\pi \cdot r \cdot dr = \rho u_m \cdot \pi r_0^2 \tag{5.200}$$

$$\therefore u_m = \frac{2}{r_0^2} \int_0^{r_0} \frac{-r_0^2}{4\mu} \frac{dp}{dx} \left(1 - \frac{r}{r_0} \right)^2 \cdot r dr \tag{5.201}$$

$$u_m = \frac{-1}{2\mu}\left(\frac{dp}{dx}\right)\left[\frac{r^2}{2} - \frac{r^4}{4r_0^2}\right]_0^{r_0} = \frac{-r_0^2}{8\mu}\frac{dp}{dx} \tag{5.202}$$

The maximum velocity u_{max} occurs at $r = r_0$.

$$u_{max} = \frac{-r_0^2}{4\mu}\frac{dp}{dx} \tag{5.203}$$

comparing Eqs. 5.202 and 5.203, we have $u_{max} = 2u_m$. Now we proceed to evaluate the shear stress, τ, and the skin friction coefficient, c_f.

$$\tau = \mu\left(\frac{\partial u}{\partial r}\right)_{r=r_0} \tag{5.204}$$

$$\tau = \mu \cdot \frac{-r_0^2}{4\mu} \cdot \frac{r}{r_0^2}\Big|_{r=r_0} \tag{5.205}$$

$$\tau = \frac{r_0}{2}\frac{dp}{dx} \tag{5.206}$$

However, from the expression for u_m we have

$$\frac{dp}{dx} = -\frac{8\mu u_m}{r_0^2} \tag{5.207}$$

$$\therefore \tau = \frac{r_0}{2} \cdot 8 \cdot \frac{\mu u_m}{r_0^2} \tag{5.208}$$

$$\tau = \frac{4\mu u_m}{r_0} \tag{5.209}$$

$$c_f = \frac{\tau}{\frac{1}{2}\rho u_m^2} = \frac{\frac{4\mu u_m}{r_0}}{\frac{1}{2}\rho u_m^2} = \frac{16}{\frac{u_m \cdot D}{\nu}} = \frac{16}{Re_D} \tag{5.210}$$

This is also known as the Fanning friction factor, f.

$$\therefore f = c_f = \frac{16}{Re_D} \tag{5.211}$$

The Darcy-Weisbach friction factor, f_D, is given by

$$f_D = 4f = \frac{64}{Re_D} \tag{5.212}$$

For turbulent flows, measurements confirm the following expressions for friction factors.

$$f_D = 0.316 Re_D^{-0.25}, Re_D \leq 2 \times 10^4 \qquad (5.213)$$

$$f_D = 0.184 Re_D^{-0.2}, Re_D \geq 2 \times 10^4 \qquad (5.214)$$

Usually $Re_D = 2300$ is considered to be the limit for transition to turbulence in internal flows. In view of this, for a smooth, circular duct, $f_D = 64/Re_D$ needs to be used for $Re_D < 2300$, except for situations where the boundary layer is "engineered" to be turbulent below this Reynolds number. Further, since $x/D = 10$ for turbulent flows, unless the tube length is very short, fully developed flow relations like the ones given above should hold. The friction factor is a strong function of the surface roughness, and the Moody's chart gives a graphical depiction of the variation of friction factor for laminar and turbulent flows. For the latter, the relative roughness is a key parameter. One can intuit that for the same Reynolds number, for a turbulent pipe flow, the higher the roughness, the higher the friction factor.

Noncircular ducts

The results we derived above for circular ducts are valid for noncircular ducts too. However, we need to use an effective diameter. This is known as the hydraulic diameter and is given by

$$D_h = \frac{4A_c}{p} \qquad (5.215)$$

where A_c is the cross-sectional area of the duct, and p is the wetted perimeter.
For a circular duct of diameter d, $A_c = \pi d^2/4$ and $p = \pi/d$ and so $D_h = d$.

Thermal considerations

Entry length: Using scale analysis, we can show that the dimensionless thermal entry length is given by $x_{th}/D \approx 0.05 Re_D Pr$ for laminar flow and approximately 10 for turbulent flow. The concept of thermal entry length is quite tricky from a conceptual viewpoint and needs an elaboration, as follows.

The mean temperature

For a duct flow, there is an absence of free stream velocity. Similarly, there is an absence of the free stream temperature. Just as we used mean velocity, here we define a mean temperature. The rate at which energy transport, E_t, occurs may be obtained by integrating the product of mass flux (ρu) and the internal energy per unit mass $(c_v T)$ over the cross-section.

$$\dot{E}_t = \int_{A_c} (\rho u)(c_v T) dA_c \qquad (5.216)$$

We define T_m such that $\dot{E}_t = \dot{m} c_v T_m$.

$$\therefore T_m = \frac{\int_{A_c} (\rho u)(c_v T) dA_c}{\dot{m} c_v} \tag{5.217}$$

For incompressible flow in a circular tube with constant c_v, it follows that

$$T_m = \frac{c_v \cdot \rho \int_0^{r_0} u \cdot T \cdot 2\pi r \, dr}{\rho \cdot \pi r_0^2 \cdot u_{mean} \cdot c_v} \tag{5.218}$$

$$T_m = \frac{2}{u_m r_0^2} \int_0^{r_0} u \cdot T \cdot r \cdot dr \tag{5.219}$$

The bulk mean temperature, T_m is also called the mixing cup temperature.

Newton's law of cooling

Once T_m is defined, we can use Newton's law of cooling for heat transfer in this geometry (tube/duct flow)

$$q = h(T_s - T_m) \tag{5.220}$$

where h is the local heat transfer coefficient, and T_s is the surface or wall temperature.

There is an essential difference between external and internal flow. While in external flow T_∞ is a constant, here T_m must vary in the flow direction. Stated explicitly, for heat transfer to occur, $dT_m/dx \neq 0$. dT_m/dx is +ve, if the fluid is getting heated and dT_m/dx is -ve, if the fluid is getting cooled.

Fully developed conditions

In the hydrodynamic case, $\partial u/\partial x = 0$ is valid for fully developed flow. But if heat transfer occurs, neither dT_m/dx nor $\partial T/\partial x$ equals zero at any radius r. $T(r)$ changes continuously with x. However, this seeming paradox may be reconciled by working with a dimensionless temperature $(T_S(x) - T(r,x))/(T_S(x) - T_m(x))$ and examining if this quantity becomes independent of x beyond a certain x. In fact, though $T(r)$ changes with x, $(T_S(x) - T(r,x))/(T_S(x) - T_m(x))$ is not a function of x in the fully developed region. Hence, the mathematical condition for fully developed flow from the viewpoint of convective heat transfer is

$$\frac{d}{dx}\left(\frac{T_S(x) - T(r,x)}{T_S(x) - T_m(x)} \right) = 0 \tag{5.221}$$

We can work a little with the dimensionless temperature considered above at $r = r_0$, as follows.

$$\frac{\partial}{\partial r}\left(\frac{T_S(x) - T(r,x)}{T_S(x) - T_m(x)} \right)_{r=r_0} = \frac{-\left(\dfrac{\partial T(r,x)}{\partial r} \right)_{r=r_0}}{T_S(x) - T_m(x)} \neq f(x) \tag{5.222}$$

We know that

$$q = -k\left(\frac{\partial T}{\partial r}\right)_{r=r_0} \tag{5.223}$$

$$\text{or } q/k = -\left(\frac{\partial T}{\partial r}\right)_{r=r_0} \tag{5.224}$$

Similarly, from Newton's law of cooling, $T_s(x) - T_m(x) = q/h$. Substituting these results in, Eq. (5.222)

$$\frac{q/k}{q/h} \neq f(x) \tag{5.225}$$

$$\therefore \frac{h}{k} \neq f(x) \tag{5.226}$$

This is indicated in Fig. 5.10 for a constant k.

Hence, in the thermally fully developed flow of a fluid with constant properties, $h \neq f(x)$ and is a constant. One way of coming to peace with these "uneasy" arguments is to first state that from measurements, we know that Eq. (5.221) is valid, and using all our definitions and ideas of convection, we get to the fact that $h \neq f(x)$ in the fully developed region. If this too causes discomfiture, one can start with the argument that measurements show that $h \neq f(x)$ beyond a certain x and we would like to know what mathematical condition $(T_s(x) - T_s(r, x))$ must satisfy in order for this to be true; this will logically lead us back to Eq. (5.221).

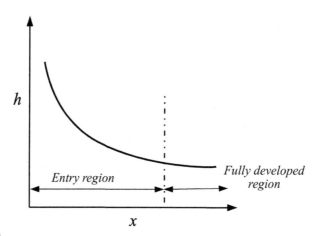

FIGURE 5.10

Variation of heat transfer coefficient, h, along the length of the tube.

Internal flow with constant heat flux, q_w

From Newton's law of cooling, $q = h\left(T_s(x) - T_m\right)$ From the earlier discussion, we concluded that for fully developed internal flow, h is constant.

As q is constant in this case, the difference $\left(T_s(x) - T_m\right)$ is also constant.

$$\therefore \frac{d}{dx}\left(T_s(x) - T_m(x)\right) = 0 \tag{5.227}$$

$$\frac{dT_s(x)}{dx} = \frac{dT_m(x)}{dx} \tag{5.228}$$

Similarly, by performing differentiation on Eq. (5.221)

$$\frac{d}{dx}\left(T_s(x) - T(r,x)\right) = 0 \tag{5.229}$$

$$\frac{dT_s(x)}{dx} = \frac{dT(r,x)}{dx} \tag{5.230}$$

$$\therefore \frac{dT_s(x)}{dx} = \frac{dT_m(x)}{dx} = \frac{dT(r,x)}{dx} \tag{5.231}$$

$\therefore$ The key result we have now obtained is that the axial gradient of temperature is not a function of of the radial location under fully developed conditions for a constant heat flux.

From the energy balance, for ideal gases and incompressible liquids,

$$dq_{conv} = \dot{m}c_p dT_m = q \cdot p \cdot dx \tag{5.232}$$

$$Q_{conv} = \dot{m}c_p\left(T_{m,o} - T_{m,i}\right) \tag{5.233}$$

$$\frac{dT_m}{dx} = \frac{qp}{\dot{m}c_p} = \frac{p \cdot h \cdot \left(T_s(x) - T_m(x)\right)}{\dot{m}c_p} \tag{5.234}$$

Integrating from $x = 0$ to x, we get

$$T_m(x) = T_{m,i} + \frac{qpx}{\dot{m}c_p} \tag{5.235}$$

From Eq. (5.235), it is clear that the mean temperature variation is linear for constant heat flux. $T_m(x)$ varies always linearly with x. $T_s(x)$ varies linearly with x (for the constant heat flux case) in the fully developed region. These are qualitatively sketched for a typical case in Fig. 5.11.

Internal flow with constant wall temperature, T_w or T_s

From Eq. (5.234), we know that $dT_m/dx = p.h\left(T_s(x) - T_m(x)\right)/\dot{m}c_p$.

Let $\Delta T = T_s(x) - T_m(x)$, where $T_s(x)$ is a constant

$$d(\Delta T) = -dT_m \tag{5.236}$$

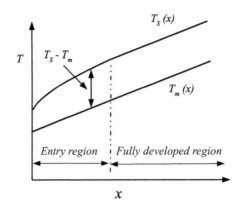

FIGURE 5.11

Variation of surface temperature and mean temperature along the length of the tube.

$$dT_m = -d(\Delta T) \tag{5.237}$$

Substituting Eq. (5.237) in Eq. (5.234)

$$\frac{dT_m}{dx} = -\frac{d(\Delta T)}{dx} = \frac{p \cdot h \cdot \Delta T}{\dot{m} c_p} \tag{5.238}$$

$$\int_{\Delta T_i}^{\Delta T_o} \frac{d(\Delta T)}{\Delta T} = -\frac{p}{\dot{m} \cdot c_p} \int_0^L h\, dx \tag{5.239}$$

$$\ln\left(\frac{\Delta T_o}{\Delta T_i}\right) = \frac{-\rho L}{\dot{m} \cdot c_p} \int_0^L \frac{1}{L} h \cdot dx \tag{5.240}$$

$$\ln\left(\frac{\Delta T_o}{\Delta T_i}\right) = -\frac{pL}{\dot{m} \cdot c_p} \bar{h} \tag{5.241}$$

$$\frac{T_s - T_{m,o}}{T_s - T_{m,i}} = \frac{\Delta T_o}{\Delta T_i} = e^{\frac{-pL\bar{h}}{\dot{m} c_p}} \tag{5.242}$$

We could have also integrated from the tube inlet to some axial position x within the tube.

$$\frac{T_s - T_m(x)}{T_s - T_{m,i}} = e^{-\frac{px\bar{h}}{\dot{m} c_p}} \tag{5.243}$$

$$Q_{conv} = \dot{m} c_p \left(T_{m,o} - T_{m,i}\right) \tag{5.244}$$

$$Q_{conv} = \dot{m} c_p \left[\left(T_s - T_{m,i}\right) - \left(T_s - T_{m,o}\right)\right] \tag{5.245}$$

$$Q_{conv} = \dot{m}c_p\left[\Delta T_i - \Delta T_0\right] \tag{5.246}$$

From Eq. (5.241)

$$ln\left(\frac{\Delta T_o}{\Delta T_i}\right) = -\frac{hpL}{\dot{m}\cdot c_p} \tag{5.247}$$

$$\therefore \dot{m}c_p = -\frac{hpL}{ln\left(\dfrac{\Delta T_o}{\Delta T_i}\right)} \tag{5.248}$$

Substituting Eq. (5.248) in Eq. (5.246)

$$Q_{conv} = \frac{\bar{h}\cdot L \cdot p\left(\Delta T_0 - \Delta T_i\right)}{ln\left(\dfrac{\Delta T_0}{\Delta T_i}\right)} \tag{5.249}$$

$$Q_{conv} = \bar{h}A_s\left(LMTD\right) \tag{5.250}$$

$$\Delta T_{lm} = LMTD = \frac{\left(\Delta T_o - \Delta T_i\right)}{ln\left(\dfrac{\Delta T_o}{\Delta T_i}\right)} \tag{5.251}$$

Given h, the area A_s and the temperatures Q_{conv} can be determined. In Eq. (5.251), LMTD is the logarithmic mean temperature difference, which physically is the mean temperature difference that is driving the convection heat transfer. Since from Eq. (5.242) it is seen that the temperature distribution is exponential, the mean temperature difference is logarithmic. A mean of the arithmetic temperature difference at the outlet and the inlet is crude and is against the physics of the problem.

Analytical solution for Nusselt number for a fully developed flow

The boundary layer energy equation for flow inside a duct or tube can be derived by modifying the energy equation given in Chapter 4 and employing boundary layer simplifications presented earlier in this chapter. The final equation turns out to be

$$u\frac{\partial T}{\partial x} + v\frac{\partial T}{\partial r} = \frac{\alpha}{r}\frac{\partial}{\partial r}\left(r\frac{\partial T}{\partial r}\right) \tag{5.252}$$

For fully developed flow, $v = 0$, $\dfrac{\partial u}{\partial x} = 0$.

For constant surface heat flux and fully developed flow, we know the following relation:

$$\frac{u}{u_m} = 2\left[1 - \left(\frac{r}{r_0}\right)^2\right] \tag{5.253}$$

$$\frac{\partial T}{\partial x} = \frac{dT_m}{dx} \tag{5.254}$$

$$\therefore \frac{1}{r}\frac{d}{dr}\left[r\frac{dT}{dr}\right] = \frac{2u_m}{\alpha}\left[1-\left(\frac{r}{r_0}\right)^2\right]\frac{dT_m}{dx} \tag{5.255}$$

Eq. (5.255) can be integrated twice in order to obtain an expression for temperature distribution as follows.

$$T = \frac{2u_m}{\alpha}\frac{dT_m}{dx}\left[\frac{r^2}{4}-\frac{r^4}{16r_0^2}\right]+A\ lnr + B \tag{5.256}$$

The two boundary conditions are

$$\text{At } r = r_0, T = T_s \tag{5.257}$$

$$\text{At } r = 0, T \text{ is finite} \tag{5.258}$$

Substituting the above boundary conditions in Eq. (5.256), we get

$$A = 0$$
$$B = T_s - \frac{2u_m}{\alpha}\frac{dT_m}{dx}\frac{3r_0^2}{16} \tag{5.259}$$

Substituting for A and B in Eq. (5.256), we have

$$T(r) = T_s - \left(\frac{2u_m}{\alpha}\cdot\frac{dT_m}{dx}\right)\left(\frac{3r_0^2}{16}+\frac{r^4}{16r_0^2}-\frac{r^2}{4}\right) \tag{5.260}$$

We know that the bulk mean temperature is given by

$$T_m = \frac{2}{u_m r_0^2}\int_0^r u\cdot T\cdot r\cdot dr \tag{5.261}$$

Substituting for $T(r)$ from Eq. (5.260) in Eq. (5.261) and completing the integration, we get the following expression for T_m

$$T_m = T_s - \frac{11}{96}\cdot\frac{2u_m}{\alpha}\cdot\frac{dT_m}{dx}\cdot r_0^2 \tag{5.262}$$

From Newton's law of cooling,

$$q = h\left(T_s - T_m\right) = h\times\frac{11}{96}\times\frac{2u_m}{\alpha}\frac{dT_m}{dx}\cdot r_0^2 \tag{5.263}$$

By working on eqn 5.263 with the understanding that dT_m/dx can be related to q through eqn.5.232, we can show that

$$T_m - T_s = -\frac{11}{48}\frac{qD}{k} \tag{5.264}$$

$$\therefore h = \frac{48}{11}\left(\frac{k}{d}\right) \tag{5.265}$$

$$Nu_D = \frac{hD}{k} = \frac{48}{11} = 4.36 \qquad (5.266)$$

The above relationship is valid for constant heat flux.

For the constant wall temperature, the analysis is more difficult and the Nusselt number turns out to be $Nu_D = 3.66$. The analytical solution can be found in advanced texts on convection.

Correlation for turbulent flow inside tubes and ducts

For fully developed (both hydrodynamically and thermally)

Colburn equation:
Recall the relation $f_D = 0.184 Re_D^{-0.2}$, $\quad Re_D \geq 2 \times 10^4$ (Eq. 5.214).

Combining this with the modified Reynolds analogy (Eq. 5.178), we get an expression for the average Nusselt number as

$$\frac{c_f}{2} = \frac{f_D}{8} = St Pr^{2/3} \left[Re_D > 10^4 \right] \qquad (5.267)$$

From the definition of Stanton number, St, from Eq. (5.267) the expression for average Nusselt number turns out to be

$$\overline{Nu_D} = 0.023 Re^{0.8} Pr^{1/3} \qquad (5.268)$$

Dittus-Boelter equation
Dittus and Boelter worked out a correlation for both heating and cooling, which is given as

$$\overline{Nu_D} = 0.023 Re^{0.8} Pr^{n} \qquad (5.269)$$

In Eq. (5.269), the exponent n = 0.4 for heating, and n = 0.3 for cooling of the fluid flowing inside the tube or duct. This equation is slightly more accurate than Eq. (5.268). Dittus-Boelter equation is the workhorse of convective heat transfer calculations in internal flows that are turbulent. These correlations can have errors of the order of ±25%. The Dittus-Boelter equation can be used for both constant wall flux and constant wall temperature conditions and can be used for turbulent flow in an annulus too. Current state of the art with three-dimensional turbulent flow and heat transfer correlations can give us Nusselt numbers far closer to truth than the above.

Example 5.5: *Consider the flow of water in a tube of diameter 25 mm. The flow rate of water in the tube is 0.16 kg/s. The tube surface is maintained at 120 °C (by, say, a vapor, like steam, condensing at its saturation pressure). Water enters the tube at 30 °C and leaves the tube at 90 °C. If the convective heat transfer coefficient associated with the flow of water inside the tube is 2250 W/m²k, determine the length of the tube required to achieve the objective.*

Solution:

Water properties are to be evaluated at a bulk mean temperature

$$\frac{T_{m,i} + T_{m,o}}{2} = \frac{30 + 90}{2} = 60 \,°C$$

$\rho = 985 \text{ kg/m}^3, c_p = 4.18 \text{ kJ/kgK}, k = 0.65 \text{ W/mK}, v = 5.1 \times 10^{-7} \text{m}^2/\text{s}$ and $Pr = 3.3$

The problem details are sketched in Fig. 5.12. Let us now draw a temperature length diagram for the heat transfer to the water in the tube. Fig. 5.13

The heat transfer rate Q in the problem under consideration is given by the following set of relations.

$$Q = \dot{m}c_p \left[T_{m,o} - T_{m,i} \right] = LMTD$$

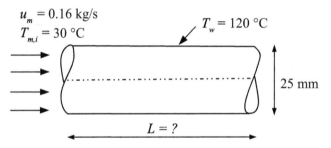

FIGURE 5.12

Flow of water in a tube.

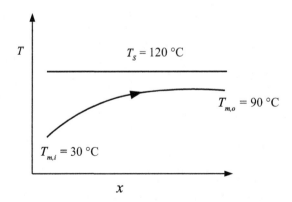

FIGURE 5.13

Temperature variation of water passing through the tube in Example 5.5.

Where LMTD is the logarithmic mean temperature difference given by

$$LMTD = \frac{\left(T_s - T_{m,o}\right) - \left(T_s - T_{m,i}\right)}{\ln\left(\dfrac{T_s - T_{m,o}}{T_s - T_{m,i}}\right)}$$

$$LMTD = \frac{30 - 90}{\ln\left(\dfrac{30}{90}\right)} = 54.61\ ^{\circ}C$$

$$Q = 0.16 \times 4.184 \times 10^3 \times (90 - 30) = 40166\,W = 40.166\ kW$$
$$Q = h \cdot A \cdot LMTD$$
$$40.166 \times 10^3 = 2250 \times \pi \times 0.025 \times L \times LMTD$$
$$40.166 \times 10^3 = 2250 \times \pi \times 0.025 \times L \times 54.61$$
$$L = 4.164\ m$$

Hence, the length of the tube required to accomplish the heat transfer is 4.164 *m*.

In this problem, there is no need for us to use an analytical result/correlation to determine the heat transfer coefficient "h," as it is already specified in the problem. However, this has been specified only for the purpose of demonstrating internal convection calculations through an example. In engineering practice, invariably what we seek is the heat transfer coefficient, which can be determined from the data given together with additional details if the flow is fully developed. Let us now solve one more example directly related to Example 5.5 to see if the heat transfer coefficient 2250 W/m²K given in the problem is reasonable.

Example 5.6: *Revisit Example 5.5. All the information given to us in Example 5.5 is available except that the heat transfer coefficient is as yet unknown. However, a key additional piece of information, namely the fact that the flow is fully developed from the inlet, is given to us. The unknown again is the length of the tube required to accomplish the heat transfer.*

Solution:
The Reynolds number of the flow is

$$Re_D = \frac{u_m d}{\nu}$$

Where the mean velocity of the water in the tube is

$$u_m = \frac{4\dot{m}}{\pi \rho d^2}$$
$$u_m = \frac{4 \times 0.16}{3.14 \times 0.025^2 \times 985} = 0.33\,m/s$$
$$\therefore Re_D = \frac{0.33 \times 0.025}{5.1 \times 10^{-7}} = 16176$$

The flow is turbulent as $Re_D > 2300$ and is fully developed as stated in the problem. Hence, the Dittus-Boelter correlation can be used to obtain h.

$$Nu_D = 0.023 Re_D^{0.8} Pr^{0.4}$$
$$Nu_D = 0.023(16176)^{0.8} (3.3)^{0.4}$$
$$Nu_D = 86.34$$

∴ The heat transfer coefficient

$$h = \frac{Nu_D k}{D} = \frac{86.34 \times 0.65}{0.025} = 2245 \ \text{W/m}^2\text{K}$$

This is bang on target and very close to the value actually specified in the previous example, that is, Discerning readers may be quick to realize that we engineered two examples to drive home the point that if we know the heat transfer coefficient a priori, we can swiftly design a heat exchanger (in this case, this was a simple tube), and even if we do not know this a priori, if we know the flow and temperature conditions, we can use convective heat transfer theory (either an analytical/numerical solution or a correlation) to obtain "h" and then complete the design. A third variant of the above problem may also be considered. If one has a tube of known diameter and length with a vapor (say, steam) condensing on it at a particular temperature, and for a given flow rate of water inside the tube, if the inlet and outlet temperatures are measured, we can use the following equation to estimate the convective heat transfer coefficient.

$$Q = \dot{m} c_p \left(T_{m,o} - T_{m,i} \right) = h \times A \times LMTD$$

This is known as the empirical approach to heat transfer and is exactly how heat transfer engineering started. Getting "h" from theory is heat transfer science morphing into heat transfer engineering. Even so, $Q = h \times A \times \Delta T$ or $h \times A \times LMTD$ is still a rate law and is not a fundamental law in the sense that it cannot be derived from first principles.

Problems

5.1   The integral form of the momentum and energy conservation for steady, laminar, incompressible, constant property two-dimensional flow and heat transfer over a flat plate can be expressed as follows.

$$\frac{\partial}{\partial x} \left[\int_0^\delta u (u_\infty - u) dy \right] = v \frac{\partial u}{\partial y} \bigg|_{y=0}$$

$$\frac{\partial}{\partial x} \left[\int_0^{\delta_T} u (T_\infty - T) dy \right] = \alpha \frac{\partial T}{\partial y} \bigg|_{y=0}$$

Starting from these equations and using linear profiles for the velocity and temperature,

    a. Obtain an expression for the velocity boundary layer thickness in terms of local Reynolds number.

    b. Derive an expression for local Nusselt number if the plate is subjected to a constant heat flux boundary condition and the Prandtl number (Pr) of the fluid is more than 1.

    c. For a given local value of Reynolds number and Prandtl number, which boundary condition type (constant T or constant q) would exhibit a higher heat transfer rate? Comment.

5.2 Use the integral form of the momentum and energy conservation for steady, laminar, incompressible, constant property two-dimensional flow and heat transfer over a flat plate given in Problem 5.1.

    a. Assuming a quadratic profile for the velocity, obtain an expression for the velocity boundary layer thickness in terms of the local Reynolds number. Also derive an expression for the average skin friction coefficient c_f.

    b. Assuming a linear temperature profile and a liquid metal $(\mathrm{Pr} \ll 1)$ as the medium, derive expressions for the local and average Nusselt numbers.

    c. If the velocity profile is assumed to be cubic, how will the expression for the Nusselt number change for $(\mathrm{Pr} \ll 1)$?

5.3 Consider a steady, turbulent boundary layer on an isothermal flat plate of length "L" at temperature T_w. The boundary layer is "tripped" at the leading edge $x = 0$ by a fine wire such that the flow is turbulent right at the leading edge of the plate. Assume constant physical properties and velocity and temperature profiles of the form

$$\frac{u}{u_\infty} = \left(\frac{y}{\delta}\right)^{1/7}$$

$$\frac{T - T_\infty}{T_w - T_\infty} = 1 - \left(\frac{y}{\delta_T}\right)^{1/7}$$

From experiments it is known that the wall shear stress is related to the boundary layer thickness by an expression of the form

$$T_w = 0.0228 \rho u_\infty^2 \left(\frac{u_\infty \delta}{\nu}\right)^{-1/4}$$

(Note that in view of the nature of the velocity profile, this expression is required for evaluating the RHS of the momentum integral equation at the wall.)

Additionally, heat flux at the wall is related to the wall temperature through the Newton's law of cooling as $Q_w = h_x(T_w - T_\infty)$

    a. By employing the momentum integral equation, derive an expression for the boundary layer thickness as a function of local Reynolds number for the turbulent boundary layer.

b. Determine the average friction coefficient over the entire plate length, $c_{f,L}$.

c. Using the energy integral equation, obtain an expression for local Nusselt number Nu_x and use this result to evaluate the average Nusselt number $\overline{Nu_L}$.

5.4 The local heat transfer coefficient for turbulent flow along a flat plate can be determined by

$$St_x Pr^{2/3} = 0.0296 Re_x^{-1/5}$$

Assume that this relation is valid from the leading edge of the flat plate. Develop an expression for the average value of the heat transfer coefficient over the length L of the flat plate.

5.5 Consider a thin horizontal flat plate of length 1 m that is maintained at a temperature of 373 K. Air at a temperature of 303 K and a velocity of 15 m/s flows over the top surface of the plate. The bottom surface of the plate is insulated. The plate is 1 m deep in the direction perpendicular to the plane of the paper. The flow is fully turbulent from the leading edge (i.e. $x = 0$). Properties of air at the film temperature of 338 K are given below for ready use, though you can actually get them from Table 5.2.

$$k = 0.03 \text{ W/mK}, \ v = 21 \times 10^{-6} \text{ m}^2/\text{s}, \ \rho = 1 \text{ kg/m}^3 \text{ and } Pr = 0.7.$$

Determine the convective heat transfer rate from the plate in the region between $x = 150$ mm and $x = 800$ mm (x is the axial distance from the leading edge). Assume the flow and heat transfer to be two-dimensional, steady, and incompressible and that air has constant properties.

5.6 Consider a laminar flow of a fluid over a flat plate maintained at a constant temperature. If the free stream velocity of the fluid is halved, determine

a. The change in the drag force on the plate.

b. The change in the rate of heat transfer between the fluid and the plate. Calculate these quantities if the velocities are tripled, assuming that the flow is still laminar.

5.7 Consider a cylindrical pin fin, 10 mm in diameter and 80 mm long made of aluminum $(k = 205 \text{ W/mK})$ The fin is placed in a cross flow of air at $T_\infty = 30 \ ^\circ\text{C}$ with $u_\infty = 5$ m/s The base of the fin is maintained at 100 °C, and the tip may be assumed to be insulated.

a. Determine the heat transfer coefficient from the fin.

b. Determine the heat transfer rate from the fin.

c. Determine the fin efficiency and fin effectiveness.

d. If the velocity of the air is increased by 50%, determine the quantities calculated in (b) and (c).

e. If the velocity of air is 5 m/s but the fin diameter is reduced to 6 mm, calculate the heat transfer rate, fin efficiency, and fin effectiveness.

f. Comment on the results obtained with a view to examining the effect of key parameters on the performance of the fin.

5.8 Consider a 60 W bulb. The surface area of the bulb is known to be $33.9 \times 10^{-3} \, \text{m}^2$. Assume the bulb to be a sphere.

   a. Determine its surface temperature when it is kept in surroundings with atmospheric air at 30 °C and air with a very mild velocity of 0.6 m/s flows over the bulb.
   b. Is the bulb safe to touch?
   c. If radiation is considered, will your estimate of the surface temperature be lower or higher?
   d. Regardless of your answer to (c), determine the surface temperature when radiation is present, with the bulb assembly emissivity assumed to be 0.2. (Please note this is a crude assumption with the assembly consisting of a filament, glass envelope, fuse, lead wires, and filament support). Assume the temperature of the surroundings in so far as radiation is concerned to be 30 °C (The problem involves iterations.)

5.9 Consider a bank of cylinders consisting of 16 rows (in the direction of the flow), with each tube having a diameter of 50 mm in an array with $S_T = 8$ mm and $S_H = 6$ mm The cylinders are heated such that their surface temperatures are maintained at 150 °C. Air flows over the cylinders with a velocity of $u_\infty = 3$ m/s at a temperature of 30 °C. Please use Table 5.2 to obtain the air properties. Determine the heat transfer coefficient for this configuration for

   a. Inline arrangement
   b. Staggered arrangement

5.10 A fluid at a temperature of 30 °C and an average velocity of 2.8 m/s is in hydrodynamically and thermally fully developed flow through a circular tube of 5 mm diameter. The fluid then enters a 6 m long heated section.
The surface of the heated section is maintained at 120 °C. Determine the fluid outlet temperature and the total heat transfer rate. Properties of the fluid may be evaluated at an estimated mean temperature of 75 °C and are as follows.
$$\left(\begin{array}{l} \rho = 1104 \, \text{kg/m}^3, c_p = 2.46 \, \text{kJ/kgK}, \ \mu = 1.07 \times 10^{-2} \, \text{Ns/m}^2, \text{k} = 0.26 \, \text{W/mK and} \\ \text{Pr} = 103 \end{array} \right)$$

5.11 Water at 30 °C is flowing in a 12 mm diameter tube with a flow rate of 0.6 kg/s. The wall temperature of the tube is 55 °C.

   a. Determine the entry length (hydrodynamic and thermal).
   b. Determine the length of the tube such that the water is heated to 45 °C
   c. Determine the additional length required for every 1 °C rise in the outlet temperature of the water.

   Use water properties given at the end of the chapter (Table 5.3) at an appropriate bulk mean temperature.

5.12 Hot water flowing in a thin-walled tube is cooled by blowing room air at 30 °C in cross flow with a velocity of $u_\infty = 10$ m/s. The inlet and outlet temperatures of water are 80 °C and 40 °C respectively. The mass flow rate of water is 0.18 kg/s and the diameter of the tube is 40 mm.

a. Determine the length of the tube required to accomplish the cooling.

b. Determine the outlet temperature of the water, if the length of the tube is doubled.

c. Determine the outlet temperature of the water for the tube length determined in (a) when the velocity of the air is doubled.

Table 5.3 Thermophysical properties of water at saturation pressure (Meyer et al., 1967; Harvey et al., 2000)

T (K)	ρ (kg/m³)	c_p (J/kg.K)	k (W/m.K)	ν (m²/s) $\times 10^6$	α (m²/s) $\times 10^8$	Pr	β (K⁻¹) $\times 10^4$
273	999.8	4220	0.5611	1.791	13.31	13.47	−0.68
275	999.9	4214	0.5645	1.682	13.40	12.55	−0.36
280	999.9	4201	0.5740	1.434	13.66	10.63	0.44
285	999.5	4193	0.5835	1.240	13.92	8.91	1.12
290	998.8	4187	0.5927	1.085	14.17	7.66	1.72
295	997.8	4183	0.6017	0.960	14.42	6.66	2.26
300	996.5	4181	0.6103	0.857	14.65	5.85	2.74
305	995.0	4180	0.6184	0.771	14.87	5.18	3.19
310	993.3	4179	0.6260	0.698	15.08	4.63	3.61
320	989.3	4181	0.6396	0.583	15.46	3.77	4.36
330	984.4	4185	0.6501	0.507	15.78	3.23	5.01
340	979.5	4189	0.6605	0.431	16.10	2.68	5.65
350	973.5	4196	0.6671	0.384	16.34	2.36	6.22
360	967.4	4202	0.6737	0.337	16.57	2.03	6.79
373	958.3	4216	0.6791	0.294	16.81	1.75	7.51
380	952.9	4227	0.6803	0.278	16.89	1.65	7.81
390	945.2	4241	0.6819	0.255	17.01	1.51	8.26
400	937.5	4256	0.6836	0.233	17.13	1.36	8.70
420	919.9	4299	0.6825	0.203	17.26	1.18	10.08
440	900.5	4357	0.6780	0.181	17.28	1.05	11.32
460	879.5	4433	0.6702	0.164	17.19	0.96	12.73
480	856.5	4533	0.6590	0.151	16.97	0.89	14.40
500	831.3	4664	0.6439	0.142	16.60	0.85	16.45
520	803.6	4838	0.6246	0.134	16.07	0.83	19.09
540	772.8	5077	0.6001	0.128	15.30	0.84	22.66
560	738.0	5423	0.5701	0.123	14.25	0.86	27.83
580	697.6	5969	0.5346	0.119	12.84	0.93	36.07
600	649.4	6953	0.4953	0.117	10.97	1.07	51.41

References

Churchill, S.W., Bernstein, M., 1977. A correlating equation for forced convection from gases and liquids to a circular cylinder in crossflow. ASME Journal of Heat Transfer 99 (1), 300–306.

Harvey, A.H., Peskin, A.P., Klein, S.A., 2000. NIST/ASME Steam Properties, NIST Standard Reference Database 10, Version 2.2. National Institute of Standard and Technology, Gaithersburg, MD.

Incropera, F.P., Lavine, A.S., Bergman, T.L., DeWitt, D.P., 2013. Principles of heat and mass transfer. Wiley, NY.

Jacobsen, R.T., Penoncello, S.G., Beyerlein, S.W., Clarke, W.P., Lemmon, E.W., 1992. A thermodynamic property formulation for air. Fluid Phase Equilibria 79, 113–124.

Kadoya, K., Matsunaga, N., Nagashima, A., 1985. Viscosity and thermal conductivity of dry air in the gaseous phase. Journal of Physical and Chemical Reference Data 14 (4), 947–970.

Lienhard V, J.H., Lienhard IV, J.H., 2020. A Heat Transfer Textbook. Courier Dover Publications, Mineola, NY.

Meyer, C.A., McClintock, R.B., Silvestri, G.J., Spencer, Jr., R.C., 1967. ASME Steam Tables, 6th edition ASME, New York.

Whitaker, S, 1972. Forced convection heat transfer correlations for flow in pipes, past flat plates, single cylinders, single spheres, and for flow in packed beds and tube bundles. AIChE Journal 18 (2), 361–371.

Zukauskas, A., 1972. Heat transfer from tubes in cross flow. Advances in Heat Transfer 8 (1), 93–160.

Natural convection

6.1 Introduction

Natural convection is a form of convection heat transfer in which the driver of the bulk motion or advection is self-induced forces. These forces may be due to temperature or concentration gradients. In this book, we focus our attention on natural convection flow and heat transfer due to temperature gradients. In the light of the above statement, it is intuitively apparent that flow and heat transfer will be strongly coupled in natural convection. Because there is bulk motion, natural convection heat transfer will be several times more than what one would obtain in molecular conduction. Even so, because no external agency such as a pump or blower is involved in natural convection, the velocities will be small, typically of the order of cm/s or tens of cm/s as opposed to several m/s, which is typically seen in forced convection. As a consequence of this, natural convection heat transfer rates will be lower than forced convection in a particular situation, if all other controlling variables like temperature difference, geometry, and medium are the same. Recall that, traditionally, convection aided by an external agency like a pump or fan or blower has come to be known as "forced" convection. In view of this, natural convection where there is no "forcing" by an external agency, so to speak, is also known as free convection.

Consider two infinitely wide horizontal parallel plates at temperatures T_1 and T_2, respectively. Let the space between the two plates be occupied by a medium like air. Two possibilities exist, as shown in Fig. 6.1.

In situation (A), the top plate is hotter than the bottom plate. In view of this, as the air (or any other medium for that matter) gets heated from the top plate, it stays at the top, as heated air is less dense. So, this represents a stable arrangement in so far as convection is concerned, meaning in this case no natural convection will occur. Even so, heat transfer will occur between T_1 and T_2 through the air via the conduction or molecular diffusion route. Radiation may also occur, but this depends on the difference between T_1 and T_2, radiative properties of the surfaces, and so on. Radiative heat transfer is the subject of chapter 8 of this book.

Suffice it to say for now that Fig. 6.1A represents a "no flow," "no natural convection" situation. The situation in Fig. 6.1B, though, is interesting. The heated plate is at the bottom, so air coming in contact with it gets heated and rises. Once the air hits the top plate, it is cooled, becomes denser, and returns to the bottom plate to get heated again and continue the cycle. This is quintessential natural convection.

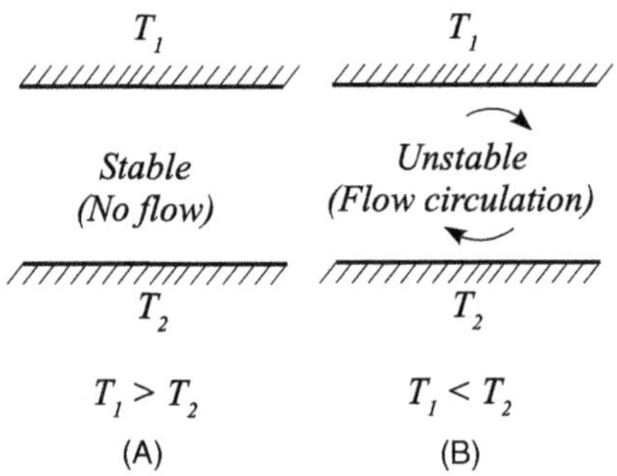

FIGURE 6.1

Medium enclosed between two parallel plates with temperatures T_1 and T_2. (A) $T_1 > T_2$ (B) $T_1 < T_2$.

Applications of this are legion, for example, cooling of electronic equipment like transformers, heat transfer in double pane windows, solar collectors, thermal hydraulics in nuclear reactors, and so on. The list is endless. Just to reinforce a point that the above is not a trite and beaten-to-death list of applications of natural convection, we would like to draw your attention to the Fukushima Daiichi nuclear disaster that happened on March 11, 2011. The disaster first started with an earthquake and as soon as the earthquake was detected, the nuclear reactors shut down. However, due to grid problems, the electricity supply failed and the emergency diesel generator sets started to ensure that the pumps circulated the coolant through the nuclear reactor cores to remove decay heat (which does not stop immediately and follows a typical $q = ae^{-bt}$ kind of distribution, with a and b being known constants, t being the time, and q being the heat decay). However, the earthquake caused a nearly 50 feet high tsunami that flooded the basement of the plant, thereby paralyzing the emergency generator. This resulted in what is known as loss of coolant accident (LOCA), reactor meltdown, and radiation release to the atmosphere. The basic problem here was that the decay heat removal system was designed only for the case of forced convection, and the system was incapable of preventing LOCA, if the emergency generators failed. A decay heat removal system that would have worked even under natural convection would have involved considerable engineering effort and inclusion of chimneys and so on, but would have saved the day. Even in the ubiquitous laptop, a heat pipe removes the heat generated by the processor but the condensation of the vapor of the heat pipe itself has to be driven by natural convection and radiation from all the surfaces of the laptop. It is worthwhile to remember that eventually any heat generated has to be released to the ambient air or a nearby lake, pond, sea, or outer space. The challenge for a heat transfer specialist is to enable and engineer this pathway that is sure-shot, safe, budget friendly, environmentally benign, and meets all guidelines and legislations.

6.2 Natural convection over a flat plate

The vertical flat plate is a frequently encountered geometry in natural convection and also serves as an excellent baseline configuration to undertake a mathematically rigorous study of natural convection, so that we eventually get results that are simple to use and are of great practical relevance in actual engineering situations.

Before getting drowned in mathematical details, let us try to intuit about what is likely to happen if a heated vertical plate of length L, at a temperature T_w is placed in quiescent (still) air at T_∞ with $T_w > T_\infty$ (see Fig. 6.2), with gravity acting downward.

Cold air coming near the plate gets heated and rises, thereby creating some sort of vacuum that is filled by fresh air rushing in towards the plate. In view of this, an upward current or convective flow is set up. Let x and y be the vertical and horizontal coordinates, respectively, and let the corresponding air velocities generated due to natural convection be u and v, respectively (These are indicated in Fig. 6.2). It is instructive to note that $u = v = 0$ everywhere if $T_w = T_\infty$. All the action is generated due to the temperature difference, $\Delta T = (T_w - T_\infty)$. The higher the ΔT, the stronger the natural convection.

Now, consider a horizontal section A-A as indicated in Fig. 6.2. At the wall (y = 0), the velocity $u = 0$. Again at a horizontal distance far away from the wall, $u = 0$ as the air is still. Even so, there is an upward motion due to buoyancy caused by ΔT, as already discussed. Hence, if we mark a point on A-A where the velocity is almost zero, say $0.01u_{max}$ with u_{max} being the maximum velocity in the air layer close to the plate, it stands to reason that one can expect a smooth velocity profile that is zero at y = 0 and $y = \delta$ where δ is the boundary layer thickness, with the maximum occurring somewhere in between. In fact, measurements by researchers have confirmed a velocity profile that looks like what is indicated in Fig. 6.2. From numerous measurements and flow visualization experiments, the boundary layer is seen to have a parabolic profile with increasing thickness as the height increases, as shown in Fig. 6.2. In continuation of the above arguments, it is not very difficult to

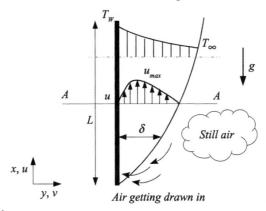

FIGURE 6.2

Natural convection from a heated vertical plate of length L.

comprehend an exponential temperature profile within the boundary layer, with T = T_w at y = 0 and T = T_∞ + 0.01 (T_w − T_∞) ≈ T_∞ at y = δ. The key differences between the forced convection and natural convection boundary layers are

1. In forced convection, the velocity boundary layer can exist independent of the thermal boundary layer, whereas in natural convection, it cannot, as the velocities themselves arise due to a temperature difference between the wall and the free stream.
2. The velocity profile is monotonic for forced convection with u = u_∞ at y = δ, while for natural convection it has to be nonmonotonic with a peak velocity occurring within the boundary layer.

6.3 Boundary layer equations and nondimensional numbers

Consider the vertical flat plate shown in Fig. 6.2 suspended in a quiescent medium at T_∞. The plate is at a uniform temperature of T_w with T_w > T_∞. Consider a two-dimensional, steady, incompressible (with density alone being a function of temperature) flow and heat transfer for a constant property flow. The governing equations are more or less the same as those we saw in Chapter 4, except for the addition of a body force in the x-momentum equation (please note that now the x-axis is along the height of the plate. i.e., vertical). The governing equations are

Continuity equation

$$\frac{\partial u}{\partial x} + \frac{\partial v}{\partial y} = 0 \tag{6.1}$$

x-momentum equation

$$\rho \left[u\frac{\partial u}{\partial x} + v\frac{\partial u}{\partial y} \right] = -\frac{\partial p}{\partial x} + \mu \left[\frac{\partial^2 u}{\partial x^2} + \frac{\partial^2 u}{\partial y^2} \right] - \rho g \tag{6.2}$$

y-momentum equation

$$\rho \left[u\frac{\partial v}{\partial x} + v\frac{\partial v}{\partial y} \right] = -\frac{\partial p}{\partial y} + \mu \left[\frac{\partial^2 v}{\partial x^2} + \frac{\partial^2 v}{\partial y^2} \right] \tag{6.3}$$

Energy equation

$$\rho C_p \left[u\frac{\partial T}{\partial x} + v\frac{\partial T}{\partial y} \right] = k \left[\frac{\partial^2 T}{\partial x^2} + \frac{\partial^2 T}{\partial y^2} \right] \tag{6.4}$$

Please note that the new term appearing in Eq. (6.2), namely, the body force term "−ρg" was not considered in forced convection. If we consider a two-dimensional control volume of size Δx.Δy.1, the term −ρ.Δx.Δy.1.g is nothing but the weight of the medium inside the control volume. Cancelling ΔxΔy and recognizing that x is positive upward, while gravity is acting downward, this term reduces to "−ρg", as shown in Eq. (6.2).

The big challenge before us is to now work out "ρg" in terms of the primary quantities of interest in our problem, those are, u, v, and T, if we choose to stay with the easier to solve incompressible flow formulation. The challenge is compounded by the fact that there is no free stream velocity u_∞ that can be used as a reference. Outside of handling the "ρg" term, we would also like to examine the possibility of using boundary layer approximations much as in the same way as we did for forced convection.

Scale for velocity

Let the scale for velocity be u_{ref} (by velocity here we mean vertical velocity). Consider Eq. (6.1). The scales for the two terms in this equation need to be of the same order, so that the two-dimensional character of the problem is intact (Bejan, 2013). Let the scale of u be u_{ref}. The scale for v is v and the length scales are L and δ for x and y, respectively. Substituting for these, in the continuity equation

$$\frac{u_{ref}}{L} \sim \frac{v}{\delta} \tag{6.5}$$

$$v \sim u_{ref}\frac{\delta}{L} \tag{6.6}$$

Since $\dfrac{\delta}{L} \ll 1$ (due to slenderness assumption of the boundary layer)

$$v \ll u \tag{6.7}$$

Also, $\dfrac{\partial}{\partial y} \gg \dfrac{\partial}{\partial x}$

$\therefore$ the y-momentum equation is $\dfrac{\delta}{L}$ times the x-momentum equation and can be "axed" from the analysis. Even so, there is a key take away from the y-momentum equation which is the following

$$-\frac{\partial p}{\partial y} = 0 \tag{6.8}$$

$$\therefore p = f(x) \text{ alone} \tag{6.9}$$

Consequent upon Eq. (6.9), the x-momentum equation reduces to

$$\rho\left[u\frac{\partial u}{\partial x} + v\frac{\partial u}{\partial y}\right] = -\frac{dp}{dx} + \mu\frac{\partial^2 u}{\partial y^2} - \rho g \tag{6.10}$$

The energy equation reduces to

$$\rho C_p\left[u\frac{\partial T}{\partial x} + v\frac{\partial T}{\partial y}\right] = \alpha\frac{\partial^2 u}{\partial y^2} \tag{6.11}$$

Applying Eq. (6.10) to the free stream, i.e., the region outside the boundary layer, we have

$$u = 0, \ v = 0, \ \frac{\partial^2 u}{\partial y^2} = 0$$

∴ Eq. (6.10) becomes

$$\frac{dp_\infty}{dx} = -\rho_\infty g \tag{6.12}$$

Because of the slenderness of the boundary layer, we then get

$$p(x, y) \sim p(x) \sim p_\infty(x) \tag{6.13}$$

Substituting for $\frac{dp}{dx}$ (or $\frac{dp_\infty}{dx}$) in Eq. (6.10)

$$\rho\left(u\frac{\partial u}{\partial x} + v\frac{\partial u}{\partial y}\right) = \left(\rho_\infty - \rho\right)g + \mu\frac{\partial^2 u}{\partial y^2} \tag{6.14}$$

The density of the medium is a function of temperature and pressure

$$\rho_\infty = \rho(T_\infty, p_{ref})$$

where p_{ref} is the reference pressure, say, at the bottom (x = 0)

$$\therefore \ \rho \approx \rho_\infty + \left(\frac{\partial\rho}{\partial T}\right)_p (T - T_\infty) + \left(\frac{\partial\rho}{\partial p}\right)_T (p - p_0) + \ldots\ldots\ldots\ldots \tag{6.15}$$

Now we make an assumption that the variation of density depends more on the temperature difference and $\left(\frac{\partial\rho}{\partial T}\right)_p$ product rather than on the pressure difference and $\left(\frac{\partial\rho}{\partial T}\right)_T$ product.

Consequent upon this assumption

$$\rho \approx \rho_\infty + \left(\frac{\partial\rho}{\partial T}\right)_p (T - T_\infty) \tag{6.16}$$

Now we introduce a quantity called the isobaric cubic expansivity denoted by β and defined as follows

$$\beta = -\frac{1}{\rho}\left(\frac{\partial\rho}{\partial T}\right)_p \tag{6.17}$$

For an ideal gas, $\beta = -\frac{1}{\rho}\left(\frac{-P}{RT^2}\right) = \frac{\rho RT}{\rho RT^2} = \frac{1}{T}$

From the definition of β it is clear that β has the units of K^{-1}. Invoking the definition of β, we now have an expression for density ρ as

$$\rho = \rho_\infty - \rho_\infty \beta(T - T_\infty) \tag{6.18}$$

$$\therefore \ \rho = \rho_\infty(1 - \beta(T - T_\infty)) \tag{6.19}$$

$\therefore$ Eq. (6.14) becomes

$$\rho_\infty \left[1 - \beta(T - T_\infty)\right]\left[u\frac{\partial u}{\partial x} + v\frac{\partial u}{\partial y}\right] = \mu\frac{\partial^2 u}{\partial y^2} + \rho_\infty g\beta(T - T_\infty) \tag{6.20}$$

The above simplification is frequently referred to as the **Boussinesq approximation**. When $\beta(T - T_\infty) \ll 1$, Eq. (6.20) simplifies to the following form

$$\left[u\frac{\partial u}{\partial x} + v\frac{\partial u}{\partial y}\right] = v\frac{\partial^2 u}{\partial y^2} + g\beta(T - T_\infty) \tag{6.21}$$

Eq. (6.21) is a key approximation in the analysis of natural convection over a flat plate.

Finally, now we are in a position to write down the governing equations for this problem, under the boundary layer simplifications, as follows

$$\frac{\partial u}{\partial x} + \frac{\partial v}{\partial y} = 0 \tag{6.22}$$

$$u\frac{\partial u}{\partial x} + v\frac{\partial u}{\partial y} = v\frac{\partial^2 u}{\partial y^2} + g\beta(T - T_\infty) \tag{6.23}$$

$$u\frac{\partial T}{\partial x} + v\frac{\partial T}{\partial y} = \alpha\frac{\partial^2 T}{\partial y^2} \tag{6.24}$$

In Eq. (6.23), the terms on the left-hand side represent inertial forces, while the first term on the right-hand represents the frictional force and the second represents the buoyancy force. In the energy equation (Eq. 6.24), the left-hand side represents the advection terms and the right-hand side represents the conduction term. Eqs. 6.22–6.24 satisfy closure, as we have three equations in three unknowns (u, v, and T). However, Eqs. 6.23 and 6.24 are coupled, consequent upon the presence of the term with the temperature difference in the momentum equation (Eq. 6.23).

Dimensionless numbers governing natural convection

We can now carry out nondimensionalization of the governing equations, with a view to obtain the pertinent dimensionless numbers governing natural convection.

In order to normalize the velocities u and v, we need a reference velocity. In the field of natural convection, a frequently used velocity scale is v/L (some researchers also have used α/L). Please note v/L has the units of velocity.

The following dimensionless quantities are introduced now

$$u^+ = \frac{u}{v/L} \tag{6.25}$$

$$v^+ = \frac{v}{v/L} \tag{6.26}$$

$$x^+ = \frac{x}{L} \tag{6.27}$$

$$y^+ = \frac{y}{L} \tag{6.28}$$

and the dimensionless temperature, ϕ, is defined as

$$\phi = \frac{(T - T_\infty)}{(T_w - T_\infty)} \tag{6.29}$$

While x^+, y^+, and ϕ all vary from 0 to 1, the same thing cannot be said about u^+ and v^+, as the normalizing velocity (v/L) is not a truly physical quantity and does not in way represent the maximum velocity in the problem. Even so, the use of (v/L) to scale velocity is useful and at the same time convenient.

Using the above framework, we now normalize Eqs. 6.22–6.24.
The continuity equation becomes

$$\frac{v}{L^2} \frac{\partial u^+}{\partial x^+} + \frac{v}{L^2} \frac{\partial v^+}{\partial y^+} = 0 \tag{6.30}$$

$$\text{or,} \quad \frac{\partial u^+}{\partial x^+} + \frac{\partial v^+}{\partial y^+} = 0 \tag{6.31}$$

The momentum equation becomes

$$\frac{v^2}{L^2.L} u^+ \frac{\partial u^+}{\partial x^+} + \frac{v^2}{L^2.L} v^+ \frac{\partial u^+}{\partial x^+} = \frac{v^2}{L^2.L} \frac{\partial^2 u}{\partial y^{+2}} + g\beta \ \phi(T_w - T_\infty) \tag{6.32}$$

On simplification, Eq. (6.32) becomes

$$u^+ \frac{\partial u^+}{\partial x^+} + v^+ \frac{\partial u^+}{\partial x^+} = \frac{\partial^2 u}{\partial y^{+2}} + \frac{g\beta\phi L^3 (T_w - T_\infty)}{v^2} \tag{6.33}$$

Please note that all the individual terms in Eq. (6.33) are dimensionless and so is ϕ.

Consequent upon this and also on its own, the group $\dfrac{g\beta L^3 \Delta T}{v^2}$ is dimensionless and a key quantity in natural convection known as the **Grashof number**, Gr_L, with $\Delta T = (T_w - T_\infty)$.

Hence Eq. (6.33) can be rewritten as

$$u^+ \frac{\partial u^+}{\partial x^+} + v^+ \frac{\partial u^+}{\partial y^+} = \frac{\partial^2 u^+}{\partial y^{+2}} + Gr_L \phi \qquad (6.34)$$

The energy equation becomes

$$u^+ \frac{\partial \phi}{\partial x^+} + v^+ \frac{\partial \phi}{\partial y^+} = \frac{\alpha}{v} \frac{\partial^2 u}{\partial y^2} \qquad (6.35)$$

Eq. (6.35) can be simplified as

$$u^+ \frac{\partial \phi}{\partial x^+} + v^+ \frac{\partial \phi}{\partial y^+} = \frac{1}{Pr} \frac{\partial^2 u}{\partial y^2} \qquad (6.36)$$

where the Prandtl number is $Pr = \dfrac{\alpha}{v}$. The importance of the Prandtl number in forced convection flows was discussed in detail in chapter 5.

In the light of the preceding arguments, it is clear that the functional dependence of u^+ and ϕ on the pertinent variables and parameters in the problem under consideration may be expressed as

$$u^+ = f_1(x^+, y^+, Gr_L, \phi) \qquad (6.37)$$

and

$$\phi = f_2(u^+, v^+, Pr) \qquad (6.38)$$

In our treatment of forced convection, we obtained a very important relation between the Nusselt number and the temperature gradient at the wall. Similarly, it is intuitive that

$$Nu_x = f_3(x^+, Gr_L, Pr) \qquad (6.39)$$

In Eq. (6.39), the y dependence is gone as $Nu_x = -\dfrac{x}{L}\left(\dfrac{\partial \phi^+}{\partial y^+}\right)_{y=0}$.

If we now seek an average Nusselt number, $\overline{Nu_L}$ as

$$\overline{Nu_L} = \frac{h_{ave}.L}{k} = \frac{\int_0^L h_x \, dx}{L} \cdot \frac{L}{k}$$

$$\overline{Nu_L} = \frac{\int_0^L \dfrac{Nu_x.k}{x} dx}{L} \cdot \frac{L}{k} = \int_0^L \frac{Nu_x dx}{x} \qquad (6.40)$$

$$\overline{Nu_L} = f_4(Gr_L, Pr)$$

In fact the product of $Gr_L.Pr$ is known as **Rayleigh number**, Ra_L.

$$Ra_L = \frac{g\beta\,\Delta TL^3}{\nu\,\alpha} \tag{6.41}$$

Hence, correlations for natural convection heat transfer coefficients are expressed in terms of Rayleigh number or Grashof number (and Prandtl number) and are typical power law forms that look like what is shown below

$$\overline{Nu_L} = aRa^b(\varsigma_1)^c(\varsigma_2)^d \tag{6.42}$$

where a, b, c, and d are constants, ς_1 may typically represent the viscosity correction term, and ς_2 may be a geometric parameter like ratio of height to width of an enclosure or ratio of inner to outer radius in a cylindrical or spherical annulus. In its simplest avatar, the Nusselt number correlations look like what is given in Eq. (6.43).

$$\overline{Nu_L} = aRa^b \tag{6.43}$$

Constants a and b need to be determined for the geometry under consideration. For simple geometries, a and b may be derived using one of the following methods.

1. Approximate methods like integral methods.
2. Analytical methods like similarity transformation.
3. Numerical methods like finite difference or finite element or finite volume method.

Of course, we should not forget that a and b can also be derived purely from experiments and the Rayleigh number, Ra_L, dependence could have also been obtained through the Buckingham pi theorem route.

Integral approach for natural convection on a vertical flat plate

The integral solution for natural convection over a flat plate has been derived by several investigators, such as Eckert and Squire (Lienhard and Lienhard, 2020). We first derive the integral forms of the momentum and energy equations that are similar to those for forced convection in the previous chapter. However, there are two key differences: (1) the presence of the buoyancy term in the x-momentum equation and (2) recognition that $\delta_t = \delta$ as the temperature difference causes the flow and, in view of this, the two boundary layers are coupled. In fact this assumption (i.e., $\delta_t \approx \delta$) is strictly valid when the Prandtl numbers are close to one. While this assumption will also work quite well for Pr $\gg$ 1, the analysis will not be accurate for Pr $\ll$ 1. In these cases, we need to rely on either experimental results or full numerical solutions to get a handle on the problem. For the sake of completeness, we present a quick overview of the derivation of the integral momentum and energy equations.

Consider the boundary layer forms of these equations (Eqs. 6.22 and 6.24)

$$u\frac{\partial u}{\partial x}+v\frac{\partial u}{\partial y}=\nu\frac{\partial^2 u}{\partial y^2}+g\beta(T-T_\infty) \tag{6.44}$$

$$u\frac{\partial T}{\partial x}+v\frac{\partial T}{\partial y}=\alpha\frac{\partial^2 T}{\partial y^2} \tag{6.45}$$

Integrating Eq. (6.44) from 0 to δ in y and Eq. (6.45) from 0 to δ_T in y, with the understanding that $\delta = \delta_T$, we have

$$\int_0^\delta u\frac{\partial u}{\partial x}dy+\int_0^\delta v\frac{\partial u}{\partial y}dy=\int_0^\delta \nu\frac{\partial^2 u}{\partial y^2}dy+\int_0^\delta g\beta(T-T_\infty)dy \tag{6.46}$$

and

$$\int_0^\delta u\frac{\partial T}{\partial x}dy+\int_0^\delta v\frac{\partial T}{\partial y}dy=\int_0^\delta \alpha\frac{\partial^2 T}{\partial y^2}dy \tag{6.47}$$

Eq. 6.46 can be written as

$$\int_0^\delta \frac{\partial u^2}{\partial x}dy-\int_0^\delta u\frac{\partial u}{\partial x}dy+\int_0^\delta \frac{\partial uv}{\partial y}dy-\int_0^\delta u\frac{\partial v}{\partial y}dy$$
$$=\nu\frac{\partial u}{\partial y}\Big|_0^\delta+\int_0^\delta g\beta(T-T_\infty)dy \tag{6.48}$$

Invoking the continuity equation for the sum of the second and fourth terms on the left-hand side and the Newton's law of viscosity for the first term on the right-hand side of the equation, we can rewrite Eq. (6.48) as

$$\frac{d}{dx}\int_0^\delta (u^2-uu_\infty)dy=-\frac{\tau_w}{\rho}+\int_0^\delta g\beta(T-T_\infty)dy \tag{6.49}$$

In arriving at Eq. (6.49), the Leibnitz rule has been used to interchange the order of differential and integral operators and $\frac{\partial}{\partial x}$ has been replaced by $\frac{d}{dx}$ as y dependence goes, upon integrating $(u^2 - uu_\infty)$ with respect to y from 0 to δ. The velocity gradient $\frac{\partial u}{\partial y}$ is also 0 at $y = \delta$. τ_w is the wall shear stress, given by $\mu(\partial u/\partial y)_{y=0}$. The integral energy equation is exactly same as the one derived in chapter 5 and is given by

$$\frac{d}{dx}\int_0^\delta u(T-T_\infty)dy=-\alpha\frac{\partial T}{\partial y}\Big|_{y=0} \tag{6.50}$$

Once these equations are in place, the integral method progresses with the usual assumptions of polynomial profiles for temperature and velocity. We have a special difficulty here, as unlike forced convection, we do not have a reference velocity. Equation 6.49 can be simplified with the understanding that u subscript infinity is 0 and this then becomes the final form of the momentum integral equation for laminar natural convection.

Assuming a linear profile for temperature (the simplest possible) of the form

$$\frac{(T - T_\infty)}{(T_w - T_\infty)} = a + b\left(\frac{y}{\delta}\right) \tag{6.51}$$

We can determine a and b from the boundary conditions

$$y = 0, \; T = T_w \tag{6.52}$$

$$y = \delta, \; T = T_\infty \tag{6.53}$$

On applying these two boundary conditions, the "trial" temperature profile turns out to be

$$\frac{(T - T_\infty)}{(T_w - T_\infty)} = 1 - \left(\frac{y}{\delta}\right) \tag{6.54}$$

From an earlier discussion on the nature of the velocity profile, it is clear that at $y = 0$, $u = 0$ and at $y = \delta$, $u = 0$ with $u = u_{max}$ somewhere in between with u_{max} itself being as yet unknown. However, a quick order of magnitude analysis will help us get a ballpark of u_{max}. Consider the x-momentum equation (Eq. 6.44), and let the scale for $u \sim u_{ref}$. Consider the two terms on the right-hand side of Eq. (6.44). The respective scales are

$$\nu \frac{u_{ref}}{\delta^2} \sim g\beta(T_w - T_\infty) \tag{6.55}$$

$$\therefore \; u_{ref} \sim \frac{\delta^2}{\nu} g\beta(T_w - T_\infty) \tag{6.56}$$

We may replace the term $\sim$ with an equality by introducing a constant "C" with the understanding that "C" will be of the order of unity.

$$\therefore \; u_{ref} = C \frac{\delta^2}{\nu} g\beta(T_w - T_\infty) \tag{6.57}$$

Now we can get back to the "trial" profile for velocity. In view of the tricky boundary conditions of $u = 0$ at $y = 0$ as well as at $y = \delta$, we assume a cubic profile for velocity as follows.

$$\frac{u}{u_{ref}} = \left[\frac{y}{\delta} + d\left(\frac{y}{\delta}\right)^2 + e\left(\frac{y}{\delta}\right)^3\right] \tag{6.58}$$

The following boundary conditions are applicable

$$\text{At, } y = 0, \; u = 0 \tag{6.59}$$

This is already satisfied by Eq. (6.58)

$$\text{At } y = \delta, \text{ u} = 0 \tag{6.60}$$

$$\therefore 1 + d + e = 0 \tag{6.61}$$

$$\text{At } y = \delta, \frac{\partial u}{\partial y} = 0 \tag{6.62}$$

$$\therefore 1 + \frac{2d}{\delta} + \frac{3e}{\delta} = 0 \tag{6.63}$$

$$\therefore 1 + 2d = -3e \tag{6.64}$$

From Eqs. 6.61 and 6.64, we have e =1 and d = −2.
∴ The velocity profile is given by

$$\frac{u}{u_{ref}} = \left[\frac{y}{\delta} - 2\left(\frac{y}{\delta}\right)^2 + \left(\frac{y}{\delta}\right)^3\right]$$
$$= \frac{y}{\delta}\left[1 - \frac{y}{\delta}\right]^2 \tag{6.65}$$

Now we will have to examine what happens if we apply the profile to the momentum equation at the wall, i.e., y = 0 for any x.

At y = 0, we know that u = 0, v = 0, and so the momentum equation reduces to

$$\frac{\partial^2 u}{\partial y^2} = -\frac{g\beta(T_w - T_\infty)}{\nu} \tag{6.66}$$

Differentiating Eq. (6.65) twice and substituting in Eq. (6.66) (at y = 0),

$$-\frac{4u_{ref}}{\delta^2} = -\frac{g\beta(T_w - T_\infty)}{\nu} \tag{6.67}$$

$$\therefore u_{ref} = \frac{\delta^2}{4\nu}g\beta(T_w - T_\infty) \tag{6.68}$$

A quick sanity check is in order. An estimate of u_{ref} from Eq. (6.57) is very similar to the one obtained in Eq. (6.68), with C = $\frac{1}{4}$. This confirms that we committed no "grave error" in assuming the cubic profile and working out the constants of the profile, as discussed above.

We now have a choice to proceed with Eqs. 6.57 or 6.68 for u_{ref}. However, if we proceed with Eq. (6.68), we are left with only one unknown which is δ and we have two Eqs. 6.49 and 6.50. This leads to a mathematical "uneasiness" (technically an overdetermined system). and so we stick with Eq. (6.57) so that we have two unknowns C and δ and two integral equations!

Substituting for the velocity profile in the momentum integral equation together with form of u_{ref}, we get the following

$$\frac{d}{dx}\int_0^\delta u_{ref}^2 \cdot \left(\frac{y}{\delta}\right)^2 \left[1-\frac{y}{\delta}\right]^4 dy = -v\frac{u_{ref}}{\delta} + g\beta(T_w - T_\infty)\int_0^\delta \left(1-\frac{y}{\delta}\right) dy \qquad (6.69)$$

On simplifying Eq. (6.69), we get

$$\frac{C^2}{v^2}g\beta(T_w - T_\infty)\frac{d}{dx}\int_0^\delta \delta^4 \cdot \left(\frac{y}{\delta}\right)^2 \left[1-\frac{y}{\delta}\right]^4 dy = \left(\frac{1}{2}-C\right)\delta \qquad (6.70)$$

We now integrate the term within the integrand on the left hand side and simplify as follows.

$$\frac{C^2}{v^2}g\beta(T_w - T_\infty)\frac{d}{dx}\int_0^\delta \delta^4 \cdot \left(\frac{y}{\delta}\right)^2 \left[1-4\frac{y}{\delta}+6\frac{y^2}{\delta^2}-4\frac{y^3}{\delta^3}+\frac{y^4}{\delta^4}\right]dy$$
$$=\left(\frac{1}{2}-C\right)\delta \qquad (6.71)$$

$$\frac{C^2}{v^2}g\beta(T_w - T_\infty)\frac{d}{dx}\left\{\delta^4\left[\frac{y^3}{3\delta^2}-\frac{y^4}{\delta^3}+\frac{6y^5}{5\delta^4}-\frac{4y^6}{6\delta^5}+\frac{y^7}{7\delta^6}\right]_0^\delta\right\}$$
$$=\left(\frac{1}{2}-C\right)\delta \qquad (6.72)$$

On simplifying, after a bit of algebra to get rid of the delta on the right hand side of the above equation, we get

$$\frac{1}{21}\frac{C^2}{v^2}g\beta(T_w - T_\infty)\delta^3\frac{d\delta}{dx}=\frac{1}{2}-C \qquad (6.73)$$

Upon integrating Eq. (6.73) and substituting for $\delta = 0$ at $x = 0$, we get

$$\therefore \delta^4 = 84\left[\frac{1}{2}-C\right]\left[\frac{C^2}{v^2}g\beta(T_w - T_\infty)\right]^{-1}x \qquad (6.74)$$

Please note that Eq. (6.74) confirms that the velocity boundary layer thickness is proportional to the buoyancy term raised to $\left(-\frac{1}{4}\right)$ and $x^{1/4}$. This revalidates the coupling between the momentum and thermal boundary layers. We can now substitute for δ in the integral energy equation and use this equation to determine the constant C. Upon doing this we can evaluate

$$q_w = -k\frac{\partial T}{\partial y}\bigg|_{y=0} = \frac{-k(T_w - T_\infty)}{\delta} \qquad (6.75)$$

with δ having been determined q_w, and hence h_x and Nu_x can be determined. One can intuit now that the presence of v in the expression for δ and that of α in the integral equation confirms that the Prandtl number Pr given by v/α must appear in some avatar in C and should eventually get into the expression for Nu_x. Let us now complete the remaining part of this fascinating journey of the integral method to solve the boundary layer equations.

Consider the integral energy equation given by Eq. (6.50)

$$\frac{d}{dx}\int_0^\delta u(T - T_\infty)\,dy = -\alpha\frac{\partial T}{\partial y}\bigg|_{y=0} \tag{6.76}$$

Substituting for the velocity and temperature profiles, we have

$$\frac{d}{dx}\int_0^1 \frac{Cg\beta(T_w - T_\infty)}{\nu}\delta^3\left(\frac{y}{\delta}\right)\left(1 - \frac{y}{\delta}\right)^3 (T_w - T_\infty)d\left(\frac{y}{\delta}\right)$$
$$= \frac{\alpha(T_w - T_\infty)}{\delta} \tag{6.77}$$

$$\frac{Cg\beta(T_w - T_\infty)}{\nu}\frac{d}{dx}\delta^3\int_0^1\left(\frac{y}{\delta}\right)\left(1 - \frac{y}{\delta}\right)^3 (T_w - T_\infty)d\left(\frac{y}{\delta}\right) = \frac{\alpha}{\delta} \tag{6.78}$$

$$\int_0^1\left(\frac{y}{\delta}\right)\left(1 - \frac{y}{\delta}\right)^3 d\left(\frac{y}{\delta}\right) = \int_0^1 x(1 - x)^3\,dx \tag{6.79}$$

where $x = \dfrac{y}{\delta}$

$$\therefore \int_0^1\left(\frac{y}{\delta}\right)\left(1 - \frac{y}{\delta}\right)^3 d\left(\frac{y}{\delta}\right) = \int_0^1 x(1 - 3x + 3x^2 - x^3)\,dx \tag{6.80}$$

On integrating and simplifying

$$\int_0^1\left(\frac{y}{\delta}\right)\left(1 - \frac{y}{\delta}\right)^3 d\left(\frac{y}{\delta}\right) = \frac{1}{20} \tag{6.81}$$

Consequent upon the evaluation of the integral we have

$$\frac{Cg\beta(T_w - T_\infty)}{\nu}\frac{d}{dx}\left(\frac{\delta^3}{20}\right) = \frac{\alpha}{\delta} \tag{6.82}$$

$$\frac{Cg\beta(T_w - T_\infty)}{\nu}\delta\frac{d}{dx}\left(\frac{\delta^3}{20}\right) = \alpha \tag{6.83}$$

$$\therefore \frac{3Cg\beta(T_w - T_\infty)}{\nu}\frac{d}{dx}\left(\frac{\delta^4}{4}\right) = 20\alpha \tag{6.84}$$

$$\frac{d}{dx}\left(\frac{\delta^4}{4}\right) = \frac{20\alpha\nu}{3Cg\beta(T_w - T_\infty)} \tag{6.85}$$

Integrating Eq. (6.85) we get

$$\delta^4 = \frac{80\alpha v}{3Cg\beta(T_w - T_\infty)} x + A \tag{6.86}$$

$$\text{At } x = 0, \ \delta = 0 \tag{6.87}$$

$$\therefore A = 0 \tag{6.88}$$

$$\therefore \delta = \left[\frac{80\alpha x v}{3Cg\beta(T_w - T_\infty)} \right]^{1/4} \tag{6.89}$$

Eqs. 6.74 and 6.89 are both expressions for the boundary layer thickness. Equating the two, we can solve for C as

$$84\left[\frac{1}{2} - C\right]\left[\frac{C^2}{v^2} g\beta(T_w - T_\infty)\right]^{-1} x = \left[\frac{80\alpha x v}{3Cg\beta(T_w - T_\infty)}\right]$$

$$84\left[\frac{1}{2} - C\right]\frac{1}{C} = \frac{80}{3}\left(\frac{1}{Pr}\right) \tag{6.90}$$

$$C = \frac{Pr}{2\left[Pr + \dfrac{20}{63}\right]}$$

Substituting in Eq. (6.74)

$$\delta^4 = \frac{80xv^2 \, 2\left[Pr + \dfrac{20}{63}\right]}{3\left[Pr \times g\beta \Delta T\right]} \tag{6.91}$$

On simplifying, we get

$$\frac{\delta}{x} = 2.7\left[Pr + 0.317\right]^{1/4} . Pr^{-1/2} Gr^{-1/4} \tag{6.92}$$

The heat transfer at the wall is given by

$$q_w = -k\frac{\partial T}{\partial y}\bigg|_{y=0} = k\frac{(T_w - T_\infty)}{\delta} \tag{6.93}$$

$$q_w = h_x.(T_w - T_\delta) = \frac{Nu_x.k}{x}(T_w - T_\infty) \tag{6.94}$$

On equating 6.93 and 6.94, we get the expression for the local Nusselt number, Nu_x as

$$Nu_x = \frac{x}{\delta} \tag{6.95}$$

Substituting for δ from Eq. (6.92)

$$Nu_x = 0.37(Ra_x)^{(1/4)}\left[\frac{Pr}{0.317 + Pr}\right]^{(1/4)} \tag{6.96}$$

Following the developments outlined in chapter 5, we can relate the average Nusselt number $\overline{Nu_L}$ from x = 0 to x = L, to the local Nusselt number, Nu_x, as

$$\overline{Nu_L} = \frac{4}{3}Nu_L \tag{6.97}$$

(The derivation of the above expression is given as an exercise in problem 6.5 at the end of this chapter).

In Eq. 6.97 Nu_L is the local Nusselt number at x = L.

$$\overline{Nu_L} = 0.49(Ra_L)^{0.25}\left[\frac{Pr}{0.317 + Pr}\right]^{0.25} \tag{6.98}$$

If we employ a quadratic profile for temperature instead of the linear temperature profile and a cubic profile for velocity, as assumed in the above development, we get the following expression for average Nusselt number as

$$\overline{Nu_L} = 0.678(Ra_L)^{0.25}\left[\frac{Pr}{0.952 + Pr}\right]^{0.25} \tag{6.99}$$

For gases, Eq. (6.98) is good enough under 20% error, though Eq. (6.99) is more accurate. The derivation of Eq. (6.99) is given as an exercise at the end of the chapter.

6.4 Empirical correlations for natural convection

The development in the previous section has historical significance, as it exemplifies our early efforts to conquer the problem of natural convection.

Churchill and Chu looked at all the experimental data on natural convection over a vertical flat plate up to that point in time and, with guidance about the form of the Nusselt number correlation from the integral solution, proposed the now widely used Churchill-Chu correlation (Churchill and Chu, 1975a) given below

$$\overline{Nu_L} = 0.68 + 0.67 Ra_L^{1/4}\left[1 + \left(\frac{0.492}{Pr}\right)^{9/16}\right]^{-4/9} \tag{6.100}$$

This works for Ra_L up to 10^{12} and is a very useful engineering result. Please see Lienhard and Lienhard (2020) for a fuller discussion of the development of a heat transfer correlation for natural convection on a vertical surface. All properties in the above correlation are to be evaluated at a mean temperature $T_{mean} = (T_w + T_\infty)/2$ except for β. For gases, $\beta = 1/T$ and the temperature T has to be T_∞. For liquids, β is available in charts or tables in books and online resources. For problems involving a large ΔT, we might face issues like errors introduced due to the Boussinesq approximation and also those introduced due to the assumptions of constant properties. However, if $(T_W - T_\infty)$ is, say, 40 or 50 °C, and we use the mean temperature of T_{mean}, we should get results that are accurate to within a few percentage points with respect to a full-blown variable property solution. In this day and age, we can actually solve the variable property case on an open-source or commercial software and check the error involved in the constant property assumption, all by ourselves. For a flat plate, $Ra_L \geq 10^9$ is taken to be the criterion for transition to turbulent natural convection.

Natural convection from vertical cylinders

If one encounters a tall vertical cylinder such that $\dfrac{\delta}{H} \ll 1$, where δ is the largest boundary layer thickness in the cylinder and H is the height of the cylinder, then the correlation for the vertical flat plate holds good. However, if this assumption does not hold, either because of a thick boundary layer or a short cylinder, correlations for the enhanced heat transfer due to curvature need to be considered. Please refer to Lienhard and Lienhard (2020) for more information on this case.

Natural convection from a heated sphere

For $Pr > 0.7$ and $Ra_D < 10^{11}$, Churchill proposed the following correlation (Churchill, 1983).

$$Nu_D = 2 + \frac{0.589\,Ra_D^{0.25}}{\left[1+\left(\dfrac{0.469}{Pr}\right)^{9/16}\right]^{4/9}} \qquad (6.101)$$

Natural convection from horizontal cylinders

Churchill and Chu have presented correlations by looking at existing data (Churchill and Chu, 1975b). The correlation is given below.

$$\overline{Nu_D} = 0.36 + \frac{0.518\,Ra_D^{1/4}}{\left[1+\left(\dfrac{0.559}{Pr}\right)^{9/16}\right]^{4/9}} \qquad (6.102)$$

This is valid for $Ra_D \leq 10^9$. For $Ra_D > 10^9$, the following correlation, applicable for turbulent flow, is useful.

$$\overline{Nu_D} = \left\{0.60 + 0.387\left[\frac{Ra_D}{\left[1+\left\{\dfrac{0.559}{Pr}\right\}^{9/16}\right]^{16/9}}\right]^{1/6}\right\}^2 \qquad (6.103)$$

Please note that the key difference between Eqs. 6.102 and 6.103 is the Rayleigh number exponents. While the Nusselt number scales as $Ra_D^{1/4}$ for laminar flow, the exponent in the case of turbulent flow is 0.33, and the usually accepted cutoff for transition to turbulence in natural convection is that Ra_L or $Ra_D \geq 10^9$, as the case may be!

Natural convection from other geometries

Correlation tables from handbooks or online resources may be referred to for correlations involving natural convection from enclosures, heated horizontal plate facing upward/downward, and for other geometries of specific interest. Properties of air are given in Table 6.3.

Example 6.1: *An aluminum plate having $\rho = 2700$ kg/m³, $k = 205$ W/mK, $C_p = 900$ J/kgK is heated to a temperature of 400 K. At this point in time ($t = 0$), the heating is stopped and the plate begins to cool in still air at 310K. The plate is 50 cm tall and 30 cm wide and has a thickness of 10 mm. The surface of the plate has an emissivity of 0.25, which is independent of the temperature.*

(a) Develop an expression for the rate of change of the plate temperature assuming spatial isothermality of the plate, neglecting surface radiation.
(b) Rework the expression obtained in (a) if radiation is considered (Temperature of the surroundings for radiation can be considered to be 37 °C).
(c) Determine initial rate of cooling when plate temperature is 400K, when (i) radiation is neglected and (ii) radiation is considered.
(d) Comment on the contribution of radiation in this problem.
(e) Is the assumption of spatial isothermality valid (i) without consideration of radiation and (ii) with consideration of radiation?

Solution
Given data
Density $\rho = 2700$ kg/m³; thermal conductivity $k = 205$ W/mK; specific heat $C_p = 900$ J/kgK; initial temperature of the plate (T_S) = 400 K; ambient temperature (T_∞) = 310 K; surface emissivity (ε) = 0.25; dimensions of the plate = 0.5 × 0.3 × 0.01 (all in m).

(a) Neglecting radiation effects, from energy balance,

$$-\rho C_p V \frac{dT}{dt} = h.(2A_s).(T - T_\infty)$$

$$\frac{dT}{dt} = -\frac{h.(2A_s).(T - T_\infty)}{\rho C_p V} \quad \frac{dT}{dt}\bigg|_{t=0} = -\frac{h.(2A_s).(T_s - T_\infty)}{\rho C_p V}$$

(b) Considering the radiation, from energy balance,

$$-\rho C_p V \frac{dT}{dt} = h(2A_s)(T - T_\infty) + \varepsilon \sigma (2A_s)(T^4 - T_\infty^4)$$

$$\frac{dT}{dt} = -\frac{(2A_s)}{\rho C_p V}\left[h(T_w - T_\infty) + \varepsilon \sigma (T^4 - T_\infty^4)\right]$$

$$\frac{dT}{dt}\bigg|_{t=0} = -\frac{(2A_s)}{\rho C_p V}\left[h(T_w - T_\infty) + \varepsilon \sigma (T_s^4 - T_\infty^4)\right]$$

(c) Film temperature, $T_f = (T_S + T_\infty)/2 = (400 + 310)/2 = 710/2 = 355$ K. Properties of air at $T_f = 355$ K:
$\nu = 2.073 \times 10^{-5}$ m²/s, $\alpha = 2.931 \times 10^{-5}$ m²/s, $k = 0.02984$ W/mK, Pr = 0.707.

$$\beta = \frac{1}{T_\infty} = 3.22 \times 10^{-3} K^{-1},$$

$$Ra_L = \frac{g\beta(T_s - T_\infty)L^3}{\nu \, \alpha}$$

$$Ra_L = \frac{9.81 \times 3.22 \times 10^{-3} \times (400 - 310) \times 0.5^3 \times 10^{10}}{2.073 \times 2.931}$$

$$Ra_L = 5.85 \times 10^8$$

From Eq. (6.100)

$$\overline{Nu_L} = 0.68 + 0.67 Ra_L^{1/4} \left[1 + \left(\frac{0.492}{Pr} \right)^{9/16} \right]^{-4/9}$$

$$\overline{Nu_L} = 0.68 + 0.67 \times (5.85 \times 10^8)^{1/4} \left[1 + \left(\frac{0.492}{0.707} \right)^{9/16} \right]^{-4/9}$$

$$\overline{Nu_L} = 80.65$$

$$\overline{h_c} = \overline{Nu_L} \times \frac{k}{L}$$

$$\overline{h_c} = 80.65 \times \frac{0.02984}{0.5}$$

$$\overline{h_c} = 4.81 \, W/m^2 K.$$

(i) Initial rate of cooling of the plate when neglecting radiation

$$\left. \frac{\partial T}{\partial t} \right|_{t=0} = -\frac{4.81 \times 2 \times 10^{-4} \times 90}{2700 \times 900 \times 10 \times 10^{-7}}$$

$$\left. \frac{\partial T}{\partial t} \right|_{t=0} = -0.0356 \, K/s$$

(ii) Initial rate of cooling of the plate when considering radiation

$$\left. \frac{\partial T}{\partial t} \right|_{t=0} = -\frac{2 \times 10^3 \times 664.87}{2700 \times 900 \times 10}$$

$$\left. \frac{\partial T}{\partial t} \right|_{t=0} = -0.055 \, K/s$$

(d) Hence, radiation plays a significant role in the heat transfer from the plate and about 35% error occurs in $\dfrac{dT}{dt}$ if radiation is not considered.

(e) Biot number $(Bi) = \dfrac{hL}{k}$

(i) Without consideration of radiation

$$Bi = \frac{4.81 \times 0.5}{205} = 0.011 < 0.1$$

Yes, the assumption of spatial isothermality is valid.

(ii) With consideration of radiation

$$q_{radiation} = \in \sigma (T_w^4 - T_\infty^4) = h_r (T_w - T_\infty)$$

with h_r being the radiative heat transfer coefficient. Introduction of this quantity is convenient but has to be used with caution as radiation is highly nonlinear.

$$h_{total} = h_c + h_r$$

$$h_{total} = h_c + \sigma \in (T_w^2 - T_\infty^2)(T_w - T_\infty)$$

$$h_{total} = 4.18 + 5.67 \times 10^{-8} \times 0.25 \times (400^2 + 310^2)(710)$$

$$h_{total} = 7.387 W/m^2 K$$

$$Bi = \frac{7.387 \times 0.5}{205} = 0.018 < 0.1$$

Yes, the assumption of spatial isothermality is valid.

Example 6.2: *Consider a cylindrical electric immersion heater with a rating of 1000W. The heater is 16 mm in diameter and 350 mm long. The heater is placed horizontally in a pool of water in a bucket at 37 °C.*

(a) Estimate the temperature of the heater under steady state. The properties of the water at temperatures 37 °C and 67 °C are tabulated below in Table 6.1 (A more comprehensive Table of properties for water was given in chapter 5). Please note that only β has to be evaluated at T_∞ and all other properties need to be evaluated at

Table 6.1 Thermophysical properties of water

S. No.	T(°C)	ρ(kg/m³)	C$_p$ (J/ kg.K)	k (W/m.K)	α(m²/s)	ν, (m²/s)	Pr
1	37	993.3	4179	0.6260	1.508×10^{-7}	6.98×10^{-7}	4.63
2	67	979.5	4189	0.6605	1.610×10^{-7}	4.31×10^{-7}	2.68

$T_f = (T_W + T_\infty)/2$, where T_W is the heater temperature. Since T_W is unknown, assume T_W initially to be 67 °C.

(b) *What will be the heater temperature be if we forget to add water in the bucket and the heater is left to operate in air?*

Solution

(a) T_w is assumed to be 67 °C as given in the problem. Film temperature, $T_f = (T_W + T_\infty)/2$ = (340 + 310)/2 = 650/2 = 325K.

Properties of water at T_f = 325K:

$v = 5.65 \times 10^{-7}$ m²/s, $\alpha = 1.56 \times 10^{-7}$ m²/s, k = 0.64325 W/mK, Pr = 3.655.
$\beta = 3.61 \times 10^{-4}$ K⁻¹ (at T_∞)

$$Ra_D = \frac{g\beta(T_w - T_\infty)L^3}{v\,\alpha}$$

$$Ra_D = \frac{9.81 \times (340 - 310) \times 0.016^3 \times 3.61 \times 10^{-4}}{5.65 \times 1.56 \times 10^{-14}}$$

$$Ra_D = 4.94 \times 10^6$$

From Eq. (6.103)

$$\overline{Nu_D} = \left(0.60 + 0.387 \left(\frac{Ra_D}{\left(1 + \left(\frac{0.559}{Pr} \right)^{9/16} \right)^{16/9}} \right)^{1/6} \right)^2$$

$$\overline{Nu_D} = \left(0.60 + 0.387 \left(\frac{4.94 \times 10^6}{\left(1 + \left(\frac{0.559}{Pr} \right)^{9/16} \right)^{16/9}} \right)^{1/6} \right)^2$$

$$\overline{Nu_D} = 27.286$$

$$h = \overline{Nu_D} \times \frac{k}{D}$$

$$h = 27.286 \times \frac{0.64325}{0.016}$$

$$h = 1097 W / m^2 K.$$

The estimated wall temperature $T_w = T_\infty + \dfrac{1000}{h \times \pi DL} = 361.8\,K$

Iteration 2

Update wall temperature from 340K to 361.8K

Film temperature, $T_f = (T_W + T_\infty)/2 = (361.8 + 310)/2 = 335.9K$.

Properties of water at $T_f = 325K$:

$v = 4.67 \times 10^{-7}$ m²/s, $\alpha = 1.6 \times 10^{-7}$ m²/s, k = 0:66 W/mK, Pr = 2.95

$$Ra_D = \frac{9.81 \times (361.8 - 310) \times 0.016^3 \times 3.61 \times 10^{-4}}{4.67 \times 1.6 \times 10^{-14}}$$

$$Ra_D = 1 \times 10^7$$

$$\overline{Nu_D} = \left(0.60 + 0.387\left(\frac{1 \times 10^7}{\left(1 + \left(\frac{0.559}{2.95}\right)^{9/16}\right)^{16/9}}\right)^{1/6}\right)^2$$

$$\overline{Nu_D} = 33.135$$

$$h = \overline{Nu_D} \times \frac{k}{D}$$

$$h = 33.135 \times \frac{0.66}{0.016}$$

$$h = 1358.15 \text{ W/m}^2\text{K}$$

Estimated wall temperature $T_w = T_\infty + \dfrac{1000}{h \times \pi DL} = 351.9$ K.

After seven iterations (see Table 6.2),

assumed T_W = Estimated T_W = 354.6 K = 81.5 °C.

(b) Now with air as the operating fluid and assuming $T_W = 360$ K

Film temperature, $T_f = (T_W + T_\infty)/2 = (360 + 310)/2 = 335$ K.

Properties of air at $T_f = 335$ K:

$v = 1.9 \times 10^{-5}$ m²/s, $\alpha = 2.72 \times 10^{-5}$ m²/s, k = 0.0285 W/mK, Pr = 0.7075

$$\beta = \frac{1}{T_\infty} = 3.2 \times 10^{-3} \text{K}^{-1}$$

Table 6.2 Iterative method to solve Example 6.2 (All temperatures in Kelvin)

S. No.	$T_{w,i}$	$T_{w,i+1}$	$(T_{w,i+1} - T_{w,i})^2$
1	340.0	361.8	475.24
2	361.8	351.8	100.00
3	351.8	355.7	15.29
4	355.7	354.1	2.59
5	354.1	354.5	0.16
6	354.5	354.6	0.01
7	354.6	354.6	0.00

$$Ra_D = \frac{9.81 \times (360-310) \times 0.016^3 \times 3.2 \times 10^{-3}}{1.9 \times 2.72 \times 10^{-10}}$$

$$Ra_L = 12540.8$$

$$\overline{Nu_D} = \left(0.60 + 0.387 \left(\frac{12540.8}{\left(1 + \left(\frac{0.559}{0.7075} \right)^{9/16} \right)^{16/9}} \right)^{1/6} \right)^2$$

$$\overline{Nu_D} = 4.6$$

$$h = \overline{Nu_D} \times \frac{k}{D}$$

$$h = 4.6 \times \frac{0.028545}{0.016}$$

$$h = 8.232 \text{ W/m}^2\text{K}.$$

Estimated wall temperature, $T_w = T_\infty + \dfrac{1000}{h \times \pi DL} = 7217.8K$

As we can see, at the end of the first iteration itself, the temperature of the heater has reached an incredibly high value far beyond the melting point of the material of the heater (what it is made of is immaterial). Hence, even by inadvertence, we can ill-afford to operate an immersion heater designed for water, with air as the medium.

Problems

6.1 A 25 cm tall vertical plate infinitely deep in the direction perpendicular to the plane of the paper and negligible thickness is maintained at a temperature of 100 °C in quiescent air at 35 °C. Get an estimate of the boundary layer thickness at
(a) 5 cm from the bottom.
(b) 15 cm from the bottom.
(c) The top of the plate.

6.2 Revisit Problem 6.1. Get an estimate of u_{ref} for the problem under question. Compare it with typical velocities encountered in forced convection, with air as the ambient medium.

6.3 A heated sphere at $T_w = 100$ °C made of copper has a diameter of 20 mm. It is kept in still air at 30 °C. The density and specific heat of copper are 8900 kg/m^3 and 0.384 kJ/kg.K, respectively.
(a) Determine the heat transfer coefficient from the sphere.
(b) At time t = 0, when the heated sphere begins to cool, what is the value of $\dfrac{dT}{dt}$ assuming the sphere to be spatially isothermal.

6.4 Consider Problem 2.4 of chapter 2 that concerned the cooling of a current carrying copper wire. In this problem, the free convection heat transfer coefficient was assumed to be 6 W/m^{2}K. Is this justified?

6.5 Consider the integral solution of natural convection over a flat plate for a fluid with Pr > 1. Considering a quadratic profile for the temperature and a cubic profile for velocity, determine the following.
(a) Expression for $\delta(x)$.
(b) Expression for Nu_x.
(c) Expression for $\overline{Nu_L}$.

6.6 Using the results from Problem 6.5, determine the location y (for any x) at which the vertical velocity is the maximum and hence determine its magnitude.

6.7 Revisit Problem 6.1. Determine the local and average Nusselt numbers at the three locations mentioned in the problem.

Table 6.3 Thermophysical properties of air at atmospheric pressure (101325 Pa) (Kadoya et al., 1985; Jacobsen et al., 1992)

T(K)	ρ(kg/m³)	c_p (J/kg.K)	μ(kg/m.s)	ν(m²/s)	k (W/m.K)	α(m²/s)	Pr
100	3.605	1039	0.711×10^{-5}	0.197×10^{-5}	0.00941	0.251×10^{-5}	0.784
150	2.368	1012	1.035	0.437	0.01406	0.587	0.745
200	1.769	1007	1.333	0.754	0.01836	1.031	0.731
250	1.412	1006	1.606	1.137	0.02241	1.578	0.721
260	1.358	1006	1.649	1.214	0.02329	1.705	0.712
270	1.308	1006	1.699	1.299	0.02400	1.824	0.712
280	1.261	1006	1.747	1.385	0.02473	1.879	0.711
290	1.217	1006	1.795	1.475	0.02544	2.078	0.710
300	1.177	1007	1.857	1.578	0.02623	2.213	0.713
310	1.139	1007	1.889	1.659	0.02684	2.340	0.709
320	1.103	1008	1.935	1.754	0.02753	2.476	0.708
330	1.070	1008	1.981	1.851	0.02821	2.616	0.708
340	1.038	1009	2.025	1.951	0.02888	2.821	0.707
350	1.008	1009	2.090	2.073	0.02984	2.931	0.707
400	0.8821	1014	2.310	2.619	0.03328	3.721	0.704
450	0.7840	1021	2.517	3.210	0.03656	4.567	0.703
500	0.7056	1030	2.713	3.845	0.03971	5.464	0.704
550	0.6414	1040	2.902	4.524	0.04277	6.412	0.706
600	0.5880	1051	3.082	5.242	0.04573	7.400	0.708
650	0.5427	1063	3.257	6.001	0.04863	8.430	0.712
700	0.5040	1075	3.425	6.796	0.05146	9.498	0.715
750	0.4704	1087	3.588	7.623	0.05425	10.61	0.719
800	0.4410	1099	3.747	8.497	0.05699	11.76	0.723

References

Bejan, A., 2013. Convection Heat Transfer. John Wiley and Sons, Hoboken, NJ.

Churchill, S.W., 1983. Comprehensive, theoretically based, correlating equations for free convection from isothermal spheres. Chem. Eng. Commun. 24 (4–6), 339–352.

Churchill, Stuart W., Chu, Humbert H.S., 1975a. Correlating equations for laminar and turbulent free convection from a vertical plate. Int. J. Heat Mass Transf. 18 (11), 1323–1329.

Churchill, S.W., Chu, H.H.S., 1975b. Correlating equations for laminar and turbulent free convection from a horizontal cylinder. Int. J. Heat Mass Transf. 18 (9), 1049–1053.

Kadoya, K., Matsunaga, N., Nagashima, A., 1985. Viscosity and thermal conductivity of dry air in the gaseous phase. J. Phys. Chem. Ref. Data 14 (4), 947–970.

Jacobsen, R.T., Penoncello, S.G., Beyerlein, S.W., Clarke, W.P., Lemmon, E.W., 1992. A thermodynamic property formulation for air. Fluid Phase Equilibria 79, 113–124.

Lienhard IV, J.H., Lienhard V, J.H., 2020. A Heat Transfer Textbook. Courier Dover Publications, Mineola, NY.

Heat exchangers

7.1 Introduction

A heat exchanger is a device that facilitates the process of heat exchange between two fluids that are at different temperatures. Heat exchangers are used in many engineering applications, such as refrigeration, heating and airconditioning systems, power plants, chemical processing systems, food processing systems, automobile radiators, and waste heat recovery units. Air preheaters, economizers, evaporators, superheaters, condensers, and cooling towers used in a power plant are a few examples of heat exchangers.

This chapter presents the parameters that influence the performance of a heat exchanger and discusses the approaches for the design of a heat exchanger or the prediction of the performance of an already existing heat exchanger.

7.2 Classification of heat exchangers

Heat exchangers can be classified based on different criteria, as listed below.

Based on the nature of the heat exchange process

1. *Direct contact–type heat exchanger*

   In this type, both fluids are from the same substance and a schematic is shown in Fig. 7.1A. For example, the hot fluid is water vapor, and the cold fluid is water.

   The limitation of a direct contact–type heat exchanger is that both the fluids must be of the same substance, like hot water and cold water, steam and water, etc.

2. *Regenerator type of heat exchanger*

   In this type of heat exchanger, the hot and cold fluids flow through the heat exchanger alternately and a schematic is shown in Fig. 7.1B. When the hot fluid flows through the heat exchanger, heat is transferred from the hot fluid to the heat exchanger wall (matrix). The hot fluid is then stopped, and the cold fluid is sent so that the heat is transferred from the heat exchanger wall (matrix) to the cold fluid. This is called a fixed matrix regenerator. A rotary regenerator employs a matrix in the form of a wheel that rotates continuously through the counterflowing streams of the hot and cold fluids.

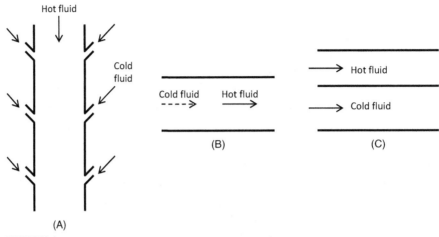

FIGURE 7.1

Schematic representation of (A) direct contact–type heat exchanger, (B) regenerator type of heat exchanger, and (C) recuperator type of heat exchanger.

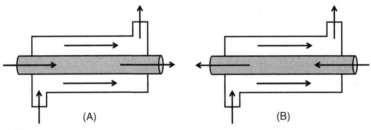

FIGURE 7.2

(A) Parallel-flow heat exchanger and (B) counterflow heat exchanger.

3. *Recuperator type of heat exchanger*

In this type of heat exchanger, both the hot and cold fluids flow through the heat exchanger simultaneously and are separated by a thin wall, as shown in Fig. 7.1C.

Based on the direction of fluid flow

1. *Parallel flow*

In a parallel-flow heat exchanger, both the hot and cold fluids move in the same direction, as shown in Fig. 7.2A.

2. *Counterflow*

In a counterflow heat exchanger, the hot and cold fluids move in opposite directions, as shown in Fig. 7.2B.

3. *Cross flow*

In a cross-flow heat exchanger, the hot and cold fluids move in perpendicular directions, as shown in Figs. 7.3A and B.

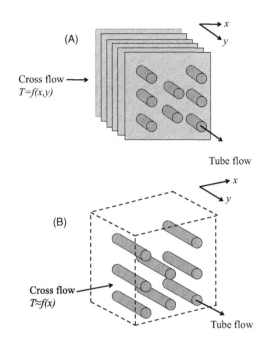

FIGURE 7.3

Cross-flow heat exchanger with (A) both fluids unmixed and (B) one fluid mixed and the other unmixed.

The difference between the heat exchangers shown in Figs. 7.3A and B is that in the former, both fluids are unmixed while in the latter, the fluid flowing outside the tubes is mixed. It is intuitively apparent that the thermal performance of the two will be different.

Based on the mechanical design

1. *Concentric tube heat exchanger*
   Also called a double pipe heat exchanger, it is one wherein one fluid flows through the inner tube, and the other fluid flows through the annulus, as shown in Figs. 7.2A and B.
2. *Shell and tube heat exchanger*
   In this type of heat exchanger, one of the fluids flows through a number of tubes stacked in a shell, and the other fluid flows outside the tubes. Depending on the requirement, there can be multiple tube or shell passes. Flow conditions in a shell and tube heat exchanger are neither parallel flow nor counter flow. Fluid flow outside the tubes is directed by separators known as baffles placed in the shell, as shown in Fig. 7.4. A concentric tube heat exchanger is the simplest form of a shell and tube heat exchanger.
3. *Multipass heat exchanger*
   Shell and tube heat exchangers and crossflow heat exchangers can be of multipass type to enhance their heat transfer capability. Multiple tube passes or shell passes

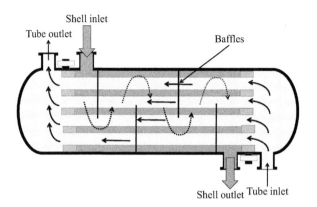

FIGURE 7.4

Shell and tube heat exchanger.

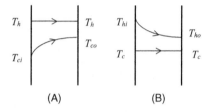

FIGURE 7.5

Temperature profiles of the hot and cold fluids in (A) a condenser and (B) an evaporator.

are chosen based on the velocity consideration, the total heat transfer area requirement, and the space (the heat exchanger length) constraints.

Based on the physical state of working fluid

1. *Condenser*

   The hot fluid condenses as the heat is transferred to the cold fluid. The temperature of the condensing (hot) fluid remains constant, as shown in Fig. 7.5A.

2. *Evaporator*

   The cold fluid evaporates due to the heat transfer from the hot fluid. The temperature of the evaporating (cold) fluid remains constant, as shown in Fig. 7.5B.

It may be noted that the convection heat transfer coefficients associated with condensation and evaporation are very high compared to single-phase heat transfer coefficients.

Based on the compactness

Compact heat exchangers pack a large amount of heat transfer surface area ($\geq 400 \text{ m}^2/\text{m}^3$) per unit volume of the heat exchanger. Gas flow is normally associated with poor heat

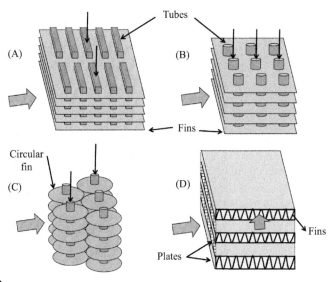

FIGURE 7.6

Compact heat exchangers: (A, B, C) fin-type heat exchanger and (D) plate-fin heat exchanger.

transfer coefficients, so compact heat exchangers are employed when the heat transfer is between two gases or between a gas and a liquid. Fin-tube heat exchangers and plate-fin heat exchangers, as shown in Fig. 7.6, are examples of compact heat exchangers.

In this chapter, the heat transfer analysis is presented only for the recuperative type of heat exchangers, which are extensively used in engineering applications.

7.3 Heat exchanger analysis

The heat transfer analysis of a heat exchanger involves relating the total heat transfer rate to variables like inlet and outlet temperatures of the hot and cold fluids, the overall heat transfer coefficient, and the overall heat transfer surface area. The analysis is essentially based on the energy balance between the heat gained by the cold fluid, the heat lost by the hot fluid, and the heat transferred through the wall that separates the two fluids. Let $\dot{m}$, c_p, and T denote the mass flow rate, the specific heat, and the temperature of a fluid, respectively. The subscripts h and c indicate the hot and cold fluids, respectively. Neglecting the heat transfer between the heat exchanger and its surroundings (assuming that heat transfer is only between the hot and cold fluids flowing through the heat exchanger) and neglecting the changes in kinetic and potential energies of the fluids and assuming a steady flow, the energy balance for a heat exchanger shown in Fig. 7.7 can be written as

$$\dot{m}_h c_{ph}(T_{hi} - T_{ho}) = \dot{m}_c c_{pc}(T_{co} - T_{ci})$$

(7.1)

FIGURE 7.7

Schematic representation of a heat exchanger.

The subscripts i and o in Eq. (7.1) indicate the inlet and the outlet, respectively. From Eq. (7.1), one can calculate the unknown temperature. However, it is not possible to determine the size of the heat exchanger required. Therefore, it is necessary to analyze the heat exchanger using the heat transfer coefficients.

$$\dot{m}_h c_{ph}(T_{hi} - T_{ho}) = \dot{m}_c c_{pc}(T_{co} - T_{ci}) = UA(\Delta T_{mean}) \tag{7.2}$$

U is the overall heat transfer coefficient in W/m² K; ΔT_{mean} is a certain mean temperature difference.

Let us consider the heat exchanger shown in Fig. 7.8.

The heat transfer from the hot fluid to the cold fluid is essentially one dimensional. Using the thermal resistance network,

$$R_{total} = R_{conv,h} + R_{cond} + R_{conv,c} \tag{7.3}$$

where R_{conv} is convective resistance and R_{cond} is conductive resistance.

$$\frac{1}{UA} = \frac{1}{h_h A_1} + \frac{\ln\left(\dfrac{r_2}{r_1}\right)}{2\pi kL} + \frac{1}{h_c A_2} \tag{7.4}$$

Here, r_1 and r_2 are the inner and outer radii of the heat exchanger tube, respectively, and A_1 and A_2 are the inner and outer surface areas of the tube, respectively. Surface area A in Eq. (7.4) can be A_1, in which case U is U_1, or A_2, in which case U is U_2.

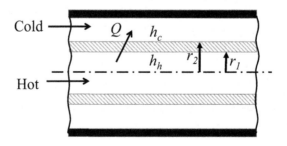

FIGURE 7.8

Schematic representation of a heat exchanger for heat transfer analysis.

For a thin wall tube, $r_2 \approx r_1$, hence $R_{cond} = 0$. Eq. (7.4) reduces to

$$\frac{1}{U} = \frac{1}{h_h} + \frac{1}{h_c} \tag{7.5}$$

[Please note that here we have used the idea $A_1 = A_2$ in Eq. (7.4), consequent upon $r_2 = r_1$]

Therefore, the overall heat transfer coefficient is purely dependent on the heat transfer coefficients of only the hot and cold fluids and does not depend on the direction of fluid flow. ΔT_{mean} is an important parameter, and its relationship with the terminal temperatures of the hot and cold fluids needs to be established.

Let $R_{scaling}$ be the resistance offered by the wall due to the formation of scales (fouling).

$$\frac{1}{UA} = \frac{1}{h_h A_1} + \frac{\ln\left(\frac{r_2}{r_1}\right)}{2\pi kL} + \frac{1}{h_c A_2} + R_{scaling} \tag{7.6}$$

Please note that Eq. (7.6) is the most general one while Eq. (7.5) is for a thin walled heat exchanger with negligible fouling resistance. Eq. (7.6) suggests that the overall heat transfer coefficient can be improved by increasing the heat transfer surface area on the side that has a lower heat transfer coefficient, and engineering efforts to improve heat exchanger performance outcomes need to be in this direction. For example, it is advisable and advantageous to add fins on the side that has a lower heat transfer coefficient, as opposed to the side with the higher coefficient, as the controlling resistance, as aforesaid, is associated with the lower heat transfer coefficient fluid or stream. Stated simply, for an air-liquid exchanger, fins or any other augmentation will be invariably done on the air side that is associated with lower heat transfer coefficients.

7.4 The LMTD method

This section presents the LMTD method of heat exchanger analysis for different types of commonly used heat exchangers.

7.4.1 The parallel-flow heat exchanger

Consider a parallel-flow heat exchanger that is insulated from its surroundings, whose overall heat transfer coefficient and the specific heat are constant. The changes in the potential and kinetic energies are assumed to be negligible. Temperature profiles for a parallel-flow heat exchanger are shown in Fig. 7.9.

For a parallel-flow heat exchanger, the difference between the temperatures of the hot and cold fluids at a section shown in the Figure is

$$\Delta T = T_h - T_c \tag{7.7}$$

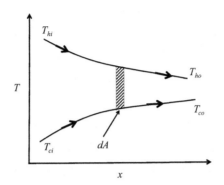

FIGURE 7.9

Schematic representation of temperature variation along the length for a parallel-flow type of heat exchanger.

$$d(\Delta T) = dT_h - dT_c \tag{7.8}$$

From energy balance,

$$dQ = -\dot{m}_h c_{ph} dT_h = \dot{m}_c c_{pc} dT_c = UdA\Delta T \tag{7.9}$$

In the above equation, the negative sign is for the heat lost.

$$d(\Delta T) = \frac{-dQ}{\dot{m}_h c_{ph}} - \frac{dQ}{\dot{m}_c c_{pc}} \tag{7.10}$$

$$d(\Delta T) = -UdA\Delta T \left(\frac{1}{\dot{m}_h c_{ph}} + \frac{1}{\dot{m}_c c_{pc}} \right) \tag{7.11}$$

Separating the variables and integrating on both sides, we have

$$\int_{inlet}^{outlet} \frac{d(\Delta T)}{\Delta T} = -\int_0^L UdA \left(\frac{1}{\dot{m}_h c_{ph}} + \frac{1}{\dot{m}_c c_{pc}} \right) \tag{7.12}$$

$$\ln(\Delta T)\Big|_{inlet}^{outlet} = -UA \left(\frac{1}{\dfrac{Q}{T_{hi} - T_{ho}}} + \frac{1}{\dfrac{Q}{T_{co} - T_{ci}}} \right) \tag{7.13}$$

[Note that in Eq. (7.13), we have used the relation $Q = \dot{m}_h c_{ph} \left(T_{hi} - T_{ho} \right) = \dot{m}_c c_{pc} \left(T_{co} - T_{ci} \right)$]

$$\ln(\Delta T_o) - \ln(\Delta T_i) = \frac{-UA}{Q} \left((T_{hi} - T_{ci}) - (T_{ho} - T_{co}) \right) \tag{7.14}$$

$$\ln\left(\frac{\Delta T_i}{\Delta T_o}\right) = \frac{UA}{Q}(\Delta T_i - \Delta T_o) \tag{7.15}$$

$$Q = UA\frac{(\Delta T_i - \Delta T_o)}{\ln\left(\dfrac{\Delta T_i}{\Delta T_o}\right)} \tag{7.16}$$

$$Q = UA(LMTD) \tag{7.17}$$

LMTD is a logarithmic mean temperature difference.

$$LMTD = \frac{\Delta T_i - \Delta T_o}{\ln\left(\dfrac{\Delta T_i}{\Delta T_o}\right)} = \frac{\theta_1 - \theta_2}{\ln\left(\dfrac{\theta_1}{\theta_2}\right)} \tag{7.18}$$

7.4.2 The counterflow heat exchanger

The assumptions for a counterflow heat exchanger are the same as those for the parallel-flow heat exchanger considered previously.

The temperature profiles for a counterflow heat exchanger are shown in Fig. 7.10.

$$dQ = -\dot{m}_h c_{ph} dT_h = -\dot{m}c_{pc} dT_c = UdA\Delta T \tag{7.19}$$

The negative sign indicates that the value of the temperature reduces along the positive x direction.

$$\Delta T = T_h - T_c \tag{7.20}$$

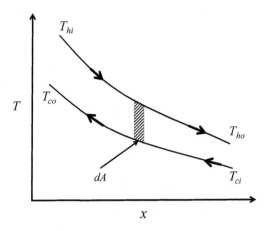

FIGURE 7.10

Schematic representation of temperature variation along the length for a counterflow type of heat exchanger.

$$d(\Delta T) = dT_h - dT_c \tag{7.21}$$

$$d(\Delta T) = \frac{-dQ}{\dot{m}_h c_{ph}} + \frac{dQ}{\dot{m}_c c_{pc}} \tag{7.22}$$

$$\int_{one\,side}^{other\,side} \frac{d(\Delta T)}{\Delta T} = \int_0^L U\,dA\left(\frac{-1}{\dot{m}_h c_{ph}} + \frac{1}{\dot{m}_c c_{pc}}\right) \tag{7.23}$$

$$\ln(\Delta T)\Big|_{one\,side}^{other\,side} = UA\left(\frac{-1}{\dfrac{Q}{T_{ho}-T_{hi}}} + \frac{1}{\dfrac{Q}{T_{ci}-T_{co}}}\right) \tag{7.24}$$

$$\ln\left(\frac{T_{hi}-T_{co}}{T_{ho}-T_{ci}}\right) = \frac{-UA}{Q}\left((T_{ho}-T_{hi})-(T_{ci}-T_{co})\right) \tag{7.25}$$

$$Q = UA\frac{(T_{hi}-T_{co})-(T_{ho}-T_{ci})}{\ln\left(\dfrac{T_{hi}-T_{co}}{T_{ho}-T_{ci}}\right)} \tag{7.26}$$

$$Q = UA(LMTD) \tag{7.27}$$

$$LMTD = \frac{\Delta T_{one\,side} - \Delta T_{other\,side}}{\ln\left(\dfrac{\Delta T_{one\,side}}{\Delta T_{other\,side}}\right)} = \frac{\theta_1 - \theta_2}{\ln\left(\dfrac{\theta_1}{\theta_2}\right)} \tag{7.28}$$

It is to be noted that while it is possible for T_{co} to be greater than T_{ho} in a counterflow heat exchanger, this is never possible in a parallel-flow heat exchanger. In a parallel-flow heat exchanger, $T_{co} \rightarrow T_{ho}$ (see Fig. 7.9. This condition happens when $x \rightarrow \infty$).

7.4.3 Heat exchangers with phase change

Let us consider a condenser for which the temperature profiles of the hot and cold fluids are as shown in Fig. 7.11.

From Fig. 7.11A,

$$\theta_1 = T_h - T_{ci} \tag{7.29}$$

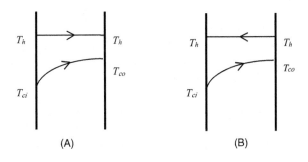

FIGURE 7.11

Schematic representation of temperature variation along the length for a condenser: (A) parallel-flow configuration and (B) counterflow configuration.

$$\theta_2 = T_h - T_{co} \tag{7.30}$$

Similarly from Fig. 7.11B,

$$\theta_1 = T_h - T_{ci} \tag{7.31}$$

$$\theta_2 = T_h - T_{co} \tag{7.32}$$

Therefore, for a condenser,

$$(\text{LMTD})_{parallel} = (\text{LMTD})_{counter} \tag{7.33}$$

Hence, the performance of parallel-flow and counterflow types of condensing heat exchanger remains the same, and the same holds true for an evaporator.

7.4.4 When is LMTD not applicable?

For a counterflow heat exchanger, the energy balance equation is given by

$$\dot{m}_h c_{ph}(T_{hi} - T_{ho}) = \dot{m}_c c_{pc}(T_{co} - T_{ci}) \tag{7.34}$$

If the heat capacities are equal, then Eq. (7.34) is reduced to

$$T_{hi} - T_{ho} = T_{co} - T_{ci} \tag{7.35}$$

$$T_{hi} - T_{co} = T_{ho} - T_{ci} \tag{7.36}$$

$$\theta_1 = \theta_2 \tag{7.37}$$

Equation 7.28 becomes

$$\text{LMTD} = \frac{0}{0} \tag{7.38}$$

The above equation is in indeterminate form. Therefore, L'Hospital's rule needs to be applied to Eq. (7.28).

$$\text{LMTD} = \lim_{\theta_1 \to \theta_2} \frac{\theta_1 - \theta_2}{\ln\left(\dfrac{\theta_1}{\theta_2}\right)} \tag{7.39}$$

Let $\dfrac{\theta_1}{\theta_2} = x$.

$$\text{LMTD} = \lim_{x \to 1} \frac{\theta_2(x-1)}{\ln(x)} \tag{7.40}$$

After applying L'Hospital's rule,

$$\text{LMTD} = \theta_2 = \theta_1 \tag{7.41}$$

Therefore, in the case of a counterflow heat exchanger, if the heat capacities are equal, then LMTD method is not applicable. LMTD remains constant along the heat exchanger length and is equal to the terminal temperature difference (θ_1 or θ_2).

The concepts and ideas introduced thus far in Section 7.4 is traditionally known as the LMTD analysis or LMTD method.

In the case of a counterflow heat exchanger, by increasing the length of the heat exchanger to infinity, the maximum possible exit temperature of the cold fluid and the minimum possible exit temperature of the hot fluid can be obtained.

$$T_{co,\,max} = T_{hi} \tag{7.42}$$

$$T_{ho,\,min} = T_{ci} \tag{7.43}$$

In the case of an infinitely long parallel-flow heat exchanger,

$$T_{co,\,max} = T_{ho,\,min} = T \tag{7.44}$$

From the energy balance equation,

$$C_h(T_{hi} - T) = C_c(T - T_{ci}) \tag{7.45}$$

From the above equation, the exit temperature (T) of the fluid(s) can be obtained.

7.4.5 Shell and tube heat exchanger

The flow configuration in a shell and tube heat exchanger is neither parallel nor counter, and hence the LMTD will be different from that of a parallel-flow or counterflow heat exchanger. LMTD analysis can still be used for a shell and tube heat exchanger making use of a correction factor F to account for the deviation from the counterflow configuration. The expression for the heat transfer rate is

$$Q = UA(\text{LMTD})_{counterflow}\, F \tag{7.46}$$

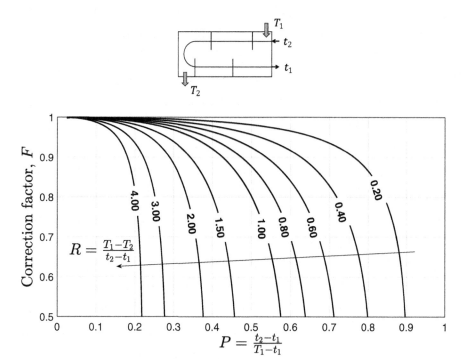

FIGURE 7.12

Correction factor for a one-shell pass and 2, 4, 6. (any multiple of 2) -tube pass shell and tube heat exchanger (Bowman et al., 1940).

The correction factor F can be obtained from the charts shown in Figs. 7.12 and 7.13 for one shell pass and two shell passes, respectively. These charts were originally developed by Bowman et al. (1940).

7.4.6 Cross-flow heat exchanger

The hot and cold fluids flow perpendicular to each other in a cross-flow heat exchanger and therefore the flow configuration is neither parallel nor counter. The LMTD approach used for a cross-flow heat exchanger is similar to that for a shell and tube heat exchanger.

$$Q = UA(LMTD)_{counterflow} F \tag{7.47}$$

The correction factor charts are available, as shown in Figs. 7.14 and 7.15, for two different configurations—one fluid mixed and the other unmixed, and both fluids unmixed, respectively.

For analyzing a heat exchanger using the LMTD method, the outlet temperature of at least one of the fluids must be known. Thus, the sizing of the heat exchanger can be done based on the overall heat transfer coefficient and the LMTD. However, in many cases, the size of a heat exchanger is specified, but the outlet temperatures are

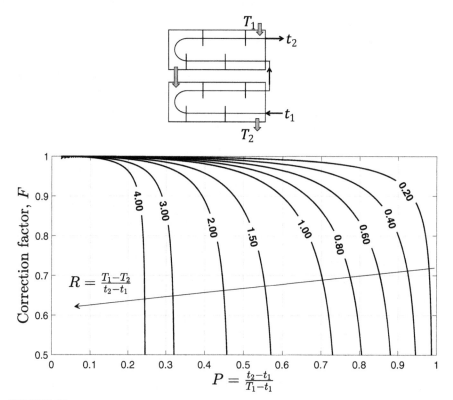

FIGURE 7.13

Correction factor for a two-shell pass and 4, 8, 12...(any multiple of 4) -tube pass shell and tube heat exchanger (Bowman et al., 1940).

unknown. Therefore, its performance needs to be predicted. In such cases, another approach called the effectiveness-NTU method is used. The NTU refers to the number of transfer units and is directly related to the size of the heat exchanger.

7.5 The effectiveness-NTU method

The heat exchanger temperature effectiveness, or simply the effectiveness, ε, is the ratio of actual heat transfer to the maximum possible heat transfer.

$$\varepsilon = \frac{Q_{act}}{Q_{max}} \tag{7.48}$$

$$\varepsilon = \frac{C_h(T_{hi} - T_{ho})}{C_{min}(T_{hi} - T_{ci})} = \frac{C_c(T_{co} - T_{ci})}{C_{min}(T_{hi} - T_{ci})} \tag{7.49}$$

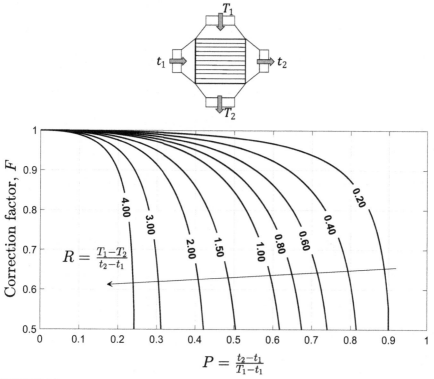

FIGURE 7.14

Correction factor for a cross-flow heat exchanger with one fluid mixed and the other unmixed (Bowman et al., 1940).

$$Q_{max} = C_{min} \Delta T_{max} \tag{7.50}$$

For any type of heat exchanger, $\Delta T_{max} = T_{hi} - T_{ci}$.

In Eq. (7.50), we have considered C_{min} and not C_{max}, as the fluid with minimum heat capacity can experience the maximum change in temperature. The NTU is given by

$$NTU = \frac{UA}{C_{min}} \tag{7.51}$$

$$C_{min} = (\dot{m}c_p)_{min} \tag{7.52}$$

$$C_{max} = (\dot{m}c_p)_{max} \tag{7.53}$$

The NTU indicates the size of the heat exchanger, as already mentioned. The capacity ratio of a heat exchanger, C, is defined as

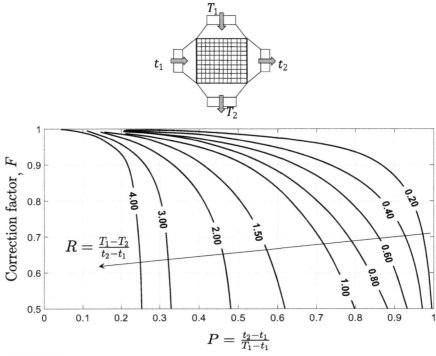

FIGURE 7.15

Correction factor for a cross-flow heat exchanger with both fluids unmixed (Bowman et al., 1940).

$$C = \frac{C_{min}}{C_{max}} \tag{7.54}$$

C varies between 0 and 1. $C = 0$ corresponds to a heat exchanger with one of the fluids undergoing phase change, condensation or evaporation, and $C = 1$ corresponds to a heat exchanger with equal heat capacity rates.

7.5.1 Effectiveness of a parallel-flow heat exchanger

The temperature profiles for the hot and cold fluids in a parallel-flow heat exchanger were shown in Fig. 7.9. Please refer to the same figure.

Let $C_c = C_{min}$

$$\Delta T = T_h - T_c \tag{7.55}$$

$$d(\Delta T) = dT_h - dT_c \tag{7.56}$$

From energy balance,

$$dQ = -\dot{m}_h c_{ph} dT_h = \dot{m}_c c_{pc} dT_c = UdA\Delta T \tag{7.57}$$

In the above equation, the negative sign is for the heat lost by the hot fluid.

$$d(\Delta T) = \frac{-dQ}{\dot{m}_h c_{ph}} - \frac{dQ}{\dot{m}_c c_{pc}} \tag{7.58}$$

$$d(\Delta T) = -UdA\Delta T \left(\frac{1}{\dot{m}_h c_{ph}} + \frac{1}{\dot{m}_c c_{pc}} \right) \tag{7.59}$$

Separating the variables and integrating on both sides

$$\int_{inlet}^{outlet} \frac{d(\Delta T)}{\Delta T} = -\int_0^L U \, dA \left(\frac{1}{\dot{m}_h c_{ph}} + \frac{1}{\dot{m}_c c_{pc}} \right) \tag{7.60}$$

$$\ln(\Delta T)\Big|_{inlet}^{outlet} = -UA \left(\frac{1}{C_h} + \frac{1}{C_c} \right) \tag{7.61}$$

$$\ln\left(\frac{\Delta T_o}{\Delta T_i} \right) = \frac{-UA}{C_c} \left(\frac{C_c}{C_h} + 1 \right) \tag{7.62}$$

$$\ln\left(\frac{\Delta T_o}{\Delta T_i} \right) = -NTU(1 + C) \tag{7.63}$$

$$\ln\left(\frac{T_{ho} - T_{co}}{T_{hi} - T_{ci}} \right) = -NTU(1 + C) \tag{7.64}$$

Here $NTU = \dfrac{UA}{C_c}$, and $C = \dfrac{C_c}{C_h}$

From energy balance,

$$\dot{m}_h c_{ph} (T_{hi} - T_{ho}) = \dot{m}_c c_{pc} (T_{co} - T_{ci}) \tag{7.65}$$

$$C_h (T_{hi} - T_{ho}) = C_c (T_{co} - T_{ci}) \tag{7.66}$$

$$T_{ho} = T_{hi} - C(T_{co} - T_{ci}) \tag{7.67}$$

From Eq. (7.64),

$$\ln\left(\frac{T_{hi} - C(T_{co} - T_{ci}) - T_{co}}{T_{hi} - T_{ci}}\right) = -NTU(1+C) \tag{7.68}$$

$$\ln\left(\frac{T_{hi} - C(T_{co} - T_{ci}) - T_{co} + T_{ci} - T_{ci}}{T_{hi} - T_{ci}}\right) = -NTU(1+C) \tag{7.69}$$

$$\ln\left(\frac{T_{hi} - T_{ci}}{T_{hi} - T_{ci}} - (1+C)\frac{T_{co} - T_{ci}}{T_{hi} - T_{ci}}\right) = -NTU(1+C) \tag{7.70}$$

$$\ln(1 - \varepsilon(1+C)) = -NTU(1+C) \tag{7.71}$$

$$1 - \varepsilon(1+C) = \exp(-NTU(1+C)) \tag{7.72}$$

$$\varepsilon = \frac{1 - \exp(-NTU(1+C))}{(1+C)} \tag{7.73}$$

7.5.2 **Effectiveness of a counterflow heat exchanger**

An expression for the effectiveness of a counterflow heat exchanger can be derived in a way similar to that of a parallel-flow heat exchanger. The temperature profiles of fluids for a counterflow heat exchanger were shown in Fig. 7.10. Please refer to the same figure.

From energy balance,

$$dQ = -\dot{m}_h c_{ph} dT_h = -\dot{m}c_{pc} dT_c = UdA\Delta T \tag{7.74}$$

The negative sign indicates that the value of temperature reduces along the positive x direction.

$$\Delta T = T_h - T_c \tag{7.75}$$

$$d(\Delta T) = dT_h - dT_c \tag{7.76}$$

$$d(\Delta T) = \frac{-dQ}{\dot{m}_h c_{ph}} + \frac{dQ}{\dot{m}_c c_{pc}} \tag{7.77}$$

$$\int_{one\,side}^{other\,side} \frac{d(\Delta T)}{\Delta T} = \int_0^L U\,dA\left(\frac{-1}{\dot{m}_h c_{ph}} + \frac{1}{\dot{m}_c c_{pc}}\right) \tag{7.78}$$

$$\ln(\Delta T)\Big|_{one\,side}^{other\,side} = UA\left(\frac{-1}{C_h} + \frac{1}{C_c}\right) \tag{7.79}$$

$$\ln\left(\frac{T_{ho} - T_{ci}}{T_{hi} - T_{co}}\right) = UA\left(\frac{-1}{C_h} + \frac{1}{C_c}\right) \tag{7.80}$$

$$\frac{T_{hi} - T_{co}}{T_{ho} - T_{ci}} = \exp(-NTU(1-C)) \tag{7.81}$$

LHS:

$$\frac{T_{hi} - T_{co}}{T_{ho} - T_{ci}} = \frac{\left(\dfrac{T_{hi} - T_{co}}{T_{hi} - T_{ci}}\right)}{\left(\dfrac{T_{ho} - T_{ci}}{T_{hi} - T_{ci}}\right)} \tag{7.82}$$

We know that

$$\varepsilon = \frac{C_h(T_{hi} - T_{ho})}{C_{min}(T_{hi} - T_{ci})} = \frac{C_c(T_{co} - T_{ci})}{C_{min}(T_{hi} - T_{ci})} \tag{7.83}$$

In this derivation, let $C_h = C_{max}, C_c = C_{min}$
Therefore,

$$\varepsilon = \frac{C_{max}(T_{hi} - T_{ho})}{C_{min}(T_{hi} - T_{ci})} = \frac{C_{min}(T_{co} - T_{ci})}{C_{min}(T_{hi} - T_{ci})} \tag{7.84}$$

From Eq. (7.84),

$$T_{ho} = T_{hi} - \varepsilon C(T_{hi} - T_{ci}) \tag{7.85}$$

$$T_{co} = T_{ci} + \varepsilon(T_{hi} - T_{ci}) \tag{7.86}$$

Substituting T_{ho} and T_{co} in Eq. (7.82),

$$\frac{\left(\dfrac{T_{hi} - (T_{ci} + \varepsilon(T_{hi} - T_{ci}))}{T_{hi} - T_{ci}}\right)}{\left(\dfrac{(T_{hi} - \varepsilon C(T_{hi} - T_{ci})) - T_{ci}}{T_{hi} - T_{ci}}\right)} = \frac{1-\varepsilon}{1-C\varepsilon} \tag{7.87}$$

From Eq. (7.81),

$$\frac{1-\varepsilon}{1-C\varepsilon} = \exp(-NTU(1-C)) \tag{7.88}$$

$$\varepsilon = \frac{1 - \exp(-NTU(1 - C))}{1 - C\exp(-NTU(1 - C))} \tag{7.89}$$

Similar expressions for the effectiveness of other types of commonly used heat exchangers were developed by Kays and London (1984) and are presented in Table 7.1. For a shell and tube heat exchanger with n shell passes, ε would be determined based on the $(NTU)_1$ of a single shell and the corresponding effectiveness (ε_1), assuming that the total NTU is equally shared between shell passes, $NTU = n(NTU)_1$. The effectiveness can also be determined from the effectiveness-NTU charts, as shown in Figs. 7.16–7.21.

Table 7.1 Heat exchanger effectiveness relations (Kays and London, 1984).

Flow configuration	Relation
Concentric tube	
Parallel flow	$\varepsilon = \dfrac{1 - \exp(-NTU(1+C))}{(1+C)}$
Counterflow	$\varepsilon = \dfrac{1 - \exp(-NTU(1-C))}{1 - C\exp(-NTU(1-C))} \quad (C < 1)$
	$\varepsilon = \dfrac{NTU}{1 + NTU} \quad (C = 1)$
Shell and tube	
One shell pass (2, 4...tube passes)	$\varepsilon_1 = 2\left\{1 + C + (1+C^2)^{1/2} \times \dfrac{1 + \exp(-(NTU)_1(1+C^2)^{1/2})}{1 - \exp(-(NTU)_1(1+C^2)^{1/2})}\right\}^{-1}$
n shell passes (2n, 4n... tube passes)	$\varepsilon = \left[\left(\dfrac{1 - \varepsilon_1 C}{1 - \varepsilon_1}\right)^n - 1\right]\left[\left(\dfrac{1 - \varepsilon_1 C}{1 - \varepsilon_1}\right)^n - C\right]^{-1}$
Cross-flow (single pass)	
Both fluids unmixed	$\varepsilon = 1 - \exp\left[\dfrac{NTU^{0.22}}{C}\left\{\exp\left(-C(NTU)^{0.78}\right) - 1\right\}\right]$
C_{max} (mixed), C_{min} (unmixed)	$\varepsilon = \dfrac{1 - \exp\{-C(1 - \exp(-NTU))\}}{C}$
C_{max} (unmixed), C_{min} (mixed)	$\varepsilon = 1 - \exp\left(\dfrac{1 - \exp(-C(NTU))}{-C}\right)$
All exchangers $(C = 0)$	$\varepsilon = 1 - \exp(-NTU)$

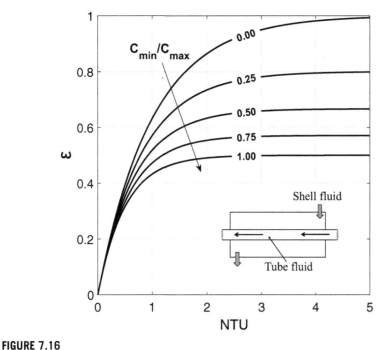

FIGURE 7.16

Effectiveness of a parallel-flow heat exchanger.

7.5.3 Comparison between parallel-flow and counterflow heat exchangers

Having seen heat exchangers with different configurations, a comparison between two simple configurations – parallel-flow and counterflow heat exchangers, is warranted. The expressions for the effectiveness of parallel-flow and counterflow configurations, as derived before, are,

$$\varepsilon_p = \frac{1 - \exp(-NTU(1+C))}{(1+C)} \tag{7.90}$$

$$\varepsilon_c = \frac{1 - \exp(-NTU(1-C))}{1 - C\exp(-NTU(1-C))} \tag{7.91}$$

Figures 7.16 and 7.17 show the effectiveness-NTU relations in graphical form for the parallel-flow and counterflow configurations, respectively. It can be seen that the effectiveness of the counterflow heat exchanger is higher than that of the parallel-flow heat exchanger for the same NTU and heat capacity ratio. This is intuitively apparent as the stream to stream temperature difference, which holds the key to the superior thermal performance of a heat exchanger, is more or less preserved in a

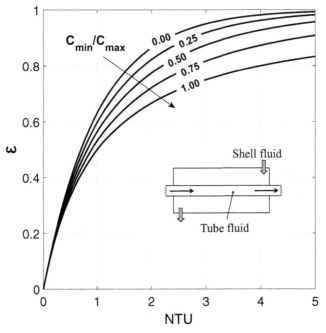

FIGURE 7.17

Effectiveness of a counterflow heat exchanger.

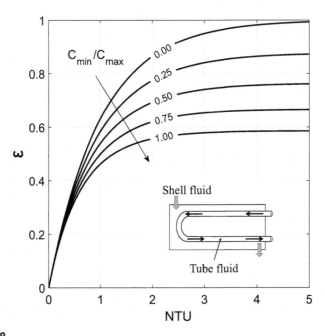

FIGURE 7.18

Effectiveness of a one-shell pass and 2, 4, 6...(any multiple of 2) -tube pass shell and tube heat exchanger.

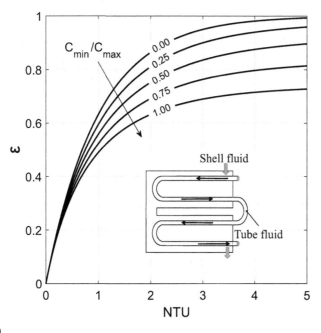

FIGURE 7.19

Effectiveness of a two-shell pass and 4, 8, 12 . . . (any multiple of 4) -tube pass shell and tube heat exchanger.

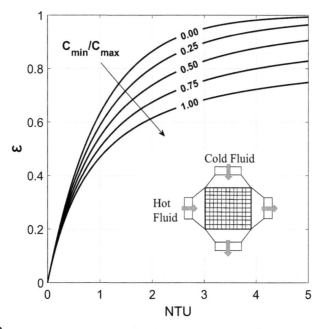

FIGURE 7.20

Effectiveness of a cross-flow heat exchanger with both fluids unmixed.

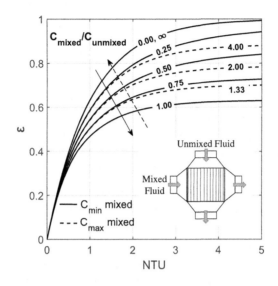

FIGURE 7.21

Effectiveness of a cross-flow heat exchanger with one fluid mixed and the other fluid unmixed.

counterflow heat exchanger from the entry to the exit, while there is a continuous decline in the stream to stream temperature difference between the hot and cold fluids from the entry to the exit in a parallel flow heat exchanger.

For a condenser,

$$Q_h = C_h(T_{h1} - T_{h2})$$
(7.92)

Since the hot fluid condenses,

$$C_h \to \infty$$
(7.93)

$$\therefore C = \frac{C_c}{C_h} = 0$$
(7.94)

If $C = 0$

$$\varepsilon_p = 1 - \exp(-NTU)$$
(7.95)

$$\varepsilon_c = 1 - \exp(-NTU)$$
(7.96)

Therefore, for a condenser, the effectiveness is the same for both the parallel-flow and counterflow configurations. Eqs. (7.95) and (7.96) are true for an evaporator as well. If $C = 1$ for a parallel-flow heat exchanger,

$$\varepsilon_p = \frac{1 - \exp(-2NTU)}{2}$$
(7.97)

If $C = 1$ for a counterflow heat exchanger,

$$\varepsilon_c = \frac{0}{0} = indeterminate\ form \tag{7.98}$$

Applying L'Hospital's rule,

$$\varepsilon_c = \frac{NTU}{1 + NTU} \tag{7.99}$$

Figures 7.18–7.21 show the effectiveness-NTU relations in graphical form for other configurations - a one-shell pass shell and tube, a two-shell pass shell and tube, a cross-flow with both fluids unmixed, and a cross-flow with one fluid mixed and the other fluid unmixed, respectively.

7.6 Comparison between the LMTD and effectiveness-NTU methods

1. If the exit temperature of at least one of the fluids is known, both the methods can be used for analyzing the heat exchanger. Out of the two, LMTD is easier and faster as compared to the effectiveness-NTU method.
2. If the exit temperatures of both fluids are unknown, the only method that can be used is the effectiveness-NTU method.

Example 7.1 *A 1 kg/s of hot water stream is cooled from 90 °C to 60 °C in a parallel-flow heat exchanger in which the cooling agent is 2 kg/s of water with inlet temperature of 40 °C. The overall heat transfer coefficient U is 1000 W/m² K. Using the LMTD method, determine the required area of the heat exchanger. Take* $c_{ph} = c_{pc} = 4182$ J/kg K .

Solution:

$$\dot{m}_h c_{ph} \left(T_{hi} - T_{ho} \right) = \dot{m}_c c_{pc} \left(T_{co} - T_{ci} \right)$$

$$1 \times 4182 \left(90 - 60 \right) = 2 \times 4182 \left(T_{co} - 40 \right)$$

$$T_{co} = 55\,°C$$

$$\Delta T_i = 90 - 40 = 50\,°C$$

$$\Delta T_o = 60 - 55 = 5\,°C$$

$$LMTD = \frac{50 - 5}{\ln\left(\frac{50}{5}\right)} = 19.54\,°C$$

$$Q = UA(\text{LMTD})$$

$$1 \times 4182 \times (90 - 60) = 1000 \times A \times 19.54$$

$$A_p = 6.42 \, \text{m}^2$$

Example 7.2 *Revisit the above problem and solve it for a counter-flow heat exchanger.*

Solution:
In the case of a counterflow–type heat exchanger,

$$\Delta T_{oneside} = 90 - 55 = 35 \,^\circ\text{C}$$

$$\Delta T_{otherside} = 60 - 40 = 20 \,^\circ\text{C}$$

$$\text{LMTD} = \frac{35 - 20}{\ln\left(\dfrac{35}{20}\right)} = 26.80 \,^\circ\text{C}$$

$$Q = UA(\text{LMTD})$$

$$1 \times 4182 \times (90 - 60) = 1000 \times A \times 26.80$$

$$A_c = 4.68 \, \text{m}^2$$

From the above two examples it is clear that $\text{LMTD}_{parallel} < \text{LMTD}_{counter}$. Therefore the performance of the counterflow heat exchanger is better than the parallel-flow type.

Example 7.3 *Revisit the above problem and solve it for a cross-flow heat exchanger with one stream mixed and the other unmixed.*

Solution:
For a cross-flow heat exchanger,

$$Q = UA(\text{LMTD})_{counter} F$$

Here F is the correction factor obtained from charts, and the procedure is as follows:

$$P = \frac{t_2 - t_1}{T_1 - t_1}$$

$$P = \frac{55 - 40}{90 - 40} = 0.3$$

$$R = \frac{T_1 - T_2}{t_2 - t_1}$$

$$R = \frac{90 - 60}{55 - 40} = 2$$

Using P and R from Fig. 7.14 we get $F = 0.9$.

$$1 \times 4182 \times (90 - 60) = 1000 \times A \times 26.80 \times 0.9$$

$$A_{cross} = 5.20 \, \text{m}^2$$

Example 7.4 *Revisit the above problem and assume that the outlet temperature of the hot fluid is not provided. Determine the outlet temperature of the hot fluid using the effectiveness-NTU method for a parallel-flow heat exchanger with A = 6.42 m².*

Solution:

$$\dot{m}_h c_{ph} = 1 \times 4182 = 4182 \, \text{W/K}$$

$$\dot{m}_c c_{pc} = 2 \times 4182 = 8364 \, \text{W/K}$$

From the above

$$C_{min} = 1 \times 4182 = 4182 \, \text{W/K}$$

$$C_{max} = 2 \times 4182 = 8364 \, \text{W/K}$$

$$C = \frac{C_{min}}{C_{max}} = 0.5$$

$$NTU = \frac{UA}{C_{min}} = \frac{1000 \times 6.42}{4182} = 1.535$$

$$\varepsilon_p = \frac{1 - \exp(-NTU(1 + C))}{(1 + C)}$$

$$\varepsilon_p = \frac{1 - \exp(-1.535(1.5))}{(1.5)} = 0.6$$

$$Q = \varepsilon_p Q_{max}$$
$$= 0.6 \times C_{min} \times (T_{hi} - T_{ci})$$
$$= 0.6 \times 4182 \times (90 - 40)$$
$$= 125460 \, \text{W}$$

$$T_{ho} = T_{hi} - \frac{Q}{\dot{m}_h c_{ph}}$$

$$= 90 - \frac{125460}{1 \times 4182}$$

$$= 60\,°C.$$

This is the same as what we started out with in Example 7.1.

Example 7.5 *A cardiac-pulmonary bypass procedure uses a cross-flow heat exchanger (both fluids unmixed) to cool the blood flowing at 5000 ml/min from a body temperature of 37 °C to 25 °C so as to induce hypothermia, which slows down metabolism and reduces oxygen demand. Iced water at 0 °C is used as the coolant, and its outlet temperature is 15 °C. The density and specific heat of blood are 1050 kg/m³ and 3740 J/kg K, respectively, and that of water are 1000 kg/m³ and 4198 J/kg K, respectively. The overall heat transfer coefficient is 750 W/m²K.*

a. Calculate the heat transfer rate for the heat exchanger.
b. Determine the coolant flow rate.
c. Calculate the heat transfer area of the heat exchanger.

Solution:

$$\dot{m}_h = \frac{5 \times 10^{-3}}{60} \times 1050 = 0.0875\,\text{kg/s}$$

From thermodynamics

$$0.0875 \times 3740(37 - 25) = \dot{m}_w \times 4198(15 - 0)$$

$$\dot{m}_w = 0.0623\,\text{kg/s}$$

$$Q = 0.0875 \times 3740(37 - 25) = 3.927\,\text{kW}$$

$$\theta_1 = 37 - 15 = 22\,°C$$

$$\theta_2 = 25 - 0 = 25\,°C$$

$$\text{LMTD} = \frac{\theta_1 - \theta_2}{ln(\frac{\theta_1}{\theta_2})}$$

$$\text{LMTD} = \frac{22 - 25}{ln\left(\frac{22}{25}\right)} = 23.46$$

$$Q = UA(LMTD)F$$

In the above equation, F is the correction factor, which can be obtained from Fig. 7.15.

$$P = \frac{t_2 - t_1}{T_1 - t_1} = \frac{18 - 0}{37 - 0} = 0.4$$

$$R = \frac{37 - 25}{15 - 0} = 0.8$$

From the figure, $F = 0.95$

$$3927 = 750 \times A \times (23.46) \times 0.95$$

$$A = 0.234 \, \text{m}^2$$

Example 7.6 *In a chemical processing plant, 10800 kg/h of water (c_p = 4.2 kJ/kg K) is to be heated from 42 °C to 62 °C with 7200 kg/h of hot water (c_p = 4.2 kJ/kg K) available at 88 °C. A shell and tube heat exchanger is proposed with cold water flowing through the tubes of 19 mm diameter at an average velocity of 0.6 m/s and hot water flowing through the shell. The space constraint limits the maximum length of the tube to 2.5 m. Assuming the overall heat transfer coefficient as 1800 W/m² K and one shell pass, determine the number of tube passes, the total number of tubes, and the length of the tubes.*

Solution:
Since the number of tube passes is not known, it can be assumed to be one. The total heat transfer rate is

$$Q_c = \dot{m}_c c_{pc} \Delta T_c = \frac{10800}{3600} \times 4.2 \times (62 - 42) = 252 \, \text{kW}$$

$$Q_h = Q_c$$

The outlet temperature of the hot fluid is

$$T_{ho} = T_{hi} - \frac{Q_h}{\dot{m}_h c_{ph}}$$

$$= 88 - \frac{252}{\frac{7200}{3600} \times 4.2}$$

$$= 58 \, °C$$

LMTD for the counterflow configuration is

$$LMTD = \frac{\theta_1 - \theta_2}{\ln\left(\frac{\theta_1}{\theta_2}\right)}$$

where,

$$\theta_1 = T_{hi} - T_{co} = 88 - 62 = 26\,°C$$

$$\theta_2 = T_{ho} - T_{ci} = 58 - 42 = 16\,°C$$

$$LMTD = \frac{26 - 16}{\ln\left(\frac{26}{16}\right)} = 20.6\,°C$$

Heat transfer area is

$$A = \frac{Q}{U \times LMTD} = \frac{252}{1.8 \times 20.6} = 6.8\,m^2$$

The mass flow rate of cold water is

$$\dot{m}_c = \rho u \left(\frac{N \pi D^2}{4}\right)$$

where N is the number of tubes.

$$N = \frac{\dot{m}_c}{\rho u \left(\frac{\pi D^2}{4}\right)} = \frac{10800/3600}{1000 \times 6 \times \frac{\pi \times 0.019^2}{4}}$$

$$= 17.63$$
$$\approx 18$$

Length of each tube can be calculated as

$$L = \frac{A}{N\pi D} = \frac{6.8}{18 \times \pi \times 0.019} = 6.33\,m$$

Since the length is larger than the maximum allowable length of 2.5 m, we need to use more than one tube pass. Next, a two-tube pass is tried.

LMTD correction factor needs to be determined from Fig. 7.12.

$$P = \frac{T_{co} - T_{ci}}{T_{ho} - T_{ci}} = \frac{62 - 42}{88 - 42} = 0.434$$

$$R = \frac{T_{hi} - T_{ho}}{T_{co} - T_{ci}} = \frac{88 - 58}{62 - 42} = 1.5$$

From Fig. 7.12, $F \approx 0.65$.
Therefore the total area required is,

$$A = \frac{Q}{U \times F \times \text{LMTD}} = \frac{252}{1.8 \times 0.65 \times 20.6} = 10.45\,\text{m}^2$$

The number of tubes per pass will still be 18, as the velocity has to be the same.
The length of the tube per pass is

$$L = \frac{A}{2N\pi D} = \frac{10.45}{2 \times 18 \times \pi \times 0.019} = 4.86\,\text{m}$$

This length is greater than the maximum allowable length of 2.5 m. Hence a two-tube pass is not sufficient. Next, a four-tube pass can be tried.

From Fig. 7.12, F factor is the same for a four-tube pass. Therefore, the area required remains the same (10.45 m²).
The length of the tube per pass is

$$L = \frac{A}{4N\pi D} = \frac{10.45}{4 \times 18 \times \pi \times 0.019} = 2.43\,\text{m}$$

This length is acceptable as it is within the permitted length of 2.5 m.
Therefore, the final parameters are as follows:
Number of tube passes = 4
Length of tube per pass = 2.43 m
Number of tubes per pass = 18
Total number of tubes = 72
Heat duty = 252 kW

For elaborate illustrations and more examples on heat exchangers, readers can refer to Lienhard IV and Lienhard V (2020), Bergman et al. (2011), Cengel (2003), and Bejan (1993).

7.7 Other considerations in the design of a heat exchanger

In addition to the heat transfer, the factors that influence the design or selection of a heat exchanger are the incurred pressure drop (or pumping power) as the fluids are forced through the heat exchanger; the construction material and its strength; the size and weight considerations; and the total cost taking into account the material, fabrication, and space requirements. These are beyond the scope of this book.

Problems

7.1. A thin-walled, double-pipe (concentric tube) heat exchanger has air flowing through the annulus and water flowing through the inner tube. The air side and water side heat transfer coefficients are 80 W/m^2 K and 4000 W/m^2 K, respectively. Determine the overall heat transfer coefficient. To enhance heat transfer, in one case, more turbulence is created on the water side, leading to an increase in the water side heat transfer coefficient by 40%, and in the other case, more turbulence is created on the air side, resulting in an increase in the air side heat transfer coefficient by 40%. Determine the increase in the overall heat transfer coefficients for both cases and comment on the results.

7.2. A heat recovery unit involves heat transfer from hot gases flowing over an aluminum tube ($k = 205$ W/m K) to water flowing through the tube. The inner and outer diameters of the tube are 19 mm and 25 mm, respectively. The outer and inner heat transfer coefficients are 100 W/m^2 K and 5000 W/m^2 K, respectively. Determine the overall heat transfer coefficient based on the inner surface area of the tube. In order to enhance the heat transfer, fins are proposed on the outer side. The scheme consists of 10 straight fins of rectangular profile provided longitudinally (evenly spaced around the circumference) over the outer surface of the tube. The fin thickness and length are 3 mm and 20 mm, respectively. Determine the enhanced overall heat transfer coefficient based on the inner surface area of the tube.

7.3. The operating condition for a certain double-pipe (concentric tube) heat exchanger is such that the LMTD is the same for both the parallel-flow and counterflow configurations. The effectiveness of the heat exchanger is 60%. Determine the required increase in the length of the heat exchanger if the effectiveness of the heat exchanger has to be increased to 80%.

7.4. A double-pipe (concentric tube) counterflow heat exchanger is designed to transfer heat from oil flowing through the annulus to water flowing through the inner tube. The inlet and outlet temperatures of water are 30 °C and 70 °C, respectively, and that of oil are 130 °C and 110 °C, respectively. The inner tube has a diameter of 25 mm, and the overall heat transfer coefficient is 400 W/m^2 K.

  a. Determine the length of the heat exchanger for a total heat transfer rate of 2500 W.

  b. Due to deterioration in the performance caused by scaling on the water side of the heat exchanger, the water outlet temperature decreases to 60 °C for the same inlet temperatures and fluid flow rates. Determine the overall heat transfer coefficient, the scaling (fouling) factor on the water side, the heat transfer rate, and the oil outlet temperature.

7.5. A one-shell and two-tube pass condenser of a steam power plant consists of brass tubes ($k = 111$ W/m K) of 19 mm outer diameter and 16 mm inner diameter. The condensate heat transfer coefficient on the outer surfaces of the tubes is 12500 W/m^2 K. Cooling water from a lake at 20 °C is pumped at 1000 kg/s through 2000 brass tubes per tube pass.

  a. For a steam condensation rate of 50 kg/s at 52 °C (latent heat of condensation, $h_{fg} = 2377$ kJ/kg), determine the inner heat transfer coefficient (based on the appropriate Nusselt number correlation), the overall heat transfer coefficient based on the outer surface area of the tube and the tube length per pass, the outlet temperature of the cooling water, and the condenser effectiveness.

b. If, after prolonged usage, scaling (fouling) causes a resistance of 8×10^{-5} m^2 K/W at the inner surface, determine the value of the overall heat transfer coefficient based on the outer surface area of the tube and the corresponding effectiveness of the condenser.

Properties of water: c_p = 4178 J/kg.K, $\mu = 0.77 \times 10^{-3}$ Pa.s, Pr = 5.2, and $k = 0.62$ W/mK

7.6. A two-shell and four-tube pass shell and tube heat exchanger is designed to cool 2.2 kg/s of oil (c_p = 2.1 kJ/kg K) from 82 °C to 55 °C with water (c_p = 4.2 kJ/kg K) at an inlet temperature of 35 °C and flowing at 3 kg/s. The overall heat transfer coefficient is 820 W/m^2 K. Determine the required heat transfer surface area. Compare the required area with that obtained from a single-shell and two-tube pass shell and tube heat exchanger for the same flow rates and temperatures.

7.7. The following are the results from the test conducted on a finned-tube, cross-flow heat exchanger used to recover the waste heat from the exhaust of a gas turbine. Gas turbine exhaust: flow rate 3 kg/s; specific heat 1100 J/kg K; inlet temperature 275 °C. Pressurized water (through the tubes): flow rate 0.75 kg/s; specific heat 4200 J/kg K; inlet temperature 40 °C and outlet temperature 130 °C. If the heat transfer surface area is 15 m^2, determine the overall heat transfer coefficient assuming (a) both fluids unmixed and (b) one fluid mixed and the other unmixed.

7.8. An intercooler of the cross-flow type (single-pass) is used for cooling 3.2 kg/s of air (c_p = 1.1 kJ/kg K) at 115 °C with water (c_p = 4.2 kJ/kg K) flowing through the tubes. The water flow rate is 2.9 kg/s and enters the intercooler at 40 °C. If the overall heat transfer coefficient based on the tube surface area of 35 m^2 is 120 W/m^2 K, determine the effectiveness of the intercooler and the outlet temperatures of air and water. Assume both fluids are unmixed.

References

Bejan, A., 1993. Heat Transfer, John Wiley & Sons, New York.

Bergman, T.L., Lavine, A.S., Incropera F.P., Dewitt, D.P., 2011. Fundamentals of Heat and Mass Transfer, seventh ed. John Wiley & Sons, NJ.

Bowman, R.A., Mueller, A.C., Nagle, W.M., 1940. Mean temperature difference in design. Trans. ASME 62 (4), 283–294.

Cengel, Y. A., 2003. Heat Transfer: A Practical Approach, second ed. McGraw-Hill, New York.

Kays, W.M., London, A.L., 1984. Compact HeatExchangers third ed. McGraw-Hill, New York.

Lienhard IV, J.H., Lienhard V, J.H., 2020. A Heat Transfer Textbook. Courier Dover Publications, Mineola, NY.

Thermal radiation

8

8.1 Introduction

Any body above a temperature of zero Kelvin emits radiation. This is a fundamental law of nature and is known as Prevost's law. Radiation occurs due to molecular activity that arises as a consequence of temperature. The molecular activity is due to the translational, rotational, and vibrational energy of matter. Radiation ceases only at absolute zero temperature. Hence, it stands to reason that any body, or anything that exists for that matter, has to and will emit radiation. It is quite intuitive to guess that radiation is a function of temperature and that it shows a monotonic increase with temperature. The challenge now is to characterize this radiation and find ways to quantify it as a mode of heat transfer in engineering problems, when several surfaces at different temperatures and radiation characteristics, whatever they may be, are present.

The theoretical framework for radiation comes from two key game changers in modern physics: (1) electromagnetic theory and (2) quantum theory.

According to electromagnetic theory, the radiation emitted by any surface travels in the form of waves with the velocity of light. In such a case, radiation can be characterized by velocity, frequency, and wavelength.
These are related by

$$c = v\lambda \tag{8.1}$$

In Eq. (8.1), c is the velocity in m/s, v is the frequency in Hz, and λ is the wavelength in μm (because meter (m) is typically too big a unit for wavelength in the study of thermal radiation).

Quantum theory, on the other hand, proposed that the radiation emitted from any surface is in the form of discrete packets called quanta.

$$E = hv \tag{8.2}$$

Here h is the Planck's constant that has a value 6.627×10^{-34} J s and E is the energy in J.

Table 8.1 shows a very basic classification of radiation based on the wavelength of the emitted radiation.

Radiation that occurs in the wavelength $0 \le \lambda \le 100\,\mu$m is frequently referred to as thermal radiation. In the study of radiation heat transfer, this wavelength range is of primary interest to us. Radiation from a surface or volume can be emitted in all

Table 8.1 A very basic classification of radiation.

Wavelength, λ, μm	Type of radiation
<0.4	Ultraviolet
0.4–0.7	Visible
>0.7	Infrared

directions. Hence, we need to understand some basic ideas in geometry to proceed further in our study of radiation.

8.2 Concepts and definitions in radiation

The elemental plane angle (see Fig. 8.1A) $d\alpha$ is defined as

$$d\alpha = \frac{dL}{r} \tag{8.3}$$

In Eq. (8.3), $d\alpha$ is in radians.

As opposed to this, the solid angle subtended by an elemental area dA_n associated with the hemisphere is given by

$$d\omega_n = \frac{dA_n}{r^2} \text{ steradians} \tag{8.4}$$

A schematic representation of solid angle is shown in Fig. 8.1B.

The solid angle subtended by an elemental area dA_n about a point on another elemental area dA_1 in the spherical coordinate system is given by

$$dA_n = (r\sin\theta)\,d\phi.rd\theta \tag{8.5}$$

$$dA_n = r^2 \sin\theta\, d\theta\, d\phi \tag{8.6}$$

where r, θ, ϕ are the radius, zenith angle, and azimuthal angle, respectively.

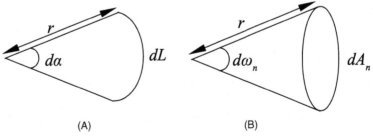

(A)　　　　　　　　(B)

FIGURE 8.1

Schematic representation of (A) a plane angle and (B) a solid angle.

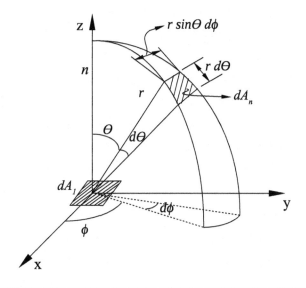

FIGURE 8.2

Solid angle subtended by an elemental area dA_n about a point dA_1 in the spherical coordinate system.

The idea of a solid angle is better understood from Fig. 8.2, which shows both the cartesian and the spherical coordinate systems.

Example 8.1: *Evaluate the total solid angle for the positive hemisphere.*
Solution:
The elemental solid angle $d\omega_n$ is given by

$$d\omega_n = \frac{dA_n}{r^2}\sin\theta d\theta d\phi \tag{8.7}$$

Integrating from 0 to $\pi/2$ for θ and 0 to 2π for ϕ, we have $\omega_n = \int_0^{2\pi}\int_0^{\frac{\pi}{2}}\sin\theta\, d\theta\, d\phi$

$$\omega_n = 2\pi \text{ steradian} \tag{8.8}$$

Hence, the solid angle associated with the hemisphere is 2π sr and that for a sphere is 4π sr, where sr is steradian.

8.3 Black body and laws of black body radiation

The study of a black body and its characteristics is central to the study of radiation heat transfer. A black body is a benchmark against which all radiative surfaces are compared.

8.3.1 **Black body**

A black body is defined as one that absorbs all the incident radiation, regardless of the direction and the wavelength.

In view of the above definition, it is evident that

1. A black body is an ideal absorber.
2. The black body is a perfect emitter. This logically follows from the definition of a black body and can be proved by a thought experiment involving a black body that is in equilibrium with its surroundings.
3. Black body radiation is independent of position (i.e., homogeneous) and direction (i.e., isotropic).
4. Consequent upon #3, the radiation from a black body is a function of only temperature.

Additionally, if all the incident radiation is absorbed and nothing is reflected, then the body appears black. However, what is visually black need not be radiatively black. Even so, all radiatively black bodies have to be visually black.

8.3.2 **Spectral directional intensity**

The spectral directional intensity of emission $I_{\lambda e}$ from a black body is given by

$$I_{\lambda e}(\lambda,\theta,\phi,T) = \frac{dQ}{dA\cos\theta d\omega d\lambda} \tag{8.9}$$

From the above mathematical definition, it is clear that the spectral direction intensity of emission $I_{\lambda e}$ is the rate at which the radiant energy emitted by a surface (dQ) in the direction of (θ,ϕ) per unit area of the surface normal to this direction $(dA\cos\theta)$ per unit solid angle $d\omega$ about (θ,ϕ) per unit wavelength $d\lambda$ about λ.

$$dQ = I_{\lambda e}(\lambda,\theta,\phi,T)dA_1\cos\theta d\omega d\lambda \tag{8.10}$$

$$\text{Let } dQ_\lambda = \frac{dQ}{d\lambda} \tag{8.11}$$

$$\text{and } dq_\lambda = \frac{dQ_\lambda}{dA_1} \tag{8.12}$$

$$dq_\lambda = I_{\lambda e}(\lambda,\theta,\phi,T)\cos\theta d\omega \tag{8.13}$$

For a black body, $I_{b,\lambda}(\lambda,\theta,\phi) \neq f(\theta,\phi)$.

Therefore, the hemispherical spectral black body emissive power of a black body at a temperature T in the wavelength interval $d\lambda$ about λ is given by

$$E_b(\lambda,T) = I_{b,\lambda}(\lambda,T)\int_0^{2\pi}\int_0^{\frac{\pi}{2}}\cos\theta\sin\theta d\theta d\phi \tag{8.14}$$

$$E_b(\lambda,T) = I_{b,\lambda}(\lambda,T) \times 2\pi \times \frac{1}{2} \qquad (8.15)$$

$$\therefore E_b(\lambda,T) = \pi I_{b,\lambda}(\lambda,T) \qquad (8.16)$$

The total hemispherical, emissive power from a black body defined as $E_b(T)$ is

$$E_b(T) = \int_0^\infty E_{b,\lambda}(\lambda,T)\,d\lambda \qquad (8.17)$$

$$E_b(T) = \int_0^\infty \int_0^{2\pi} \int_0^{\frac{\pi}{2}} I_{b,\lambda}(\lambda,T)\cos\theta\sin\theta\,d\theta\,d\phi\,d\lambda \qquad (8.18)$$

Eq. (8.18) is a powerful equation that connects $I_{b,\lambda}(\lambda,T)$ to $E_b(T)$. Hence, if $I_{b,\lambda}(\lambda,T)$ is known, $E_b(T)$ can be determined.

8.3.3 Planck's distribution

After many unsuccessful attempts by many researchers to get the correct distribution, Planck proposed the following distribution for the spectral intensity of emission from a black body at a wavelength λ and temperature T as

$$I_{b,\lambda}(\lambda,T) = \frac{C_1' \lambda^{-5}}{e^{(C_2/\lambda T)} - 1} \qquad (8.19)$$

where C_1' and C_2 are the first and the second radiation constants respectively and are given by

$$C_1' = 2hc_0^2 = 1.192 \times 10^8\ \frac{\text{W}\mu\text{m}^4}{\text{m}^2\text{sr}}$$

$$C_2 = \frac{hc_0}{k} = 1.4380 \times 10^4\ \mu\text{mK}$$

$$k = 1.38 \times 10^{-23}\ \text{J/K (Boltzmann's constant)}$$

$$h = 6.627 \times 10^{-34}\ \text{Js (Planck's constant)}$$

$$c_0 = 3 \times 10^8\ \text{m/s (velocity of radiation in vacuum)}$$

The hemispherical, spectral emissive power $E_{b,\lambda}(\lambda,T)$ given by $\pi I_{b,\lambda}(\lambda,T)$ with $I_{b,\lambda}(\lambda,T)$ from the above distribution is plotted against λ for a few temperatures and is shown in Fig. 8.3. In this case, the constant C_1' gets changed to C_1, where $C_1 = \pi C_1'$.

From the plots, the following are evident.

1. For every temperature T, $E_{b,\lambda}$ continuously varies with λ.
2. At a given λ, $E_{b,\lambda}$ is higher at a higher temperature.
3. For every temperature T, there exists a wavelength, λ at which $E_{b,\lambda}$ has a peak.
4. This peak shifts to the left side at higher temperatures, which means that at higher temperatures, the peak occurs at lower wavelengths.

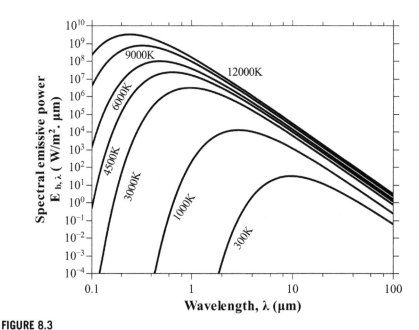

FIGURE 8.3

Planck's distribution for a few representative temperatures.

The following two laws emerge from the foundations laid by Planck's distribution:

1. Wien's displacement law
2. Stefan-Boltzmann law

8.3.4 Wien's displacement law

Let us now try to determine where the peak $I_{b,\lambda}$ occurs at a given temperature T. For maxima or minima of $I_{b,\lambda}$, the condition given below has to be satisfied.

$$\frac{\partial I_{b,\lambda}}{\partial \lambda} = 0 \tag{8.20}$$

$$\frac{\partial}{\partial \lambda}\left(\frac{C_1' \lambda^{-5}}{\left(e^{C_2/\lambda T} - 1\right)}\right) = 0 \tag{8.21}$$

$$\lambda_{max} T = 2898 \ \mu mK \tag{8.22}$$

(Details of the full solution to Eq. (8.21) are presented in Example 8.3).

8.3.5 Stefan-Boltzmann law

$$E_b(T) = \int_{\lambda=0}^{\infty} E_{b,\lambda} \, d\lambda = \sigma T^4$$
(8.23)

In fact the, Stefan Boltzmann law was obtained before the discovery of Planck's distribution using the thermodynamic route with the constant σ that has a numerical value of 5.67×10^{-8} W/m^{2}K^4 (known as the Stefan-Boltzmann constant), coming from a matching of experimental data of $E_b(T)$ against T^4. Hence, it was an absolute necessity for any proposal of $I_{b,\lambda}(\lambda T)$ to satisfy the Stefan Boltzmann law when $\pi I_{b,\lambda}(\lambda, T)$ is integrated from $\lambda = 0$ to $\lambda = \infty$.

Example 8.2: *Consider the Planck's distribution for the spectral, hemispherical emissive power of a black body given by*

$$E_{b,\lambda} = \frac{C_1 \lambda^{-5}}{e^{\frac{C_2}{\lambda T}} - 1}$$

where $C_1 = \pi C_1'$

1. In the limit of $\dfrac{C_2}{\lambda T} \to 0$, what form will the Planck's distribution take?

2. Compute the % error between the values of $E_{b,\lambda}$ (Planck's) with the $E_{b,\lambda}$ calculated from the simplified expression obtained in (1) for (a) $\dfrac{C_2}{\lambda T} = 1$, (b) $\dfrac{C_2}{\lambda T} = 0.1$, and (c) $\dfrac{C_2}{\lambda T} = 0.01$. Comment on the results.

3. In the limit of $\dfrac{C_2}{\lambda T} \to \infty$, what form will the Planck's distribution take?

4. Compute the % error between the values of $E_{b,\lambda}$(Planck's), with the $E_{b,\lambda}$ calculated from the simplified expression obtained in (3) for (a) $\dfrac{C_2}{\lambda T} = 1$, (b) $\dfrac{C_2}{\lambda T} = 10$, and (c) $\dfrac{C_2}{\lambda T} = 50$. Comment on the results.

Solution:

$$E_{b,\lambda} = \frac{C_1 \lambda^{-5}}{e^{\frac{C_2}{\lambda T}} - 1}$$

1. When $\dfrac{C_2}{\lambda T} \to 0$

We have $e^x = 1 + x + \dfrac{x^2}{2!} + \dfrac{x^3}{3!} + \ldots$

Here, $x = \dfrac{C_2}{\lambda T};$ for $x \to 0,$ $e^x = 1 + x,$ $e^{\frac{C_2}{\lambda T}} = 1 + \dfrac{C_2}{\lambda T}$

$$\therefore E_{b,\lambda} = \frac{C_1 \lambda^{-5}}{\frac{C_2}{\lambda T}}$$

This is known as the Rayleigh-Jeans distribution; it holds good for longer wavelengths and it fails in the asymptotic limit of $\lambda \to 0$.

2.
$$\% \text{ error } = \frac{E_{b,\lambda}(Planck's) - E_{b,\lambda}(Rayleigh\ Jeans)}{E_{b,\lambda}(Planck's)} \times 100$$

$$= \frac{\dfrac{C_1 \lambda^{-5}}{\left(e^{\left(\frac{C_2}{\lambda T}\right)} - 1\right)} - \dfrac{C_1 \lambda^{-5}}{\left(\dfrac{C_2}{\lambda T}\right)}}{\dfrac{C_1 \lambda^{-5}}{\left(e^{\left(\frac{C_2}{\lambda T}\right)} - 1\right)}} \times 100 = \left(1 - \frac{\left(e^{\frac{C_2}{\lambda T}} - 1\right)}{\dfrac{C_2}{\lambda T}}\right) \times 100$$

(a) $\dfrac{C_2}{\lambda T} = 1$

% error = −71.82%

(b) $\dfrac{C_2}{\lambda T} = 0.1$

% error = −5.17%

(c) $\dfrac{C_2}{\lambda T} = 0.01$

% error = −0.502%

As the value of $\dfrac{C_2}{\lambda T}$ decreases, the simplified expression approaches Planck's values of $E_{b,\lambda}$. Hence, the Rayleigh-Jeans approximation, works well for longer wavelengths (and so low values of $\dfrac{C_2}{\lambda T}$)

3. When $\dfrac{C_2}{\lambda T} \to \infty$

$$E_{b,\lambda} = \frac{C_1 \lambda^{-5}}{e^{\frac{C_2}{\lambda T}} - 1}$$

$$e^{\frac{C_2}{\lambda T}} \gg 1$$

$$\therefore E_{b,\lambda} = \frac{C_1 \lambda^{-5}}{e^{\frac{C_2}{\lambda T}}}$$

The above equation is known as the Wien's distribution; while this works well for shorter λs, at higher λs it tends to saturate.

4.
$$\% \text{ error } = \frac{E_{b,\lambda}(Planck) - E_{b,\lambda}(Wien)}{E_{b,\lambda}(Planck)} \times 100$$

$$= \frac{\dfrac{C_1 \lambda^{-5}}{e^{\frac{C_2}{\lambda T}} - 1} - \dfrac{C_1 \lambda^{-5}}{e^{\frac{C_2}{\lambda T}}}}{\dfrac{C_1 \lambda^{-5}}{e^{\frac{C_2}{\lambda T}} - 1}} \times 100 = \left(1 - \frac{\left(e^{\frac{C_2}{\lambda T}} - 1\right)}{e^{\frac{C_2}{\lambda T}}}\right) \times 100 = e^{-\frac{C_2}{\lambda T}} \times 100$$

(a) $\dfrac{C_2}{\lambda T} = 1$

% error = 36.79%

(b) $\dfrac{C_2}{\lambda T} = 10$

% error = 0.0045%

(c) $\dfrac{C_2}{\lambda T} = 50$

% error = 1.93×10^{-20}%

As the value of $\dfrac{C_2}{\lambda T}$ increases, the simplified expression approaches Planck's values of $E_{b,\lambda}$.

Example 8.3:

Consider $\dfrac{C_1 \lambda^{-5}}{\left[e^{\frac{C_2}{\lambda T}} - 1 \right] \sigma T^5}$.

1. What are the units of the above quantity?
2. Determine the value of λT at which the value of this function is the maximum.
3. Hence determine the maximum value of this function.
4. What potential use do you see in introducing such a function?
Solution:

Consider,

$$\frac{C_1 \lambda^{-5}}{\left[e^{\frac{C_2}{\lambda T}} - 1 \right] \sigma T^5}$$

1. Let us first work out the units of the above quantity.

$$C_1 = 2\pi h c_0^2 = 3.742 \times 10^8 \ W \mu m^4/m^2, \ C_2 = \frac{h c_0}{k} = 1.439 \times 10^4 \ \mu m K$$

Hence, the units of given quantity are $\dfrac{W \mu m^4/m^2 \times (\mu m)^{-5}}{W/m^2/K^4 \times K^5} = \dfrac{1}{\mu m K}$

2. $\dfrac{d\left(\dfrac{C_1 \lambda^{-5}}{\left(e^{\frac{C_2}{\lambda T}} - 1 \right) \sigma T^5} \right)}{d(\lambda T)} = 0$

$$\frac{d\left(\frac{C_1}{\left(e^{\frac{C_2}{\lambda T}}-1\right)\sigma(\lambda T)^5}\right)}{d(\lambda T)} = 0$$

$$\frac{5\times(\lambda T)^4 \times \left(e^{\frac{C_2}{\lambda T}}-1\right)+(\lambda T)^5 \times e^{\frac{C_2}{\lambda T}} \times -\frac{C_2}{\lambda T}}{\left((\lambda T)^5 \times \left(e^{\frac{C_2}{\lambda T}}-1\right)\right)^2} = 0$$

$$5\left(e^{\frac{C_2}{\lambda T}}-1\right)-\frac{C_2}{\lambda T}\times e^{\frac{C_2}{\lambda T}} = 0$$

$$\frac{\frac{C_2}{\lambda T}\times e^{\frac{C_2}{\lambda T}}}{-\left(1-e^{\frac{C_2}{\lambda T}}\right)} = 5$$

$$Let\ \frac{C_2}{\lambda T}=x\frac{xe^x}{e^x-1} = 5$$

$$x = 5\left(\frac{e^x-1}{e^x}\right)$$

Writing this in algorithmic form and solving by the successive substitution method details of which are given in Table 8.2, we have

$$x_{i+1} = 5\left(\frac{e^{x_i}-1}{e^{x_i}}\right)$$

In the above expression, i refers to the iteration number.

$$x^* = \frac{C_2}{\lambda T} = 4.965$$

Table 8.2 A successive substitution method for example 8.3.

Iteration	x_i	x_{i+1}	Residual $(x_{i+1}-x_i)^2$
1	3	4.75	3.06
2	4.75	4.95	0.044
3	4.95	4.964	1.96×10^{-4}
4	4.964	4.965	1.96×10^{-6}

$$\lambda_{max} T = \frac{1.439 \times 10^4}{4.965}$$

$$\boxed{\therefore \lambda_{max} T = 2898 \, \mu mK}$$

3. This expression is actually the Wien's displacement law and confirms that λ_{max} decreases when the temperature T increases.

$$\text{Maximum value of the quantity} = \frac{C_1 \lambda^{-5}}{\left(e^{\frac{C_2}{\lambda T}} - 1\right) \sigma T^5}$$

$$= \frac{C_1}{\left(e^{\frac{C_2}{\lambda T}} - 1\right) \sigma (\lambda T)^5}$$

$$= \frac{3.742 \times 10^8}{\left(e^{\frac{1.439 \times 10^4}{2898}} - 1\right) 5.67 \times 10^{-8} (2898)^5}$$

$$\text{Maximum value} = 2.267 \times 10^{-4} \, \mu m^{-1} K^{-1}$$

4. Through the introduction of this quantity, one can obtain a universal black body curve, wherein black body curves for different temperatures merge into one curve. This quantity is a function only of (λT) product. The resulting curve has a peak value as shown in (3), and this happens at $\lambda T = 2898 \, \mu mK$. More about this in the ensuing section.

8.3.6 Universal black body curve

Consider the function $\frac{E_{b,\lambda}(\lambda T)}{\sigma T^5}$. This can be written as

$$\frac{E_{b,\lambda}}{\sigma T^5} = \frac{C_1}{\sigma (\lambda T)^5 \left(e^{\frac{C_2}{\lambda T}} - 1\right)} \tag{8.24}$$

$\therefore \frac{E_{b,\lambda}}{\sigma T^5} = f(\lambda T)$ alone. Hence, if we plot the quantity $\frac{E_{b,\lambda}}{\sigma T^5}$ against λT, we get one curve. This is known as the universal black body curve shown in Fig. 8.4.

This curve is useful in situations where we want to determine the fraction of the radiation emitted in a wavelength interval, say, for example, $(\lambda_2 - \lambda_1)$ for a temperature T. This can be depicted as shown in Fig. 8.5.

The fraction or F function is then defined as

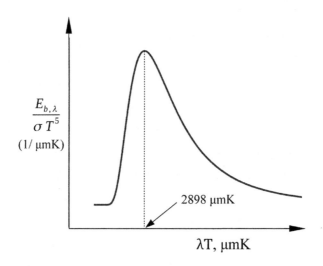

FIGURE 8.4

Universal black body curve.

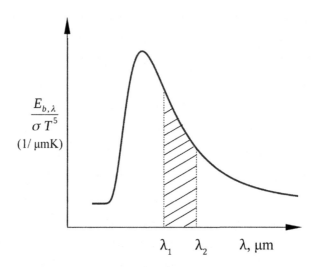

FIGURE 8.5

The universal black body curve and its application in radiation problems.

$$F_{0-\lambda T} = \frac{\int_0^{\lambda T} f(\lambda T)d(\lambda T)}{\int_0^{\infty} f(\lambda T)d(\lambda T)} \tag{8.25}$$

$$E_{0-\lambda T} = F_{0-\lambda T} \times \sigma T^4 \tag{8.26}$$

$$F_{\lambda_1 T - \lambda_2 T} = F_{0-\lambda_2 T} - F_{0-\lambda_1 T} \tag{8.27}$$

From this fraction, the hemispherical emissive power in the wavelength interval λ_1 to λ_2 can be determined as $F_{\lambda_1 T - \lambda_2 T}(\sigma T^4)$. The F function chart can be tabulated and is presented in Table 8.3.

Table 8.3 Black body radiation functions (F-function) Table.

λT (μmK)	$F_{(0 \rightarrow \lambda)}$	$I_{\lambda,b}(\lambda T)/\sigma T^5 (\mu m.K.sr)^{-1}$	$\dfrac{I_{\lambda,b}(\lambda,T)}{I_{\lambda,b}(\lambda_{max},T)}$
200	0.000000	3.711772×10^{-28}	0.000000
400	0.000000	4.877254×10^{-14}	0.000000
600	0.000000	1.036654×10^{-9}	0.000014
800	0.000016	9.883195×10^{-8}	0.001370
1000	0.000320	1.182284×10^{-6}	0.016385
1200	0.002130	5.228789×10^{-6}	0.072464
1400	0.007778	1.341736×10^{-5}	0.185946
1600	0.019691	2.487352×10^{-5}	0.344712
1800	0.039292	3.750250×10^{-5}	0.519732
2000	0.066653	4.927725×10^{-5}	0.682914
2200	0.100782	5.889147×10^{-5}	0.816153
2400	0.140119	6.580943×10^{-5}	0.912027
2600	0.182951	7.005151×10^{-5}	0.970816
2800	0.227691	7.194804×10^{-5}	0.997099
2898	0.249913	7.215735×10^{-5}	1.000000
3000	0.273004	7.195287×10^{-5}	0.997166
3200	0.317847	7.052908×10^{-5}	0.977434
3400	0.361457	6.809082×10^{-5}	0.943644
3600	0.403307	6.498093×10^{-5}	0.900545
3800	0.443063	6.146872×10^{-5}	0.851870
4000	0.480541	5.775734×10^{-5}	0.800436
4200	0.515662	5.399468×10^{-5}	0.748291
4400	0.548431	5.028467×10^{-5}	0.696875
4600	0.578903	4.669735×10^{-5}	0.647160
4800	0.607171	4.327738×10^{-5}	0.599764
5000	0.633350	4.005083×10^{-5}	0.555048
5200	0.657564	3.703038×10^{-5}	0.513189
5400	0.679946	3.421938×10^{-5}	0.474233
5600	0.700626	3.161478×10^{-5}	0.438137
5800	0.719732	2.920932×10^{-5}	0.404800
6000	0.737386	2.699316×10^{-5}	0.374087

(Continued)

Table 8.3 Black body radiation functions (F-function) Table. (*Cont.*)

λT (µmK)	$F_{(0 \to \lambda)}$	$I_{\lambda,b}(\lambda T)/\sigma T^5 (\mu m.K.sr)^{-1}$	$\dfrac{I_{\lambda,b}(\lambda,T)}{I_{\lambda,b}(\lambda_{max},T)}$
6200	0.753704	2.495496×10^{-5}	0.345841
6400	0.768793	2.308268×10^{-5}	0.319894
6600	0.782754	2.136417×10^{-5}	0.296077
6800	0.795680	1.978750×10^{-5}	0.274227
7000	0.807657	1.834121×10^{-5}	0.254183
7200	0.818763	1.701444×10^{-5}	0.235796
7400	0.829070	1.579705×10^{-5}	0.218925
7600	0.838643	1.467960×10^{-5}	0.203439
7800	0.847543	1.365338×10^{-5}	0.189217
8000	0.855825	1.271039×10^{-5}	0.176148
8200	0.863538	1.184331×10^{-5}	0.164132
8400	0.870728	1.104546×10^{-5}	0.153075
8600	0.877437	1.031075×10^{-5}	0.142893
8800	0.883702	9.633637×10^{-6}	0.133509
9000	0.889559	9.009093×10^{-6}	0.124853
9200	0.895038	8.432545×10^{-6}	0.116863
9400	0.900169	7.899844×10^{-6}	0.109481
9600	0.904977	7.407225×10^{-6}	0.102654
9800	0.909488	6.951272×10^{-6}	0.096335
10000	0.913723	6.528882×10^{-6}	0.090481
10500	0.923232	5.601903×10^{-6}	0.077635
11000	0.931410	4.830388×10^{-6}	0.066942
11500	0.938479	4.184824×10^{-6}	0.057996
12000	0.944616	3.641843×10^{-6}	0.050471
12500	0.949969	3.182853×10^{-6}	0.044110
13000	0.954656	2.792992×10^{-6}	0.038707
13500	0.958777	2.460320×10^{-6}	0.034097
14000	0.962413	2.175193×10^{-6}	0.030145
14500	0.965634	1.929783×10^{-6}	0.026744
15000	0.968496	1.717707×10^{-6}	0.023805
15500	0.971047	1.533730×10^{-6}	0.021255
16000	0.973328	1.373542×10^{-6}	0.019035
16500	0.975374	1.233578×10^{-6}	0.017096
17000	0.977214	1.110874×10^{-6}	0.015395
17500	0.978873	1.002956×10^{-6}	0.013900
18000	0.980373	9.077520×10^{-7}	0.012580
18500	0.981732	8.235169×10^{-7}	0.011413
19000	0.982966	7.487779×10^{-7}	0.010377
19500	0.984090	6.822862×10^{-7}	0.009456

Table 8.3 Black body radiation functions (F-function) Table. (*Cont.*)

λT (µmK)	$F_{(0 \to \lambda)}$	$I_{\lambda,b}(\lambda T)/\sigma T^5$ (µm.K.sr)$^{-1}$	$\dfrac{I_{\lambda,b}(\lambda,T)}{I_{\lambda,b}(\lambda_{max},T)}$
20000	0.985114	6.229790×10^{-7}	0.008634
25000	0.991726	2.763310×10^{-7}	0.003830
30000	0.994851	1.403976×10^{-7}	0.001946
35000	0.996514	7.862366×10^{-8}	0.001090
40000	0.997478	4.736531×10^{-8}	0.000656
45000	0.998075	3.020096×10^{-8}	0.000419
50000	0.998464	2.015049×10^{-8}	0.000279
55000	0.998728	1.395263×10^{-8}	0.000193
60000	0.998914	9.964006×10^{-9}	0.000138
65000	0.999048	7.303743×10^{-9}	0.000101
70000	0.999148	5.474729×10^{-9}	0.000076
75000	0.999223	4.183928×10^{-9}	0.000058
80000	0.999281	3.252025×10^{-9}	0.000045
85000	0.999327	2.565680×10^{-9}	0.000036
90000	0.999363	2.051193×10^{-9}	0.000028
95000	0.999392	1.659422×10^{-9}	0.000023
100000	0.999415	1.356864×10^{-9}	0.000019

Example 8.4: *If the Sun's equivalent black body temperature is 5800 K, determine the wavelength corresponding to maximum emission. Also determine the fraction of the emission that is emitted in the following spectral regions (1) ultraviolet (UV), (2) visible, and (3) IR.*

Solution:

If the Sun's equivalent body temperature = 5800 K, then,

$$\lambda_{max} = \frac{2898}{5800} \approx 0.5 \, \mu m$$

Solar emission: thermal emission range 0.1–100 µm, T_{Sun} = 5800 K

UV:0.1–0.4µm

$$\lambda_1 T = 0.1 \times 5800 = 580 \ \mu mK : F(0 \to \lambda_1) = 0$$

$$\lambda_2 T = 0.4 \times 5800 = 2320 \ \mu mK : F(0 \to \lambda_2) = 0.1245$$

$$F_{uv} = \frac{\int_{0.1}^{0.4} E_{b,\lambda} \, d\lambda}{\int_0^\infty E_{b,\lambda} \, d\lambda} = F(0 \to \lambda_2) - F(0 \to \lambda_1) = 0.1245$$

Visible:0.4–0.7μm

$$\lambda_1 T = 0.4 \times 5800 = 2320 \,\mu mK : F(0 \rightarrow \lambda_1) = 0.1245$$

$$\lambda_2 T = 0.7 \times 5800 = 4060 \,\mu mK : F(0 \rightarrow \lambda_2) = 0.4914$$

$$F_{vis} = \frac{\int_{0.4}^{0.7} E_{b,\lambda} \, d\lambda}{\int_0^\infty E_{b,\lambda} \, d\lambda} = F(0 \rightarrow \lambda_2) - F(0 \rightarrow \lambda_1) = 0.3669$$

Infrared:0.7–100μm

$$\lambda_1 T = 0.7 \times 5800 = 4060 \,\mu mK : F(0 \rightarrow \lambda_1) = 0.4914$$

$$\lambda_2 T = 100 \times 5800 = 580000 \,\mu mK : F(0 \rightarrow \lambda_2) = 1$$

$$F_{ir} = \frac{\int_{0.7}^{100} E_{b,\lambda} \, d\lambda}{\int_0^\infty E_{b,\lambda} \, d\lambda} = F(0 \rightarrow \lambda_2) - F(0 \rightarrow \lambda_1) = 0.5086$$

The above example clearly shows that more than 1/3 of the Sun's radiation is contained in an extremely narrow wavelength interval of 0.4–0.7 μm, that is, in the visible part of the spectrum. So, every bulb manufacturer's dream is to reproduce this day light!

8.4 Properties of real surfaces

A black body is an ideal one that is hard to realize in engineering practice. Hence, there is a need to characterize the properties of real surfaces. The primary purpose of the black body idealization is to serve as the theoretical ideal surface against which real surfaces can be characterized.

8.4.1 Emissivity (ε)

The emissivity (ε) of a surface is defined as the ratio of emission from a real surface in general to the emission from a black body at the same temperature.

$$\varepsilon = \frac{\text{Intensity of emission from a real surface}(I_{\lambda,e})}{\text{Intensity of emission from a black body}(I_{b,\lambda})} \tag{8.28}$$

Emissivity has no units and ranges from 0 to 1.

However, the problem is that intensity of emission from a real body ($I_{\lambda,e}$) is a function of wavelength, temperature, and direction, whereas the intensity of emission for a black body ($I_{b,\lambda}$) is a function of wavelength and temperature alone.

$$I_{\lambda,e} = f(\lambda, T, \theta, \phi) \tag{8.29}$$

$$I_{b,\lambda} = I_{b,\lambda}(\lambda,T) \tag{8.30}$$

The spectral directional emissivity $\varepsilon'_\lambda(\lambda,T,\theta,\phi)$ of a real surface is then given by

$$\varepsilon'_\lambda(\lambda,T,\theta,\phi) = \frac{I_{\lambda,e}(\lambda,T,\theta,\phi)}{I_{b,\lambda}(\lambda,T)} \tag{8.31}$$

The nomenclature is ε'_λ, where λ indicates that it is the spectral quantity and $'$ indicates that it is a directional quantity.

The hemispherical spectral emissivity, $\varepsilon_\lambda(\lambda,T)$ is defined as

$$\varepsilon_\lambda(\lambda,T) = \frac{\text{Spectral emissive power from a real surface}}{\text{Spectral emissive power from a black body}} \tag{8.32}$$

$$\varepsilon_\lambda(\lambda,T) = \frac{\int_{\phi=0}^{2\pi}\int_{\theta=0}^{\pi/2} I_{\lambda,e}(\lambda,T,\theta,\phi)\cos\theta\,\sin\theta\,d\theta\,d\phi}{\int_{\phi=0}^{2\pi}\int_{\theta=0}^{\pi/2} I_{b,\lambda}(\lambda,T)\cos\theta\,\sin\theta\,d\theta\,d\phi} \tag{8.33}$$

$$\varepsilon_\lambda(\lambda,T) = \frac{\int_{\phi=0}^{2\pi}\int_{\theta=0}^{\pi/2} I_{\lambda,e}(\lambda,T,\theta,\phi)\cos\theta\,\sin\theta\,d\theta\,d\phi}{I_{b,\lambda}(\lambda,T)\times\pi} \tag{8.34}$$

Substituting Eq. (8.31) in Eq. (8.34),

$$\varepsilon_\lambda(\lambda,T) = \frac{\int_{\phi=0}^{2\pi}\int_{\theta=0}^{\pi/2} \varepsilon'_\lambda(\lambda,T,\theta,\phi)I_{\lambda,b}(\lambda,T)\cos\theta\,\sin\theta\,d\theta\,d\phi}{I_{b,\lambda}(\lambda,T)\times\pi} \tag{8.35}$$

$$\varepsilon_\lambda(\lambda,T) = \frac{\int_{\phi=0}^{2\pi}\int_{\theta=0}^{\pi/2} \varepsilon'_\lambda(\lambda,T,\theta,\phi)\cos\theta\,\sin\theta\,d\theta\,d\phi}{\pi} \tag{8.36}$$

$$\varepsilon_\lambda(\lambda,T) = \frac{1}{\pi}\int_{\phi=0}^{2\pi}\int_{\theta=0}^{\pi/2} \varepsilon'_\lambda(\lambda,T,\theta,\phi)\cos\theta\,\sin\theta\,d\theta\,d\phi \tag{8.37}$$

The hemispherical total emissivity $\varepsilon(T)$, which is often the key engineering property, is given by

$$\varepsilon(T) = E(T)/E_b(T) = \frac{\text{Total hemispherical emissive power of a real surface}}{\text{Total hemispherical emissive power of a black body}} \tag{8.38}$$

$$\varepsilon(T) = \frac{\int_{\lambda=0}^{\infty}\int_{\phi=0}^{2\pi}\int_{\theta=0}^{\pi/2} I_{\lambda,e}(\lambda,T,\theta,\phi)\cos\theta\,\sin\theta\,d\theta\,d\phi\,d\lambda}{\int_{\lambda=0}^{\infty}\int_{\phi=0}^{2\pi}\int_{\theta=0}^{\pi/2} I_{b,\lambda}(\lambda,T)\cos\theta\,\sin\theta\,d\theta\,d\phi\,d\lambda} \tag{8.39}$$

$$\varepsilon(T) = \frac{\int_{\lambda=0}^{\infty}\int_{\phi=0}^{2\pi}\int_{\theta=0}^{\pi/2} \varepsilon'_\lambda(\lambda,T,\theta,\phi)I_{b,\lambda}(\lambda,T)\cos\theta\,\sin\theta\,d\theta\,d\phi\,d\lambda}{\int_{\lambda=0}^{\infty}\int_{\phi=0}^{2\pi}\int_{\theta=0}^{\pi/2} I_{b,\lambda}(\lambda,T)\cos\theta\,\sin\theta\,d\theta\,d\phi\,d\lambda} \tag{8.40}$$

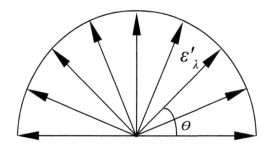

FIGURE 8.6

Variation of emissivity with θ for a diffuse surface.

$$\varepsilon(T) = \frac{\int_{\lambda=0}^{\infty}\int_{\phi=0}^{2\pi}\int_{\theta=0}^{\pi/2}\varepsilon_\lambda'(\lambda,T,\theta,\phi)I_{b,\lambda}(\lambda,T)\cos\theta\,\sin\theta\,d\theta\,d\phi\,d\lambda}{\int_{\lambda=0}^{\infty}I_{b,\lambda}(\lambda,T)\int_{\phi=0}^{2\pi}\int_{\theta=0}^{\pi/2}\cos\theta\,\sin\theta\,d\theta\,d\phi\,d\lambda} \tag{8.41}$$

$$\varepsilon(T) = \frac{\int_{\lambda=0}^{\infty}I_{b,\lambda}(\lambda,T)\int_{\phi=0}^{2\pi}\int_{\theta=0}^{\pi/2}\varepsilon_\lambda'(\lambda,T,\theta,\phi)\cos\theta\,\sin\theta\,d\theta\,d\phi\,d\lambda}{\int_{\lambda=0}^{\infty}\pi\,I_{b,\lambda}(\lambda,T)d\lambda} \tag{8.42}$$

$$\therefore \varepsilon(T) = \frac{\int_{\lambda=0}^{\infty}I_{b,\lambda}(\lambda,T)\pi\varepsilon_\lambda\,d\lambda}{E_b(T)} \text{ [follows from equation 8.34]} \tag{8.43}$$

$$\varepsilon(T) = \frac{\int_{\lambda=0}^{\infty}E_{b,\lambda}(\lambda,T)\varepsilon_\lambda(\lambda,T)d\lambda}{E_b(T)} \tag{8.44}$$

Diffuse surface: If radiation emitted by a surface is independent of direction, such a surface is called a diffuse surface. The emissivity of a diffuse surface is the same in all directions and is schematically shown in Fig. 8.6.

For a diffuse surface, $\varepsilon_\lambda'(\lambda,T,\theta,\phi) \neq f(\theta,\phi)$. Hence, the hemispherical total emissivity $\varepsilon(T)$ is given by

$$\varepsilon(T) = \frac{\pi\int_{\lambda=0}^{\infty}\varepsilon_\lambda'I_{b,\lambda}(\lambda,T)d\lambda}{\pi\int_{\lambda=0}^{\infty}I_{b,\lambda}\,d\lambda} \tag{8.45}$$

$$\varepsilon(T) = \frac{\int_{\lambda=0}^{\infty}\varepsilon_\lambda'I_{b,\lambda}(\lambda,T)d\lambda}{\int_{\lambda=0}^{\infty}I_{b,\lambda}\,d\lambda} \tag{8.46}$$

$$\varepsilon(T) = \frac{\int_{\lambda=0}^{\infty}\varepsilon_\lambda'I_{b,\lambda}(\lambda,T)d\lambda}{\sigma T^4} \tag{8.47}$$

For example, consider the variation of the spectral emissivity of a diffuse surface as shown in Fig. 8.7

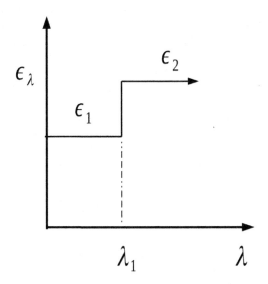

FIGURE 8.7

Representative variation of emissivity of a diffuse surface with wavelength.

The hemispherical total emissivity of a surface having spectral variation, as shown in Fig. 8.7, can be calculated as below

$$\varepsilon(T) = \frac{\int_{\lambda=0}^{\infty} \varepsilon_\lambda E_{b,\lambda}(\lambda,T)d\lambda}{\sigma T^4} \tag{8.48}$$

$$= \frac{\varepsilon_1 \int_{\lambda=0}^{\lambda_1} E_{b,\lambda}(\lambda,T)d\lambda}{\sigma T^4} + \frac{\varepsilon_2 \int_{\lambda_1}^{\infty} E_{b,\lambda}(\lambda,T)d\lambda}{\sigma T^4} \tag{8.49}$$

$$= \varepsilon_1 F_{0-\lambda_1} + \varepsilon_2(1 - F_{0-\lambda_1}) \tag{8.50}$$

Gray surface: If the radiation emitted by a surface is independent of wavelength, such a surface is called as a gray surface.

For a gray surface, $\varepsilon'_\lambda(\lambda,T,\theta,\phi) \neq f(\lambda)$. The emissivity of a gray surface is independent of wavelength.

If a surface is both gray and diffuse, its directional spectral emissivity is equal to hemispherical total emissivity. The emissivity of a gray and diffuse surface is independent of both direction and wavelength.

Example 8.5: *The spectral emissivity distribution of a diffuse surface at a temperature of 1000 K is shown in Fig. 8.8.*

1. Calculate the hemispherical total emissivity of the surface.
2. Determine the hemispherical total emissive power of the surface.

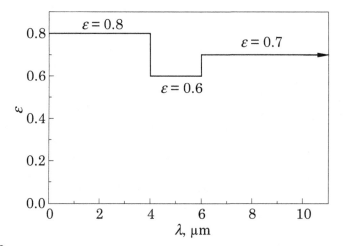

FIGURE 8.8

Variation of emissivity with wavelength.

3. What fraction of (2) is in the region $0 \leq \lambda \leq 6\ \mu m$?
4. What should be the emissivity of a gray body at 1200 K to have the same hemispherical, total emissive power as this body as calculated in (2)?

Solution:

1. $$\varepsilon = \frac{\int_0^\infty E_\lambda\, d\lambda}{\int_0^\infty E_{b,\lambda}\, d\lambda} = \frac{\int_0^\infty \varepsilon_\lambda E_{b,\lambda}\, d\lambda}{\int_0^\infty E_{b,\lambda}\, d\lambda}$$

$$\varepsilon = \frac{\int_0^{\lambda_1} \varepsilon_\lambda E_{b,\lambda}\, d\lambda}{\int_0^\infty E_{b,\lambda}\, d\lambda} + \frac{\int_{\lambda_1}^{\lambda_2} \varepsilon_\lambda E_{b,\lambda}\, d\lambda}{\int_0^\infty E_{b,\lambda}\, d\lambda} + \frac{\int_{\lambda_2}^\infty \varepsilon_\lambda E_{b,\lambda}\, d\lambda}{\int_0^\infty E_{b,\lambda}\, d\lambda}$$

$$\varepsilon = \varepsilon_{\lambda_1} \cdot F(0 \to \lambda_1) + \varepsilon_{\lambda_2} \cdot F(\lambda_1 \to \lambda_2) + \varepsilon_{\lambda_3} \cdot F(\lambda_\infty \to \lambda_2)$$

$$\varepsilon = \varepsilon_{\lambda_1} F_{0-\lambda_1} + \varepsilon_{\lambda_2}(F_{0-\lambda_2} - F_{0-\lambda_1}) + \varepsilon_{\lambda_3}(F_{0-\lambda_\infty} - F_{0-\lambda_2})$$

The values of the fraction F can be obtained from the F charts.

$$\lambda_1 T = 4000\ \mu mK : F(0 \to \lambda_1) = 0.4805$$

$$\lambda_2 T = 6000\ \mu mK : F(0 \to \lambda_2) = 0.7374$$

$$\varepsilon = 0.8 \times 0.4805 + 0.6(0.7374 - 0.4804) + 0.7(1 - 0.7374)$$

$$\therefore \varepsilon = 0.7224$$

2. Hemispherical total emissive power

$$E_b = \varepsilon \sigma T^4$$
$$= 0.7224 \times 5.67 \times 10^{-8} \times 1000^4$$
$$\therefore E_b = 40.964 \text{ kW/m}^2$$

3. Fraction of emissive power in the region $0 \le \lambda \le 6 \text{ μm}$

$$\lambda_1 T = 4000 \text{ μmK} : F(0 \to \lambda_1) = 0.4805$$

$$\lambda_2 T = 6000 \text{ μmK} : F(0 \to \lambda_1) = 0.7374$$

$$
\begin{aligned}
Fraction &= \frac{\int_0^6 E_\lambda \, d\lambda}{\int_0^\infty \varepsilon_\lambda E_{b,\lambda} \, d\lambda} = \frac{\int_0^6 \varepsilon_\lambda E_{b,\lambda} \, d\lambda}{\int_0^\infty \varepsilon_\lambda E_{b,\lambda} \, d\lambda} \\[2mm]
&= \frac{\varepsilon_{\lambda_1} \cdot F(0 \to \lambda_1) + \varepsilon_{\lambda_2} \cdot F(\lambda_1 \to \lambda_2)}{\varepsilon} \\[2mm]
&= \frac{\varepsilon_{\lambda_1} F_{0-\lambda_1} + \varepsilon_{\lambda_2} [F_{0-\lambda_2} - F_{0-\lambda_1}]}{\varepsilon} \\[2mm]
&= \frac{0.8 \times 0.4805 + 0.6 \times (0.7374 - 0.4805)}{0.7224}
\end{aligned}
$$

$$Fraction = 0.7396$$

4. For the emissivity of a gray body at 1200 K to have the same hemispherical, total emissive power as this body as calculated in (2).

$$E_b = 40.964 \times 10^3 \text{ W/m}^2 = \varepsilon \sigma T_b^4$$

here, $T_b = 1200 \text{ K}$

$$\varepsilon = \frac{40.964 \times 10^3}{5.67 \times 10^{-8} \times 1200^4}$$

$$\boxed{\varepsilon = 0.348}$$

Though we have given an example with an unrealistic, piecewise constant, spectral emissivity distribution, the same technique can be used for more complex, continuous emissivity distributions as well.

Hence, a smooth distibution can be considered to be made up of a large number of infinitesimal "boxes" using constant emissivities.

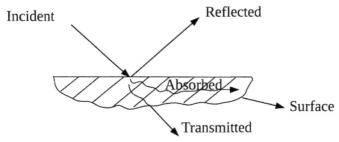

FIGURE 8.9

Absorption, reflection, and transmission processes associated with a semitransparent medium.

8.4.2 Apportioning of radiation falling on a surface

So far, we considered how real surface emits radiation at a given temperature. Let us now consider a surface that is receiving incident radiation as shown in Fig. 8.9. At steady state, using energy balance we have,

$$G_{inc} = G_{ref} + G_{abs} + G_{tran} \tag{8.51}$$

where G_{inc} is the incident radiation on the surface, and G_{ref}, G_{abs}, and G_{tran} are the radiation reflected, absorbed, and transmitted by the surface, respectively. Please note that all these are in W/m^2, if we consider total quantities for $\lambda = 0$ and $\lambda = \infty$ and the units will be W/m^{2}μm if our consideration is spectral, i.e. in a wavelength interval dλ about λ.

Dividing this equation by G_{inc} throughout we get,

$$\frac{G_{inc}}{G_{inc}} = \frac{G_{ref}}{G_{inc}} + \frac{G_{abs}}{G_{inc}} + \frac{G_{tran}}{G_{inc}} \tag{8.52}$$

Reflectivity(ρ): The ratio of radiation reflected by the surface to the total radiation falling on the surface.

Absorptivity(α): The ratio of radiation absorbed by the surface to the total radiation falling on the surface.

Transmissivity(τ): The ratio of radiation transmitted by the surface to the total radiation falling on the surface.

From Eq. (8.52), we have,

$$1 = \text{reflectivity}(\rho) + \text{absorptivity}(\alpha) + \text{transimissivity}(\tau) \tag{8.53}$$

$$\rho + \alpha + \tau = 1 \tag{8.54}$$

For an opaque surface, the transmissivity is $\tau = 0$.

Therefore, for an opaque surface, we have

$$\alpha + \rho = 1 \tag{8.55}$$

or

$$\rho = 1 - \alpha \tag{8.56}$$

Eq. (8.56) is a key engineering result that relates the absorptivity and reflectivity of an opaque surface. Please note that the reflectivity, absorptivity and transmissivity are dimensionless quantities, having the range 0 to 1.

8.4.3 Spectral directional absorptivity

The quantity α appearing in Eq. (8.56) is an average quantity over an hemisphere and over the wavelength 0 to ∞. Now, we need to get down to the most basic definition for a particular angle (θ, ϕ) and a wavelength interval $d\lambda$ about λ. This is denoted as spectral, directional absorptivity. The spectral directional absorptivity gives the ratio of radiation absorbed to that incident in a given direction and wavelength. Mathematically, it is given by

$$\alpha'_\lambda(\lambda_i, T_i, \theta_i, \phi_i) = \frac{dQ_{absorbed}}{I_{\lambda,i}\, dA \cos(\theta_i) d\omega\, d\lambda} \tag{8.57}$$

For a diffuse surface,

$$\alpha'_\lambda \neq f(\theta, \phi) \tag{8.58}$$

For a gray surface,

$$\alpha'_\lambda \neq f(\lambda) \tag{8.59}$$

For a gray diffuse surface $\alpha'_\lambda = \alpha$. This is a major engineering simplification and is often not a bad assumption.

The relation between emissivity (ε) and absorptivity (α) cannot be obtained by theory because they come from completely different processes. Absorption is the capacity of the body to absorb, whereas emission is related to translation, rotation, and vibration of molecules. So, any relation between α and ε should come empirically from the experiments. Kirchoff was the first scientist who performed those experiments and proposed the famous Kirchoff's law that relates emissivity and absorptivity.

8.5 Kirchoff's law

Kirchoff's law states that, the spectral directional absorptivity is equal to the spectral directional emissivity.

$$\alpha'_\lambda(\lambda, T, \theta, \phi) = \varepsilon'_\lambda(\lambda, T, \theta, \phi) \tag{8.60}$$

For a diffuse surface,

$$\varepsilon'_\lambda, \alpha'_\lambda \neq f(\theta, \phi) \tag{8.61}$$

$$\therefore \varepsilon_\lambda = \alpha_\lambda \tag{8.62}$$

For a gray and diffuse surface,

$$\varepsilon = \alpha \tag{8.63}$$

Eq. (8.63) is one of the most important simplifications in radiation heat transfer. For an opaque and diffuse surface, if we know ε, then from the above equation we know α then and from Eq. (8.56) we know ρ. Hence we know every property that is required to perform radiation calculations involving surface to surface radiation heat transfer.

8.6 Net radiative heat transfer from a surface

Consider a thin, opaque surface exposed to an incident radiation of G (W/m^2) as shown in Fig. 8.10.

At steady state, from energy balance, the net radiation from a surface is given as

$$Q_{radiation} = Outgoing\ radiation\ -\ Incoming\ radiation \tag{8.64}$$

$$Q_{radiation} = Q_{emitted} + Q_{reflected} - Q_{incident} \tag{8.65}$$

$$Q_{radiation} = (\varepsilon \sigma T^4 + \rho G - G)A \tag{8.66}$$

The radiosity (usually denoted by J) of a surface is the leaving radiation from a surface, consisting of emission and reflection from a surface. The incoming radiation is also known as irradiation and is denoted by G.

$$Q_{radiation} = (Radiosity - Irradiation)A \tag{8.67}$$

$$Q_{radiation} = (\varepsilon \sigma T^4 + (\rho - 1)G)A \tag{8.68}$$

$$Q_{radiation} = (\varepsilon \sigma T^4 - (1 - \rho)G)A \tag{8.69}$$

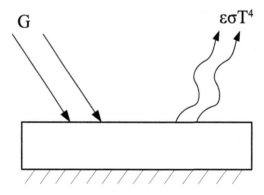

FIGURE 8.10

Depiction of net radiation from an opaque surface.

For an opaque and diffuse surface, $1 - \rho = \alpha = \alpha_\lambda$

$$Q_{radiation} = (\varepsilon \sigma T^4 - \alpha G)A \tag{8.70}$$

where 'α' is the total hemispherical absorptivity or simply absorptivity and is given by

$$\alpha = \frac{\int_{\lambda=0}^{\infty} \alpha_\lambda G_{\lambda,i} \, d\lambda}{\int_{\lambda=0}^{\infty} G_{\lambda,i} \, d\lambda} \tag{8.71}$$

In Eq. (8.71), the numerator represents the radiation absorbed by the surface in the wavelength interval 0 to ∞. The denominator shows the total incident radiation in the same wavelength range. It is straight forward so that, $\alpha_\lambda = \alpha_\lambda$ for a gray surface for which $\alpha_\lambda \neq f(\lambda)$.

Example 8.6: *For an opaque diffuse surface ε_λ varies with λ as shown in Fig. 8.11. Plot α_λ vs. λ and ρ_λ vs. λ for this surface.*
Solution:
From Kirchoff's law, $\alpha_\lambda = \varepsilon_\lambda$, and for an opaque and diffuse surface,

$$\rho_\lambda = 1 - \alpha_\lambda$$

Hence α_λ and ρ_λ will vary with λ as shown in Fig. 8.12.

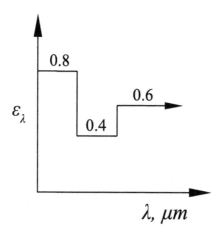

FIGURE 8.11

Spectral emissivity distribution for example 8.6.

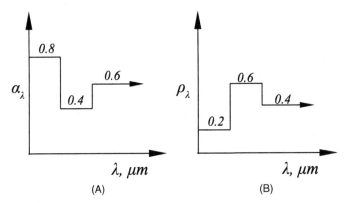

FIGURE 8.12

Spectral distribution of (A) absorptivity and (B) reflectivity for example 8.6.

Example 8.7: *Consider an opaque surface whose spectral hemispherical absorptivity varies with wavelength, as shown in Fig. 8.13(A). The spectral distribution of incident radiation is also shown in Fig. 8.13(B).*

1. Determine the hemispherical total absorptivity of the surface.
2. If this surface is diffuse and is at 1200 K, what is its total hemispherical emissivity?
3. Determine the net radiation heat transfer from the surface.

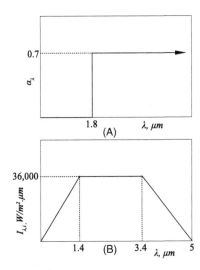

FIGURE 8.13

Variation of (A) hemispherical spectral absorptivity and (B) spectral irradiance with wavelength.

Solution:

1. Hemispherical total absorptivity of the surface.

$$\alpha = \frac{\int_{\lambda=0}^{\infty} \alpha_\lambda G_{\lambda_i} \, d\lambda}{\int_{\lambda=0}^{\infty} G_{\lambda_i} \, d\lambda}$$

Total irradiance or irradiation G in W/m² is given by

$$G = \int_{\lambda_i}^{\infty} G_{\lambda_i} \, d\lambda = \frac{1}{2} \times 1.4 \times 36000 + 2 \times 36000 + \frac{1}{2} \times 1.6 \times 36000$$

$$G = 126000 \text{ W/m}^2$$

$$\alpha = \frac{0 \times \int_0^{1.8} G_{\lambda_i} \, d\lambda}{126000} + \frac{0.7 \times \int_{1.8}^{3.4} G_{\lambda_i} \, d\lambda}{126000} + \frac{0.7 \times \int_{3.4}^{5} G_{\lambda_i} \, d\lambda}{126000}$$

$$\alpha = \frac{0.7 \times 36000 \times 1.6 + 0.7 \times 36000 \times \frac{1}{2} \times 1.6}{126000} = \frac{60.48}{126}$$

$$\alpha = 0.48$$

2. Total hemispherical emissivity.

For a diffuse surface, $\alpha_\lambda = \varepsilon_\lambda$

$\lambda_1 = 1.8 \, \mu\text{m}$, $\lambda_1 T = 2160 \, \mu\text{mK}$

From the F function chart, we get $F_{0 \to \lambda,T} = 0.094$

$$\varepsilon(T) = \frac{\int_0^{1.8} 0 \times \varepsilon_\lambda E_{b,\lambda} \, d\lambda}{E_b(T)} + \frac{\int_{1.8}^{\infty} 0.7 \times \varepsilon_\lambda E_{b,\lambda} \, d\lambda}{E_b(T)} = 0 + 0.7(1 - 0.094)$$

$$\varepsilon = 0.6342$$

3. Net radiation heat transfer from the surface.

$$Q = \varepsilon \sigma T^4 + \rho G - G = \varepsilon \sigma T^4 - \alpha G = 0.6342 \times 5.67 \times 10^{-8} \times 1200^4 - 0.48 \times 12600$$

$$Q = 14.09 \text{ kW/m}^2$$

Example 8.8: *Solar flux of 975 W/m² is incident on the top surface of a plate whose solar absorptivity is 0.9 and emissivity is 0.1. The air and surroundings are at 30 °C, and the convection heat transfer coefficient between the plate and air is 25 W/m² K. Assuming that the bottom side of the plate is perfectly insulated, determine the steady state temperature of the plate. (Refer to Fig. 8.14)*

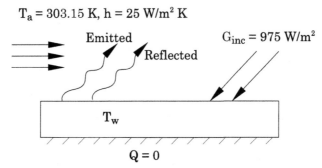

FIGURE 8.14

Various heat transfer process associated with example 8.8.

Solution:

Net emission + reflection + convection = Incident

From the energy balance, we have (since ρ_{zinc} is there σT^4_∞ need not be taken into account.)

$$\varepsilon\sigma\left(T^4\right)+\rho G_{inc}+h\left(T-T_\infty\right)=G_{inc}$$

$$\varepsilon\sigma\left(T^4\right)+h\left(T-T_\infty\right)=\alpha G_{inc}$$

$$0.1\times5.67\times10^{-8}\times\left(T^4\right)+25\left(T-303.15\right)=0.9\times975 \qquad (8.72)$$

This is a nonlinear equation and needs an iterative solution. We rearranging the Eq. (8.74)

$$T=\frac{0.9\times975-0.1\times5.67\times10^{-8}\times\left(T^4\right)}{25}+303.15$$

We carryout iterations with the following "algorithm"

$$T_{i+1}=\frac{0.9\times975-0.1\times5.67\times10^{-8}\times\left(T_i^4\right)}{25}+303.15$$

The results of the iteration process are shown in Table 8.4.

Table 8.4 A successive substitution method for example 8.8.

Iteration	T_i, K	T_{i+1}, K	Residual $(T_{i+1}-T_i)^2$, K^2
1	350	334.84	229.62
2	334.84	335.39	0.3050
3	335.39	335.38	3.56×10^{-4}
	$T=335.38$ K		

The steady state temperature of plate is 335.38K

8.7 Radiation heat transfer between surfaces

There are two ways of dealing with radiation heat transfer between multiple surfaces.

1. Network method
2. Enclosure theory

The network method is hard to scale up for a large number of surfaces. In this book, our focus is on the enclosure theory. The enclosure theory was developed by Prof. E. M. Sparrow and his colleagues at the University of Minnesota, Minneapolis, in the United States in the early 1960s and was originally developed for gray diffuse evacuated surfaces. Let us now see how this method works.

Consider an N surface evacuated enclosure as shown in Fig. 8.15

According to the enclosure theory, if we account for all the radiation coming out of every surface (J_i) and find out what is the radiation falling on that surface (G_i) from the other surfaces, then the net radiative heat transfer from any surface is given by Eq. (8.73).

$$q_i = J_i - G_i \qquad (8.73)$$

In Eq. (8.73), J_i is called the radiosity or outgoing radiation and G_i is called irradiation or incoming radiation both in W/m^2 as already mentioned.

For example, consider a three-surface enclosure with the top surface exposed to atmosphere, as shown in Fig. 8.16A This problem can be handled using the enclosure theory by closing the open cavity with an imaginary surface having an emissivity $(\varepsilon) = 1$ at temperature T_∞. Similarly, a single surface too can be handled by closing the surface with a hemispherical black body at temperature T_∞ as shown in Fig. 8.16B.

This way, even open surfaces can be handled effectively. So the enclosure theory can be applied from $N = 1$ to $N \to \infty$, where N is the number of surfaces in the enclosure.

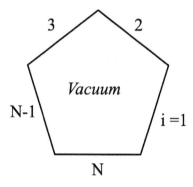

FIGURE 8.15

A typical N surface evacuated enclosure.

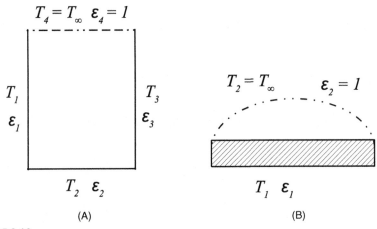

FIGURE 8.16

(A) Three surfaces enclosure, and (B) single surface exposed to atmosphere at T_∞.

8.8 Radiation view factor and its determination

From the preceding discussion, it is clear that for the ith surface in the enclosure, the net radiative flux depends on J_i and G_i. One can easily intuit that the orientations of the surface with respect to each other will play a critical part in the determination of G_i, J_i, and hence q_i. To be able to quantify this, we again take recourse to solid geometry and introduce a key quantity called the view factor or shape factor.

The view factor F_{ij} between two surface areas A_i and A_j is the fraction of radiation leaving the surface i that is intercepted by the surface j. From the foregoing discussion, the following points are evident.

1. The view factor is dimensionless.
2. The value of the view factor lies between 0 to 1, with the minimum being 0 and the maximum being 1.

Consider two surfaces in space with an inclined orientation as shown in Fig. 8.17.

1. The areas of the two surfaces are A_i and A_j.
2. The temperatures of the two surfaces are T_i and T_j.
3. The emissivities of the two surfaces are ε_i and ε_j.

Consider elemental areas dA_i and dA_j on the surfaces i and j respectively with the normal unit vectors n_i and n_j as shown. Let the intensity from the ith and j^{th} surfaces be I_i and I_j, respectively.

From the definition of solid angle, the amount of radiation that originates from dA_i and falling on dA_j can be written in terms of intensity as

$$dQ_{dA_i - dA_j} = I_i \, dA_i \cos\theta_i \, d\omega_{j-i} \qquad (8.74)$$

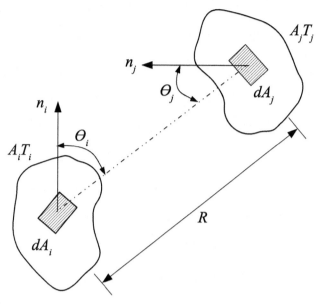

FIGURE 8.17

Figure showing two surfaces A_i and A_j with respective orientations in space, with the goal being the determination of the view factor from i^{th} surface to the jth surface, denoted by F_{ij}.

where

$$I_i = I_{i,emitted+reflected} \tag{8.75}$$

$$d\omega_{j-i} = \frac{dA_j cos\theta_j}{R^2} \tag{8.76}$$

For a diffuse surface

$$I_i = \frac{J_i}{\pi} \tag{8.77}$$

$$dQ_{dA_i-dA_j} = \frac{J_i cos\theta_i cos\theta_j dA_i dA_j}{\pi R^2} \tag{8.78}$$

The radiation leaving dA_i is $J_i.dA_i$.

∴ The fraction of the radiation leaving dA_i that is intercepted by dA_j, known as the elemental view factor $dF_{dA_i-dA_j}$, is given by

$$dF_{dA_i-dA_j} = \frac{dQ_{dA_i-dA_j}}{J_i dA_i} = \frac{J_i \cos\theta_i \cos\theta_j dA_i dA_j}{J_i dA_i \pi R^2} \tag{8.79}$$

$$dF_{dA_i-dA_j} = \frac{\cos\theta_i \cos\theta_j dA_j}{\pi R^2} \tag{8.80}$$

The view factor from A_i to A_j or from i to j is then given by

$$F_{A_i-A_j} = F_{i-j} = \frac{\displaystyle\iint_{A_j A_i} dQ_{dA_i-dA_j}}{\displaystyle\int J_i \, dA_i} \qquad (8.81)$$

For uniform radiosity

$$F_{A_i-A_j} = \frac{1}{A_i}\left(\frac{\displaystyle\iint_{A_j A_i} \cos\theta_i \cos\theta_j \, dA_i \, dA_j}{\pi R^2}\right) \qquad (8.82)$$

Similarly,

$$F_{A_j-A_i} = \frac{1}{A_j}\left(\frac{\displaystyle\iint_{A_i A_j} \cos\theta_i \cos\theta_j \, dA_i \, dA_j}{\pi R^2}\right) \qquad (8.83)$$

From Eqs. 8.82 and 8.83, it is evident that

$$A_i F_{ij} = A_j F_{ji} \qquad (8.84)$$

Eq. (8.84) is frequently referred to as the reciprocity or reciprocal rule.

8.8.1 View factor algebra

Consider an evacuated enclosure with N surfaces, as shown in Fig. 8.18. There will be N^2 view factors for this problem arranged in the matrix form below. Unless we calculate N^2 view factors, we will not be able to solve the radiation problem of finding the net radiation from any given surface. The view factor matrix for this enclosure is given as,

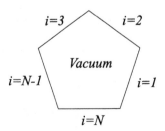

FIGURE 8.18

A depiction of an evacuated enclosure with N surfaces.

$$\begin{bmatrix} F_{11} & F_{11} & \cdots & \cdots & \cdots & F_{1N} \\ F_{21} & F_{22} & \cdots & \cdots & \cdots & F_{2N} \\ \vdots & \vdots & & & & \vdots \\ F_{N1} & F_{N2} & \cdots & \cdots & \cdots & F_{NN} \end{bmatrix}$$

For any surface, it is intuitively apparent that the sum of the view factors should be 1.

$$\sum_{j=1}^{N} F_{ij} = 1, \quad 1 \le i \le N \tag{8.85}$$

In a N surface enclosure there are N such sum rules.

The number of reciprocal rules Eq. (8.84) available for N surface enclosure is $^{N}C_2$.

∴ The total number of view factors to be independently determined are

$$= N^2 - {}^{N}C_2 - N \tag{8.86}$$

$$= N^2 - \frac{N(N-1)}{2} - N \tag{8.87}$$

$$= \frac{2N^2 - N^2 + N - 2N}{2} \tag{8.88}$$

$$= \frac{N(N-1)}{2} \tag{8.89}$$

∴ The total number of view factors to be independently determined are $^{N}C_2$. In the view of the preceding discussion, for a N surface enclosure, the number of independent view factors that need to be obtained by using the tedious view factor expression (Eq. 8.82) is $^{N}C_2$ while this number for a N surface enclosure with all plane or convex surface is $^{N}C_2$-N. The self-view factor for any plane/convex surface is $F_{ii} = 0$. Suppose all the surfaces in the N-surface enclosure are plane or convex, then N view factor values are zero.

For example, consider a triangular enclosure with all the surfaces being plane, as shown in Fig. 8.19. According to Eq. (8.89), the number of independent view factors to be determined from the fundamental view factor is three.

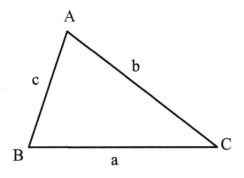

FIGURE 8.19

Triangular enclosure.

Total number of view factors $= N^2 = 3^2 = 9$.

Total number of view factors that can be determined from reciprocal rules $= {}^N C_2 = {}^3 C_2 = 3$.

Total number of view factors that can be determined from summation rule $= N = 3$.

Total number of view factors need to be independently determined $= 9 - 3 - 3 = 3$.

Total number of self-view factors $= 3$.

Therefore, the total number of view factors to be independently determined is $3 - 3 = 0$.

So, we should be able to get all the view factors by just using view factors algebra. Assuming unit depth in the direction perpendicular to the plane of the paper, we have the following:

Area of the surface $AB = c$.

Area of the surface $BC = a$.

Area of the surface $AC = b$.

From the summation rule

$$F_{aa} + F_{ab} + F_{ac} = 1 \tag{8.90}$$

$$F_{ba} + F_{bb} + F_{bc} = 1 \tag{8.91}$$

$$F_{ca} + F_{cb} + F_{cc} = 1 \tag{8.92}$$

Since all the surfaces are plane, the self-view factors for all the surfaces are zero. That is,

$$F_{aa} = F_{bb} = F_{bb} = 0$$

Eqs. (8.90–8.92) can now be written as

$$F_{ab} + F_{ac} = 1 \tag{8.93}$$

$$F_{ba} + F_{bc} = 1 \tag{8.94}$$

$$F_{cb} + F_{ca} = 1 \tag{8.95}$$

From the reciprocal rule,

$$aF_{ab} = bF_{ba} \tag{8.96}$$

$$aF_{ac} = cF_{ca} \tag{8.97}$$

$$bF_{bc} = cF_{cb} \tag{8.98}$$

Multiplying Eq. (8.93) by a, Eq. (8.94) by b, and Eq. (8.96) by c, we have

$$aF_{ab} + aF_{ac} = a \tag{8.99}$$

$$bF_{ba} + bF_{bc} = b \tag{8.100}$$

$$cF_{ca} + cF_{cb} = c \tag{8.101}$$

Adding Eqs. (8.99) and (8.100) and subtracting Eq. (8.101) from the sum, we have

$$aF_{ab} + aF_{ac} + bF_{ba} + bF_{bc} - cF_{ca} - cF_{cb} = a + b - c \tag{8.102}$$

Now, using Eqs. (8.97 and 8.98) rearranging Eq. (8.102), we have

$$aF_{ab} + (aF_{ac} - cF_{ca}) + bF_{ba} + (bF_{bc} - cF_{cb}) = a + b - c \tag{8.103}$$

However, $aF_{ac} = cF_{ca}$ and $bF_{bc} = cF_{cb}$ by the reciprocal rule.
∴ Eq. (8.103) simplifies to

$$aF_{ab} + bF_{ba} = a + b - c \tag{8.104}$$

Again applying Eq. (8.96), i.e., the reciprocal rule on (8.104), we have

$$2aF_{ab} = a + b + c \tag{8.105}$$

$$\therefore F_{ab} = \frac{a + b - c}{2a} \tag{8.106}$$

By induction,

$$F_{bc} = \frac{b + c - a}{2b} \tag{8.107}$$

$$F_{ca} = \frac{c + a - b}{2c} \tag{8.108}$$

$$F_{ba} = \frac{a + b - c}{2b} \tag{8.109}$$

$$F_{cb} = \frac{b + c - a}{2c} \tag{8.110}$$

$$F_{ac} = \frac{c + a - b}{2a} \tag{8.111}$$

Example 8.9: *Determine all the view factors for the following two dimensional (i.e., very long) surfaces (refer to Fig. 8.20). The enclosed spaces in all the problems are evacuated.*
Solution:
(1)

$$F_{11} = 0 \text{ and } F_{12} = 1$$

$$A_1 = 2rL \quad and \quad A_2 = 2\pi\left(1 - \frac{80}{360}\right)rL = \frac{7}{9} \times 2\pi rL$$

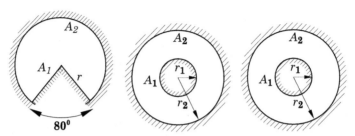

FIGURE 8.20

(A) Long duct of radius r, (B) long cylindrical annulus, and (C) sphere within a sphere under consideration in example 8.9.

$$A_1 F_{12} = A_2 F_{21}; \ F_{21} = \frac{A_1}{A_2} = \frac{2rL}{\frac{7}{9} 2\pi rL} = \frac{9}{7\pi}$$

$$F_{21} = 0.409 \ and \ F_{22} = 0.591$$

$$\textbf{View factor matrix} = \begin{bmatrix} 0 & 1 \\ 0.409 & 0.591 \end{bmatrix}$$

(2)

$$F_{11} = 0 \quad and \quad F_{12} = 1$$

$$A_1 = 2\pi r_1 L \quad and \quad A_2 = 2\pi r_2 L$$

$$A_1 F_{12} = A_2 F_{21}; \ F_{21} = \frac{A_1}{A_2} = \frac{2\pi r_1 L}{2\pi r_2 L} = \frac{r_1}{r_2}$$

$$\therefore F_{22} = 1 - \frac{r_1}{r_2}$$

$$\textbf{View factor matrix} = \begin{bmatrix} 0 & 1 \\ \dfrac{r_1}{r_2} & 1 - \dfrac{r_1}{r_2} \end{bmatrix}$$

(3)

$$F_{11} = 0 \quad and \quad F_{12} = 1$$

$$A_1 = 4\pi r_1^2 \quad and \quad A_2 = 4\pi r_2^2$$

$$A_1 F_{12} = A_2 F_{21}; \ F_{21} = \frac{A_1}{A_2} = \frac{4\pi r_1^2}{4\pi r_2^2} = \frac{r_1^2}{r_2^2}$$

$$\therefore F_{22} = 1 - \frac{r_1^2}{r_2^2}$$

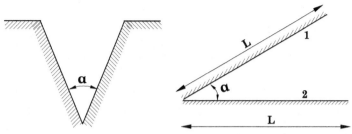

FIGURE 8.21

Infinitely deep wedge under consideration in example 8.10.

$$\text{View factor matrix} = \begin{bmatrix} 0 & 1 \\ \dfrac{r_1^2}{r_2^2} & 1 - \dfrac{r_1^2}{r_2^2} \end{bmatrix}$$

Example 8.10: *Consider an infinitely deep wedge with surfaces 1 and 2, as shown in Fig. 8.21.*

1. Determine F_{12}.

2. What will be the value of F_{12} when $\alpha = 90°$?

3. What is the value of F_{12} when $\alpha = 180°$? What do you infer from this result?

Solution:

(1) We know that, for a triangular enclosure with sides a, b, and c

$$F_{ab} = \frac{a + b - c}{2a}$$

$$\therefore F_{12} = \frac{L_1 + L_2 - L_3}{2L_1}$$

$$L_1 = L_2 = L \quad and \quad L_3 = x$$

$$\therefore F_{12} = \frac{L + L - x}{2L} = \frac{2L - x}{2L} = 1 - \frac{x}{2L}$$

From Fig 8.22

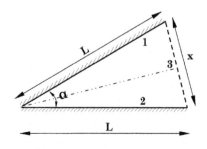

FIGURE 8.22

View factors for an infinitely deep wedge under consideration in example 8.10.

$$sin\frac{\alpha}{2} = \frac{x/2}{L} = \frac{x}{2L}$$

$$\therefore F_{12} = 1 - sin\left(\frac{\alpha}{2}\right)$$

(2)

When $\alpha = 90°$

$$F_{12} = 1 - sin45$$

$$= 1 - \frac{1}{\sqrt{2}}$$

$$\therefore F_{12} = 0.293$$

(3)

When $\alpha = 180°$

$$F_{12} = 1 - sin90°$$

$$= 1 - 1 = 0$$

$$\therefore F_{12} = 0$$

As $\alpha = 180°$, both the surfaces are on the same plane, and they do not see each other. Hence, it stands to reason that the resulting view factor equals zero.

Example 8.11: *Consider an evacuated two-dimensional quadrilateral enclosure, as shown in Fig. 8.23.*
1. Determine F_{ac}.
2. If all the four sides of a quadrilateral are equal, that is, it becomes a square, what will be the value of F_{ac}?

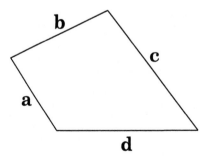

FIGURE 8.23

Two-dimensional quadrilateral enclosure under consideration in example' 8.11.

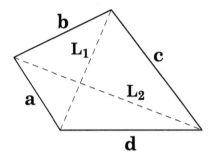

FIGURE 8.24

Two-dimensional quadrilateral enclosure under consideration in example 8.11.

Solution:

(1) Let us join the opposite vertices so that we now have two diagonals with lengths L_1 and L_2 as shown in Fig. 8.24.

From the sum rule for the quadrilateral, we have,

$$F_{aa} + F_{ab} + F_{ac} + F_{ad} = 1$$

As 'a' is a plane surface $F_{aa} = 0$

$$F_{ac} = 1 - (F_{ab} + F_{ad})$$

From the view factor relations for a triangle, we have

$$F_{ab} = \frac{a + b - L_2}{2a}$$

$$F_{ad} = \frac{a + d - L_1}{2a}$$

Substuting for F_{ab} and F_{ad} in the expression for F_{ac} we get the following,

$$F_{ac} = 1 - \frac{a + d - L_1}{2a} - \frac{a + b - L_2}{2a}$$

$$F_{ac} = \frac{(L_1 + L_2) - (b + d)}{2a}$$

This is known as the Hottel's crossed string method. The view factor between the opposite sides of a two-dimensional quadrilateral enclosure is given by the difference between the sum of the crossed strings and the sum of the uncrossed strings divided by twice the length of the side of the enclosure from which the view factor to the opposite side is sought.

$$F_{ac} = \frac{(Sum\ of\ the\ crossed\ strings) - (Sum\ of\ the\ uncrossed\ strings)}{2a}$$

(2) For a square $a = b = c = d$ and $L_1 = L_2 = \sqrt{2}a$

$$F_{ac} = \frac{(\sqrt{2}a + \sqrt{2}a) - (a + a)}{2a} = \frac{2\sqrt{2} - 2}{2} = \sqrt{2} - 1$$

$$\therefore F_{ac} = 0.414$$

Evaluation of the view factors for three-dimensional surfaces using the Eq. (8.82) can be very tedious. So, researchers over the years developed many techniques to reduce the complexity of the view factor evaluation (see Balaji, 2014) for a full discussion on this. For regular three-dimensional objects like cylinders and cuboidal surfaces, charts are available to determine the view factors. Many techniques like contour integration have also been developed to aid in the evaluation of view factors from three dimensional surfaces.

8.9 The radiosity-irradiation method

Now, we will use all of what we have studied so far in our pursuit of radiation and employ the radiosity irradiation method to determine the net radiative heat transfer from a surface.

Consider an evacuated enclosure with N surfaces, as shown in Fig. 8.25.
The radiation leaving from the ith surface of the enclosure is given by

$$A_i J_i = \text{Emitted radiation} + \text{Reflected radiation} \qquad (8.112)$$

$$A_i J_i = A_i \varepsilon_i \sigma T_i^4 + A_i \rho_i G_i \qquad (8.113)$$

In the above equation J_i is the leaving radiation or radiosity and G_i is the incoming radiation or irradiation, as already introduced.
Let all surfaces be opaque.

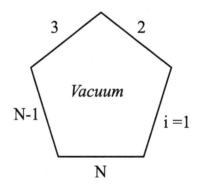

FIGURE 8.25

Depiction of an evacuated enclosure with N surfaces.

$$\tau_i = 0, 1 \leq i \leq N \tag{8.114}$$

Let all the surfaces be gray and diffuse, which is a reasonable engineering approximation. In the view of this for all i,

$$\varepsilon_i = \alpha_i \tag{8.115}$$

We know that

$$\alpha_i + \rho_i = 1 \tag{8.116}$$

$$\therefore \rho_i = 1 - \alpha_i = 1 - \varepsilon_i \tag{8.117}$$

Substituting for ρ_i in Eq. (8.118), we have

$$A_i J_i = A_i \varepsilon_i \sigma T_i^4 + A_i (1 - \varepsilon_i) G_i \tag{8.118}$$

It is also evident that if we consider the radiation falling on the ith surface, given by $A_i G_i$, we get

$$A_i G_i = \sum_{j=1}^{N} A_j F_{ji} J_j = \sum_{j=1}^{N} A_i F_{ij} J_j \tag{8.119}$$

The irradiation G_i on the ith surface is given by

$$G_i = \sum_{j=1}^{N} F_{ij} J_j \tag{8.120}$$

Substituting for G_i in Eq. (8.118), we have

$$J_i = \varepsilon_i \sigma T_i^4 + (1 - \varepsilon_i) \sum_{j=1}^{N} F_{ij} J_j \tag{8.121}$$

Eq. (8.121) is a powerful one and gives us a system of N equations in N unknowns. By solving the system of equations, we can obtain $J_1, J_2, J_3, \ldots, J_N$.

Once we know J_i, we can calculate G_i. From G_i and J_i, we can calculate net radiative heat transfer from any surface by Eq. (8.122)

$$q_i = J_i - G_i \tag{8.122}$$

or, from Eq. (8.122), we can write q_i in terms of G_i alone

$$q_i = \varepsilon_i \sigma T_i^4 + (1 - \varepsilon_i) G_i - G_i \tag{8.123}$$

$$q_i = \varepsilon_i \sigma T_i^4 + G_i - \varepsilon_i G_i - G_i \tag{8.124}$$

$$q_i = \varepsilon_i \sigma T_i^4 - \varepsilon_i G_i \tag{8.125}$$

$$q_i = \varepsilon_i(\sigma T_i^4 - G_i)$$ (8.126)

or, from Eq. (8.122), we can write q_i in terms of J_i alone

$$J_i = \varepsilon_i \sigma T_i^4 + (1 - \varepsilon_i)G_j$$ (8.127)

$$G_i = \frac{J_i - \varepsilon_i \sigma T_i^4}{(1 - \varepsilon_i)}$$ (8.128)

$$q_i = J_i - \frac{J_i - \varepsilon_i \sigma T_i^4}{(1 - \varepsilon_i)}$$ (8.129)

$$q_i = \frac{J_i - \varepsilon_i J_i - J_i + \varepsilon_i \sigma T_i^4}{(1 - \varepsilon_i)}$$ (8.130)

$$q_i = \frac{\varepsilon_i[\sigma T_i^4 - J_i]}{(1 - \varepsilon_i)}$$ (8.131)

Eq. (8.131) requires some caution in its use, as it is not applicable for a black surface.

So, if we know G_i or J_i alone, we can calculate the net radiative heat transfer from any surface.

For a black body, $\varepsilon_i = 1$, and so J_i is given by

$$J_i = \varepsilon_i \sigma T_i^4 + (1 - \varepsilon_i)G_i$$ (8.132)

$$J_i = \sigma T_i^4$$ (8.133)

Another surface of interest in the radiation problem is a reradiating surface, which is the radiation equivalent of an insulated surface.

∴ For a reradiating surface, $q_i = 0$.

$$q_i = J_i - G_i = 0$$ (8.134)

$$\therefore J_i = G_i$$ (8.135)

Additionally,

$$\frac{\varepsilon_i}{1 - \varepsilon_i}[\sigma T_i^4 - J_i] = 0$$ (8.136)

$$J_i = \sigma T_i^4$$ (8.137)

In the view of the above the key result is,

$$J_i \neq f(\varepsilon_i)$$ (8.138)

The key question to be answered now is: How do we engineer a reradiating surface? A very thick wall/surface made of a low thermal conductivity medium and with a vacuum or near vacuum will come close to a re-radiating condition. Else, conjugate heat transfer has to be considered in a wall, that is, $q_{conduction} = q_{radiation}$ and we need to write out the expressions for $q_{conduction}$ and $q_{radiation}$ and we will obtain an equation in which unknown will be the surface temperature.

Example 8.12: *Consider two infinitely long parallel plates that are separated by a distance L, as shown in Fig. 8.26. The plates are very deep in the direction perpendicular to the plane of the paper. The space between the two plates is evacuated. The two plates are maintained at temperatures of 810 K and 490 K, respectively. The two plates have emissivities of 0.95 and 0.45 respectively. Determine the net radiation heat transfer between the two plates.*

Solution:

$F_{11} = 0$ and $F_{12} = 1$

$F_{21} = 1$ and $F_{22} = 0$

View factor matrix =

$$\begin{bmatrix} 0 & 1 \\ 1 & 0 \end{bmatrix}$$

(8.139)

We can solve this problem using the radiosity-irradiation method.

$$J_i = \varepsilon_i \sigma T_i^4 + (1 - \varepsilon_i) G_i$$

(8.140)

$$G_i = \sum_{j=1}^{N} F_{ji} J_j$$

(8.141)

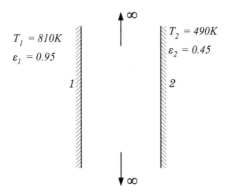

$T_1 = 810K$

$\varepsilon_1 = 0.95$

$T_2 = 490K$

$\varepsilon_2 = 0.45$

1

2

FIGURE 8.26

Two infinitely long parallel plates undergoing surface radiation exchange and under consideration in example 8.12.

$$q_i = J_i - G_i \tag{8.142}$$

Now, we write the radiosity and irradiation expressions for the two surfaces 1 and 2.

$$G_1 = F_{21}J_2^{\cdot 1} + F_{11}J_1^{\cdot 0} \implies G_1 = J_2 \tag{8.143}$$

$$G_2 = F_{12}J_1^{\cdot 1} + F_{22}J_2^{\cdot 0} \implies G_2 = J_1 \tag{8.144}$$

$$J_1 = \varepsilon_1 \sigma T_1^4 + (1 - \varepsilon_1) J_2 \tag{8.145}$$

$$J_2 = \varepsilon_2 \sigma T_2^4 + (1 - \varepsilon_2) J_1 \tag{8.146}$$

Substituting J_2 from Eq. (8.146) in Eq. (8.145), we have

$$J_1 = \varepsilon_1 \sigma T_1^4 + (1 - \varepsilon_1) \left[\varepsilon_2 \sigma T_2^4 + (1 - \varepsilon_2) J_1 \right]$$

$$J_1 \left[1 - (1 - \varepsilon_1)(1 - \varepsilon_2) \right] = \varepsilon_1 \sigma T_1^4 + (1 - \varepsilon_1) \varepsilon_2 \sigma T_2^4$$

$$J_1 = \frac{\varepsilon_1 \sigma T_1^4 + \varepsilon_2 \sigma T_2^4 - \varepsilon_1 \varepsilon_2 \sigma T_2^4}{1 - (1 - \varepsilon_1)(1 - \varepsilon_2)}$$

$$q_1 = J_1 - G_1 \implies q_1 = J_1 - J_2$$

$$q_1 = \frac{\varepsilon_1 \varepsilon_2 \left[\sigma T_1^4 - \sigma T_2^4 \right]}{1 - (1 - \varepsilon_1)(1 - \varepsilon_2)}$$

$$q_1 = \frac{\sigma \left[T_1^4 - T_2^4 \right]}{\dfrac{1}{\varepsilon_1} + \dfrac{1}{\varepsilon_2} - 1}$$

$$q_1 = \frac{5.67 \times 10^{-8} \left[810^4 - 490^4 \right]}{\dfrac{1}{0.95} + \dfrac{1}{0.45} - 1}$$

$$\therefore q_1 = 9292.4 \, \text{W/m}^2$$

The expression we derived above is a powerful engineering formula known as the parallel plate formula. This can be used to very swiftly determine the net radiative heat transfer between two infinitely long parallel plates. Double pane windows, radiation shields and several engineering applications find this formula handy.

Example 8.13: *Revisit example 8.12. Now a radiation shield with an emissivity of 0.05 on both sides is placed between plates 1 and 2. The details are provided in Fig. 8.27. Determine*

1. The net radiation heat transfer in the presence of the shield.
2. The equilibrium temperature of the shield.

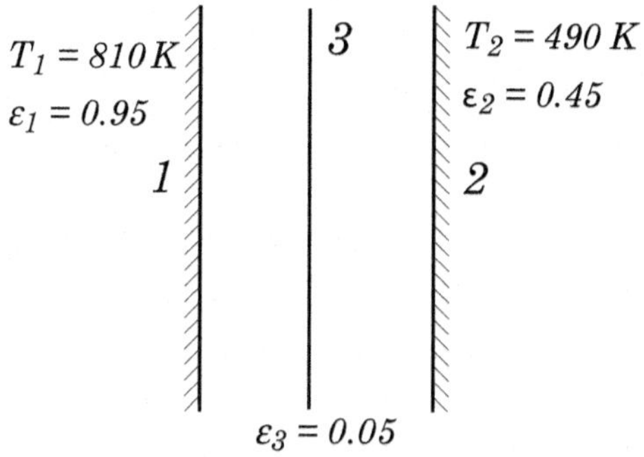

$$T_1 = 810\,K$$
$$\varepsilon_1 = 0.95$$

3

$$T_2 = 490\,K$$
$$\varepsilon_2 = 0.45$$

1

2

$$\varepsilon_3 = 0.05$$

FIGURE 8.27

Two infinitely long parallel plates with shield in between under consideration in example 8.13.

Solution:

From the above problem, we have

$$q_{13} = \frac{\sigma\left[T_1^4 - T_3^4\right]}{\dfrac{1}{\varepsilon_1} + \dfrac{1}{\varepsilon_3} - 1}$$

$$q_{32} = \frac{\sigma\left[T_3^4 - T_2^4\right]}{\dfrac{1}{\varepsilon_3} + \dfrac{1}{\varepsilon_2} - 1}$$

At a steady state, $q_{13} = q_{32} = q_{12,shield}$

$$\therefore \frac{\sigma\left[T_1^4 - T_3^4\right]}{\dfrac{1}{\varepsilon_1} + \dfrac{1}{\varepsilon_3} - 1} = \frac{\sigma\left[T_3^4 - T_2^4\right]}{\dfrac{1}{\varepsilon_3} + \dfrac{1}{\varepsilon_2} - 1}$$

$$\frac{810^4 - T_3^4}{\dfrac{1}{0.95} + \dfrac{1}{0.05} - 1} = \frac{T_3^4 - 490^4}{\dfrac{1}{0.05} + \dfrac{1}{0.45} - 1}$$

$$\frac{810^4 - T_3^4}{20.05} = \frac{T_3^4 - 490^4}{21.22}$$

$$T_3 = \left[\frac{21.22 \times 810^4 + 20.05 \times 490^4}{21.22 + 20.05}\right]^{\frac{1}{4}}$$

$$\therefore T_3 \approx 706.64\,K$$

$$q = q_{13} = \frac{\sigma\left[T_1^4 - T_3^4\right]}{\dfrac{1}{\varepsilon_1} + \dfrac{1}{\varepsilon_3} - 1}$$

$$= \frac{5.67 \times 10^{-8}\left(810^4 - 706.64^4\right)}{\dfrac{1}{0.95} + \dfrac{1}{0.05} - 1}$$

$$\therefore q = 512.15\,W/m^2$$

We see that q has drastically reduced (with respect to the solution in example 8.12) by placing just one low emissivity shield. If more are placed, this could help us obtain a "super insulation". If all the surfaces and "N" shields placed in between all have the same emissivity, it can be shown that, $q_{Nshields} = q_{noshield}/(N + 1)$.

8.10 Introduction to gas radiation

The study of heat transfer through media, which can absorb, emit, and scatter radiation, has been receiving increasing attention in the last few decades. Such media are called radiatively participating media or simply participating media. Applications include emission from rocket nozzles, combustion chambers, radiation to and from the Earth and other planets, glass making, IC engines so on.

Consider solar radiation incident on the Earth's surface, as shown in Fig. 8.28. The curve with the dotted line corresponds to the intensity of radiation emitted by a black body at 5800 K. This corresponds to the temperature of the photosphere or the outer layer of the Sun. The actual solar irradiation at the top of the atmosphere is shown as a solid line. The inner curve shows the solar irradiation after passing through the Earth's atmosphere. Here, we can see regions having lower irradiation

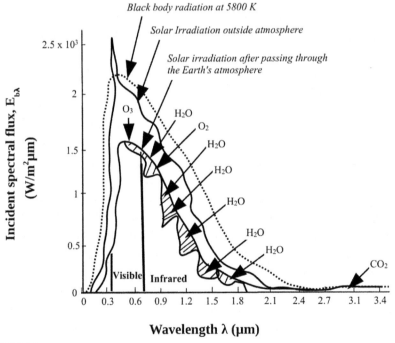

FIGURE 8.28

Approximation of variation of incident solar radiation and attenuation by the atmosphere (Adapted from Balaji C., 2014).

due to high absorption by CO_2 and H_2O. This figure clearly shows that gas absorption is spectrally dependent and that the atmosphere is truly a participating medium. There are also transparent windows where there is little absorption.

8.11 Equation of transfer or radiative transfer equation (RTE)

Consider a gas volume with a cross sectional area dA and thickness dx as shown in Fig. 8.29. The incoming spectral radiation in the direction x is $I_{\lambda x}$ while $I_{\lambda,x+dx}$ is the spectral radiation exiting the gas volume. The area dA is normal to the direction x such that the radiation is traveling in a direction normal to the cross-sectional area. When we neglect scattering, the only two other phenomena involved in this problem are emission and absorption.

The energy absorbed by the gas in the interval $d\lambda$ is given by

$$\kappa_\lambda I_\lambda dA dx \tag{8.147}$$

where k_λ is the monochromatic or spectral absorption coefficient (m^{-1}).
The energy emitted by the gas volume is given by

$$\varepsilon_\lambda I_{b,\lambda}(T_g)dAdx \tag{8.148}$$

where ε_λ is the monochromatic or spectral emission coefficient (m^{-1}).
By energy balance, we have

$$\frac{dI_\lambda}{dx} = \varepsilon_\lambda I_{b\lambda}(T_g) - \kappa_\lambda I_\lambda \tag{8.149}$$

On rearranging, we get the following equation

$$\frac{dI_\lambda}{dx} + \kappa_\lambda I_\lambda = \varepsilon_\lambda I_{b,\lambda}(T_g) \tag{8.150}$$

Eq. (8.150) is known as the **RTE, or radiative transfer equation**. It is deceptively innocuous as ε_λ and α_λ invariably have strong spectral dependence. Additionally, if "reflection" or scattering from gas molecules is concerned, the problem gets even more formidable, with scattering being a function of wavelength and angle in many cases.

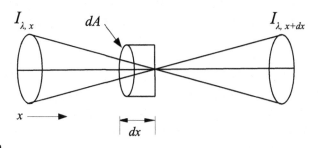

FIGURE 8.29

Gas volume used in the derivation of the RTE.

Consider the gas under question in an isothermal cavity at T_g. Let the temperature of the gas be T_g. For this situation we have isotropic radiation; that is, radiation intensity is the same everywhere and so dI_λ/dx is 0 everywhere within the isothermal cavity. This can be written as

$$I_\lambda = I_{b,\lambda}(T_g) \tag{8.151}$$

Substituting in the above Eq. (8.150)

$$0 + \kappa_\lambda I_{b,\lambda}(T_g) = \varepsilon_\lambda I_{b,\lambda}(T_g) \tag{8.152}$$

$$\therefore \kappa_\lambda = \varepsilon_\lambda \tag{8.153}$$

The above is actually Kirchhoff's law.
Substituting for ε_λ in Eq. (8.150)

$$\frac{dI_\lambda}{dx} + \kappa_\lambda I_\lambda = \kappa_\lambda I_{b,\lambda}(T_g) \tag{8.154}$$

Let us consider a situation where the absorption is much larger compared to emission. Consider a plane wall at temperature T_w (corresponding to $x = 0$) surrounded by the gas that is emitting and absorbing. In view of the foregoing assumption, the RTE becomes

$$\frac{dI_\lambda}{dx} + \kappa_\lambda I_\lambda = 0 \tag{8.155}$$

or

$$\frac{dI_\lambda}{I_\lambda} = -\kappa_\lambda dx \tag{8.156}$$

If $k_\lambda \neq f(\lambda)$, that is, the gas is gray,

$$\frac{dI}{I} = -\kappa dx \tag{8.157}$$

If we integrate this, we get $I = Ce^{-kx}$.

At $x = 0, I = I_w$, i.e. the wall intensity.

$\therefore C = I_w$, Now the solution becomes $I = I_w e^{-kx}$.

The above solution clearly shows that radiation decreases exponentially with x. This is known as **Beer-Lambert's law**. A typical example is solar radiation decreasing exponentially from the surface as we go deeper into the ocean waters.

8.11.1 Determination of heat fluxes

The heat flux q is of fundamental importance to engineers. Hence, the intensity I or I_λ, to be more general, must be converted into a flux form.

Now let us consider the full spectrum of electromagnetic radiation, that is, $\lambda = 0$ to $\lambda = \infty$. The RTE becomes

$$\frac{dI}{dx} + kI_\lambda = kI_b(T_g) \tag{8.158}$$

$$Now \quad I_b = \frac{E_b T}{\pi}$$

$$I_b = \frac{\sigma T_g^4}{\pi}$$

The solution to Eq. (8.158) is given by

$$I^+\left(\frac{k_x}{\cos\theta}\right) = \frac{\sigma T_w^4}{\pi} e^{\frac{-k_x}{\cos\theta}} + \frac{\sigma T_g^4}{\pi}\left(1 - e^{\frac{-k_x}{\cos\theta}}\right) \tag{8.159}$$

(See Balaji, 2014), for a derivation of the above result. In the above equation θ is the angle between a straight path of radiation and a path.) The heat flux at $x = 0$, going out in the positive direction of x is given by

$$q_L^+ = \int_{\phi=0}^{2\pi}\int_{\theta=0}^{\frac{\pi}{2}} I_L^+(\cos\theta)\cos\theta\sin\theta\,d\theta\,d\phi \tag{8.160}$$

The integral appearing on the right hand side of Eq. (8.160) is hard to integrate. We can, of course, easily do it on a computer this day and age. Historically, this was solved by an introducing an exponential integral $E_n(t)$, as defined below. We can introduce the exponential function $E_n(t)$ here as

$$E_n(t) = \int_0^1 \mu^{n-2}exp\left(-\frac{t}{\mu}\right)d\mu \tag{8.161}$$

The exponential integral of third order is given by

$$E_3(t) = \int_0^1 \mu^1 exp\left(-\frac{t}{\mu}\right)d\mu \tag{8.162}$$

In Eq. (8.162) μ is basically a dummy variable.
$E_3(x)$ satisfies the following asymptotic properties.

$$\lim_{t\to0} E_3(x) = \left(\frac{1}{2} - x\right) \tag{8.163}$$

$E_3(0) = 0.5$. The x in Eq. (8.162) corresponds to $k.x$. The product of κ and x is known as the optical depth and is given by τ or τ_x.

When the optical depth is very small, $E_3(x)$ approaches 0. In gas radiation, this is known as the **optically thin limit**.
q_L^+ can now be written as

$$q_L^+ = \sigma T_g^4 + 2\sigma(T_w^4 - T_g^4)E_3(\tau_L) \tag{8.164}$$

$$q_L^+ = 2E_3(\tau_L)\sigma T_w^4 + \sigma T_g^4[1 - 2E_3(\tau_L)] \tag{8.165}$$

For an optically thin gas,

$$E_3(k_L) = \frac{1}{2} - k_L \tag{8.166}$$

$$and \quad 1 - 2E_3(k_L) = \tau_L \tag{8.167}$$

$$\therefore q_L^+ = 2\tau_L \sigma T_g^4 + (1 - 2\tau_L)\sigma T_w^4 \tag{8.168}$$

Therefore the radiation arriving in a plane gas layer of thickness L with the wall at $x = 0$ being at a temperature T_w at $x = L$ consists of two parts. These are

1. Radiation coming directly from the gas (first term).
2. Radiation coming from the wall and that is attenuated by the gas (second term).

We can now introduce two key quantities, namely (1) gas emissivity ε_g and (2) gas absorptivity τ_g as

$$\varepsilon_g = 2\tau_L = (2L)(K) \tag{8.169}$$

$$\tau_g = 1 - 2\tau_L = (1 - \varepsilon_g) \tag{8.170}$$

$$\therefore q_L^+ = \varepsilon_g \sigma T_g^4 + \tau_g \sigma T_w^4 \tag{8.171}$$

Finally we are in a position to define the gas emissivity as $2.L.k$, where k is the gas absorptivity and $2L$ represents a kind of mean length traveled by the radiation. This mean length is frequently referred to as the mean beam length.

Consider a hemispherical gas volume whose radius is R_o (see Fig. 8.30). Consider an elemental area dA at the center of the bottom surface. Let the gas volume emit and absorb with no scattering.

Consider an optically thin gas at a temperature T_g. For this situation,

$$I(x) = \frac{\sigma T_W^4}{\pi} e^{(-\kappa x)} + \frac{\sigma T_g^4}{\pi}(1 - e^{(-\kappa x)}) \tag{8.172}$$

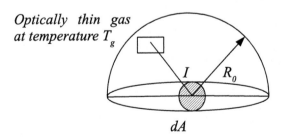

Optically thin gas at temperature T_g

I R_0

dA

FIGURE 8.30

Hemispherical gas volume used for elucidating the concept of mean beam length.

Converting I to q with R replacing x, we have

$$q(R) \approx \pi \left[\frac{\sigma T_w^4}{\pi} e^{(-\kappa R)} + \frac{\sigma T_g^4}{\pi} (1 - e^{(-\kappa R)}) \right] \tag{8.173}$$

If wall radiation is neglected, as for example when $T_w < T_g$, we have

$$q(R) = \sigma T_g^4 (1 - e^{(-\kappa R)}) \tag{8.174}$$

For optically thin gases, $\kappa R \ll 1$

$$\therefore q(R) = \sigma T_g^4 \kappa R \tag{8.175}$$

$q(R)$ can be written as $\varepsilon_g \sigma T_g^4$ where ε_g can be defined as the gas emissivity. From the above definition, it is clear that $\varepsilon_g = \kappa R$.

So, a gas emissivity can be defined as the product of two quantities. The first one κ is the thermal part, and the second one R is the geometric part. R is, in fact, the mean beam length L_m for the gas. The key achievement thus far has been our ability to separate out the geometric part from the thermal part. That in gas radiation, the gas emissivity ε_g is a function of the geometry is unsettling for us. However, the thin gas approximation and the accompanying simplifications lets us introduce emissivity ε_g and proceed with an engineering treatment of gas radiation which is often adequate if our problem of interest is a rocket nozzle or a combustion chamber.

For a few commonly encountered geometries, the mean beam lengths are given in Table 8.5. The product of κ and L_m will then give ε_g, which gives us a handle in solving the problem of gas radiation.

The foregoing framework presented basic ideas in gas radiation with a "clever" strategy to obtain gas emissivity. Please note that in this development, both ε_g and α_g

Table 8.5 Mean beam lengths L_m for different gas geometries.

Sl. No.	Geometry	Characteristic length	L_e
1.	Hemisphere radiating to an element at the center of the base	Radius R	R
2.	Sphere (radiation to surface)	Diameter D	0.65D
3.	Infinite circular cylinder (radiation to the curved surface)	Diameter D	0.95D
4.	Circular cylinder of equal height and diameter (radiation to the entire surface)	Diameter D	0.60D
5.	Infinite parallel planes (radiation to planes)	Spacing between planes L	1.80L
6.	Cube (radiation to any surface)	Side L	0.66L
7.	Arbitrary shape of volume V (radiation to the surface of area A)	Volume to area ratio V/A	3.6V/A

Adapted from Howell et al. (2011).

are constructs we introduced to make the problem more tractable. The basic properties in gas radiation are the monochromatic emission coefficient, ε_λ, and monochromatic absorption coefficients α_λ, both of which are not dimensionless and have units of m^{-1}.

8.11.2 Enclosure analysis in the presence of an absorbing or emitting gas

We now extend the theory of evacuated enclosures to problems involving radiation in optically thin gases. Consider two walls of an enclosure with areas A_1 and A_2 (see Fig. 8.31). Consider two elemental areas in them. Let the distance between them be R. Let the unit vectors be n_1 and n_2, and the angles subtended by them be θ_1 and θ_2, respectively.

Let all the surfaces in the enclosure be gray and diffuse. Now let us consider an optically thin gas filling the enclosure.

The radiation leaving dA_1 that falls on dA_2 is given by

$$\frac{J_1 dA_1 dA_2 cos(\theta_1)cos(\theta_2)}{\pi R^2}e^{(-\kappa R)} \tag{8.176}$$

$$e^{(-\kappa R)} \approx 1 - \kappa R \tag{8.177}$$

using Eq. (8.177), Eq. (8.176) can be expressed as

$$\frac{J_1 dA_1 dA_2 cos(\theta_1)cos(\theta_2)}{\pi R^2} - \kappa \frac{J_1 dA_1 dA_2 cos(\theta_1)cos(\theta_2)}{\pi R} \tag{8.178}$$

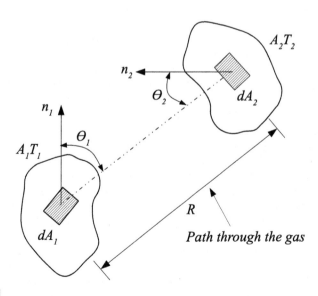

FIGURE 8.31

Enclosure analysis in the presence of an absorbing and emitting gas.

Assuming a uniform radiosity of J_1, the irradiation of A_2 due to radiation from A_1 is

$$J_1\kappa \int_{A_1}\int_{A_2} \frac{dA_1 dA_2 \cos(\theta_1)\cos(\theta_2)}{\pi R^2} - J_1\kappa \int_{A_1}\int_{A_2} \frac{dA_1 dA_2 \cos(\theta_1)\cos(\theta_2)}{\pi R} \qquad (8.179)$$

The first term within the integral is $A_1 F_{12}$, which is also $A_2 F_{21}$ (Please recall the definition of view factor F_{ij}). In the second term the integrand should also be something similar, and let us say it can be represented as $A_1 L_{12} = A_2 L_{21}$.
Let us say it is $A_1 L_{12}$, where L_{12} is the mean beam length.
Therefore, the irradiation on A_2 due to radiation coming from A_1 is given by

$$J_1 A_1 F_{12} - \kappa J_1 A_1 L_{12} = A_2 G_2^+ \qquad (8.180)$$

Using the reciprocal rule the above equation can be written

$$A_2 G_2^+ = J_1 A_2 F_{21} - \kappa J_1 A_2 L_{21} \qquad (8.181)$$

$$G_2^+ = J_1 F_{21}\left(1 - \frac{\kappa L_{21}}{F_{21}}\right) = J_1 F_{21}\left(1 - \frac{\kappa L_{12}}{F_{12}}\right) \qquad (8.182)$$

By induction,

$$G_1^+ = J_2 F_{12}\left(1 - \frac{\kappa L_{21}}{F_{21}}\right) \qquad (8.183)$$

Substituting for $1 - \dfrac{\kappa L_{21}}{F_{21}}$, which can be considered as $\tau(g)$ or the transmissivity of the gas, we have

$$G_1^+ = J_2 F_{12}\tau_{12} \qquad (8.184)$$

L_{12} and L_{21} represent the mean beam length

L_{ij} also follows the reciprocal rule, as in Eq. (8.185)

$$\tau_{12} = 1 - \frac{\kappa L_{21}}{F_{21}} = 1 - \frac{\kappa L_{12}}{F_{12}} \qquad (8.185)$$

The irradiation falling on the ith surface in the enclosure can then be written as

$$G_i = \varepsilon_g \sigma T_g^4 + \sum_{j=1}^{N} F_{ij} J_j \tau_{ij} \qquad (8.186)$$

$$\text{where } \tau_{ij} = \left(1 - \frac{\kappa L_{ij}}{F_{ij}}\right) = \left(1 - \frac{\kappa L_{ji}}{F_{ji}}\right) \qquad (8.187)$$

Please note that the irradiation G_i on the ith surface of the enclosure includes the gas radiation (first term) and the radiation falling from the other surfaces that is attenuated by the gas as seen by the presence of the term τ_{ij}. The radiosity of the ith surface is given by

$$J_i = \varepsilon_i \sigma T_i^4 + (1 - \varepsilon_i)G_i \qquad (8.188)$$

Finally, the net radiation from the ith surface is given by

$$q_i = J_i - G_i \tag{8.189}$$

The simplification afforded by the above development is too profound to be missed. The above development clearly shows that with a workaround, the original theory of radiosity for evacuated enclosures can be applied for an optically thin, gray case. One can easily see that for $\kappa = 0$, that is, the gas is transparent $\varepsilon_g = 0$ and $\tau_g = 1$ and Eq. (8.190) reduces to $G_i = \sum_{J=1}^{N} F_{ij} J_j$, which is nothing but Eq. (8.120) that gave an expression for irradiation, G_i on the ith surface in a N surface evacuated enclosure.

8.11.3 Calculation of emissivities and absorptivities for a mixture of gases

Mixtures of gases are encountered in furnaces and in internal combustion chambers. Burning hydrocarbons results in water vapor and carbon dioxide. Both are radiatively participative and they interact such that they have overlapping spectral bands.

A detailed method of solving would involve solving the RTE at millions of lines in the EM spectrum. This is very cumbersome.

For a quick engineering solution, we need a simplified approach.

The gas emissivity ε_g is a function of the following five quantities.

L_m = mean beam length
P_g = partial pressure of the gas
P = total pressure of the gas
C = concentration of other gases
T_g = temperature of the gas

This can be written as $\varepsilon_g = f(L_m, P_g, P, C, T_g)$. It is intuitive to see that emissivity is a function of temperature. The concentration of other gases needs to be included as the gas emissivity can depend on the interaction of gases and overlap of absorption bands. Hottel prepared charts for a total mixture pressure of 1 atmosphere, and a correction factor is to be applied for other pressures (Hottel and Hoyt, 1954).

The total emissivity of CO_2, ε_c, is a function of the product of the partial pressure of CO_2 and mean beam length and its gas temperature.

$$\varepsilon_c = f_1(P_c L_m, T_g)$$

The total emissivity of water vapor, ε_w, is a function of the partial pressure of water vapor, mean beam length, and gas temperature.

$$\varepsilon_w = f_2(P_w L_m, T_g)$$

A mixture of carbon dioxide and water vapor is a nongray gas. In view of this, $\varepsilon_g \neq \alpha_g$.

Leckner correlations:

Leckner gave correlations for calculation of the total emissivity and absorptivity of a mixture of CO_2 and H_2O (Leckner, 1972).

The emissivity for zero partial pressure condition is given by

$$\varepsilon_0(p_aL, p=1bar, T_g) = exp\left[\sum_{i=0}^{M}\sum_{j=0}^{N}c_{ji}\left(\frac{T_g}{T_0}\right)^i\left(\log\frac{p_aL}{(p_aL)_0}\right)^j\right] \tag{8.190}$$

where $T_0 = 1000$ K and $(p_aL)_0 = 1$ bar cm.
The emissivity for different pressure conditions is given by

$$\frac{\varepsilon(p_aL, p, T_g)}{\varepsilon_0(p_aL, 1bar, T_g)} = 1 - \frac{(a-1)(1-P_E)}{a+b-1+P_E}exp\left(-c\left[\log_{10}\frac{(p_aL)_m}{p_aL}\right]^2\right) \tag{8.191}$$

In Eq. (8.191), P_E is an effective pressure and a, b, c, and $(p_aL)_m$ are correlation parameters.

As aforementioned, a correction factor must be added to take care of the overlap of both carbon dioxide and water vapor. The constants in the above equation are given in Tables 8.6 and 8.7. The correction factor for emissivity is given by

$$\Delta\varepsilon = \left[\frac{\zeta}{10.7+101\zeta} - 0.0089\zeta^1 0.4\right]\left[\log_{10}\frac{(p_{H_2O}+p_{CO_2})}{L}(p_aL)_0\right]^{2.76} \tag{8.192}$$

$$\zeta = \frac{p_{H_2O}}{p_{H_2O}+p_{CO_2}} \tag{8.193}$$

Finally, to calculate the emissivity and absorptivity of the gas mixture containing H_2O and CO_2, the following equations can be used.

$$\varepsilon_i(p_iL, p, T_g) = \varepsilon_{0,i}(p_iL, 1bar, T_g)\left(\frac{\varepsilon}{\varepsilon_0}\right)_i(p_iL, p, T_g), \quad i = CO_2 \text{ or } H_2O \tag{8.194}$$

Table 8.6 Correlation constants for the determination of the total emissivity for water vapor.

M,N	2,2		
$C_{00}\ldots C_{N0}$	−2.2118	−1.1987	0.035596
...	0.85667	0.93048	−0.4391
$C_{0M}\ldots C_{NM}$	−0.10838	−0.17156	0.045915
P_E	$(p + 2.56p_a/\sqrt{t})/p_0$		
$(p_aL)_m/(p_aL)_0$	$13.2\,t^2$		
A	2.144	$t < 0.75$	
	$1.888-2.053\log_{10}t,$	$t > 0.75$	
B	$1.1/t^{1.4}$		
C	0.5		

$T_0 = 1000$K, $p_0 = 1$ bar, $t = T/T_0$, $(p_aL)_0 = 1$ bar cm

Table 8.7 Correlation constants for the determination of the total emissivity for carbon dioxide.

M,N	2,3			
$C_{00}...C_{N0}$	−3.9893	2.7669	−2.1081	0.39163
...	1.271	−1.1090	1.0195	−0.21897
$C_{0M}...C_{NM}$	−0.23678	−0.19731	−0.19544	0.044644
P_E	$(p + 0.28p_a)/p_o$			
$(p_aL)_0$	0.054/t^2,	$t < 0.75$		
	0.225 t^2,	$t > 0.75$		
A	$1 + 0.1 / t^{1}.45$			
B	0.23			
C	1.47			

$T_0 = 1000$ K, $p_0 = 1$ bar, $t = T/T_0$, $(p_aL)_0 = 1$ bar cm

$$\alpha_i(p_iL,p,T_g,T_s) = \left(\frac{T_g}{T_s}\right)^{(1/2)} \varepsilon_i\left(p_iL\frac{T_s}{T_g},p,T_s\right) \quad i = CO_2 \text{ or } H_2O \tag{8.195}$$

$$\varepsilon_{CO_2+H_2O} = \varepsilon_{CO_2} + \varepsilon_{H_2O} - \Delta\varepsilon(p_{H_2O}L,p_{CO_2}L), \tag{8.196}$$

$$\alpha_{CO_2+H_2O} = \alpha_{CO_2} + \alpha_{H_2O} - \Delta\varepsilon(p_{H_2O}L\frac{T_s}{T_g}, p_{CO_2}L\frac{T_s}{T_g}) \tag{8.197}$$

Example 8.14: *A furnace is in the shape of a sphere and is 0.6 m in diameter. It is filled with a gas mixture at a temperature of 1500 K and a pressure of 1.8 atmosphere. The gas mixture consists of H_2O at a partial pressure of 0.8 atm and CO_2 with a partial pressure of 0.4 atm. The cavity wall is black and is maintained at 600 K. Determine the heat transfer to the wall.*

Solution:

Given that:

Diameter of the spherical cavity = 0.6 m

Pressure of the gas mixture = 1.8 atm

Temperature of the gas mixture = 1500 K

Partial pressure of carbon dioxide = 0.4 atm

Partial pressure of water vapour = 0.8 atm

Wall temperature of the cavity = 600 K

$$L_m = \text{Mean beam length} = 0.65D = 0.39 \text{ m} \tag{8.198}$$

Emissivity calculation of CO_2

$$p_aL = 0.4 \times 0.39 = 0.156 \text{ atm m} \tag{8.199}$$

From Table 8.7,

$$P_E = 1.912, t = 1.5, a = 1.056 b = 0.23, c = 1.47$$
$$\text{and } (p_a L)_m = 0.506 \text{ bar cm}$$

Substituting the above values in Eq. (8.190), we get $\varepsilon_0 = 0.11$.
From Eq. (8.191), we get $\varepsilon_{CO_2} = 0.111$.
Emissivity calculation of H_2O:

$$p_a L = 0.8 \times 0.39 = 0.312 \text{ atm m}$$

From Table 8.6,

$$P_E = 3.47, \ t = 1.5, a = 1.53 b = 0.62, c = 0.5$$
$$\text{and } (p_a L)_m = 29.7 \text{ bar cm}$$

Substituting the above values in Eq, 8.190, we get $\varepsilon_0 = 0.195$.
From Eq. (8.191), we get $\varepsilon_{H_2O} = 0.25$.

$$\zeta = \frac{p_{H_2O}}{p_{H_2O} + p_{CO_2}} = \frac{0.44}{0.22 + 0.44} = 0.67$$

From Eq. (8.192), $\Delta\varepsilon = 0.035$

$$\begin{aligned}\varepsilon_{CO_2 + H_2O} &= \varepsilon_{CO_2} + \varepsilon_{H_2O} - \Delta\varepsilon \\ &= 0.111 + 0.250 - 0.035 \\ &= 0.326\end{aligned}$$

Total emissivity of CO_2 and $H_2O = 0.326$
In a similar fashion, absorptivity can be calculated using Eq. (8.196).

$$\text{Absorptivity of } CO_2 = 0.131$$

$$\text{Absorptivity of } H_2O = 0.297$$

From Eq. (8.197),
Total absorptivity of CO_2 and $H_2O = 0.131 + 0.297 - 0.035 = 0.393$
Heat transfer to the wall having a temperature of 600 K,

$$q_1 = J_1 - G_1 \tag{8.200}$$

$$Q_1 = (J_1 - G_1)4\pi R^2 \tag{8.201}$$

$$\varepsilon_1 = 1 \tag{8.202}$$

$$J_1 = \varepsilon_1 \sigma T_1^4 + (1 - \varepsilon_1)G_1 = \sigma T_1^4 \tag{8.203}$$

$$G_1 = \varepsilon_g \sigma T_g^4 + \tau_g(\sigma T_1^4) \tag{8.204}$$

$$\alpha_g + \tau_g + \rho_g = 1; \; \rho_g = 0 \tag{8.205}$$

$$\alpha_g + \tau_g = 1; \; \tau_g = 1 - \alpha_g \tag{8.206}$$

$$G_1 = \varepsilon_g \sigma T_g^4 + (1 - \alpha_g)(\sigma T_1^4) \tag{8.207}$$

$$q_1 = J_1 - G_1 = \sigma T_1^4 - \varepsilon_g \sigma T_g^4 - \sigma T_1^4 + \alpha_g \sigma T_1^4 \tag{8.208}$$

$$q_1 = \alpha_g \sigma T_1^4 - \varepsilon_g \sigma T_g^4 \tag{8.209}$$

For the problem under consideration,

$$\begin{aligned} q_1 &= 0.393 \times 5.67 \times 10^{-8} \times 600^4 \\ &\quad -0.326 \times 5.67 \times 10^{-8} \times 1500^4 \end{aligned} \tag{8.210}$$

$$q_1 = -90.69 \frac{kW}{m^2} \tag{8.211}$$

$$Q_1 = q_1 4\pi R^2 = -102.57 \, kW \tag{8.212}$$

$$\boxed{Q_1 = -102.57 \, kW}$$

Problems

8.1 Determine the wavelength that corresponds to the maximum emission from the following surfaces.
  a. Tungsten filament at 2900 K.
  b. Heated metal at 1100 K.
  c. Human skin at 306 K.
  d. A metal surface cooled cryogenically and maintained at 90 K.

8.2 The spectral, hemispherical emissivity of tungsten may be approximated by the distribution shown in Fig. 8.32. Consider a cylindrical tungsten filament that is of diameter $D = 0.8$ mm and length $L = 20$ mm. The filament is enclosed in an evacuated bulb and is heated by an electrical current to a steady-state temperature of 2900 K. What is the total, hemispherical emissivity when the filament temperature is 2900 K?

8.3 An opaque surface has a hemispherical spectral reflectivity, as shown in Fig. 8.33A. It is subjected to spectral irradiation, as shown in Fig. 8.33B.
  a. Sketch the spectral hemispherical absorptivity distribution.
  b. Determine the total irradiation on the surface.
  c. Determine the radiant flux that is absorbed by the surface.
  d. Determine the total hemispherical absorptivity of the surface.

8.4 An opaque surface 2 m × 2 m is maintained at 400 K and is simultaneously exposed to solar irradiation with $G = 1200$ W/m^2. The surface is diffuse and its spectral absorptivity is $\alpha_\lambda = 0$, 0.8, 0, and 0.9 for $0 \, \mu m \leq \lambda \leq 0.5 \, \mu m, 0.5 \, \mu m \leq \lambda \leq 1 \, \mu m, 0 \, \mu m \leq \lambda \leq 0.5 \, \mu m, \lambda \geq 2 \, \mu m$, respectively.

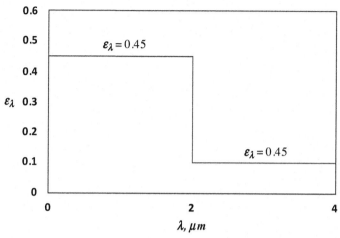

FIGURE 8.32

Spectral emissivity.

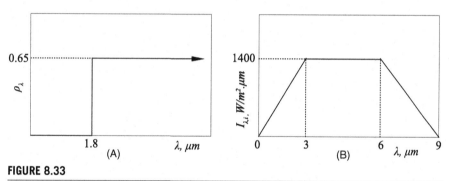

FIGURE 8.33

Variation of (A) hemispherical spectral reflectivity and (B) spectral irradiance with wavelength.

Determine the absorbed irradiation, emissive power, radiosity, and net radiation heat transfer from the surface.

8.5 Consider a two-dimensional evacuated rectangular enclosure shown in Fig. 8.34. Determine all the view factors.

8.6 Consider a two-dimensional evacuated regular hexagonal duct of side **a**, as shown in Fig. 8.35. Determine all the view factors.

8.7 Consider two very large parallel plates with diffuse gray surfaces as shown in Fig. 8.36. Details are given in the figure. Determine the net radiation heat transfer between the surfaces.

8.8 An oven is in the shape of a triangle, which is infinitely deep in the direction perpendicular to the paper. The heated surface is maintained at 1200 K, and

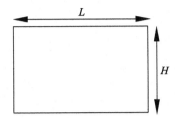

FIGURE 8.34

Rectangular enclosure.

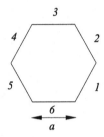

FIGURE 8.35

Two-dimensional regular hexagonal duct of side **a**.

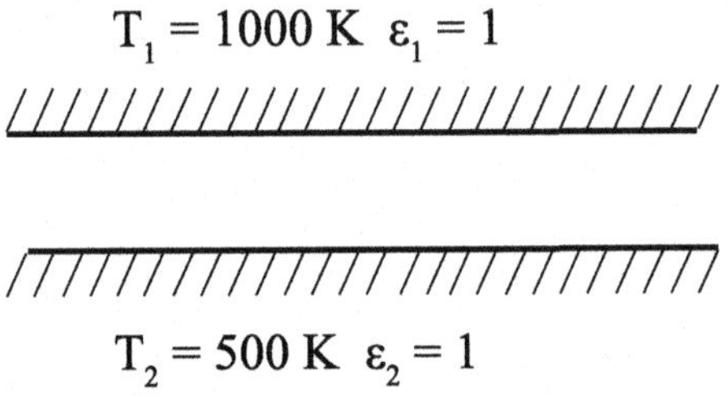

$$T_1 = 1000 \text{ K} \quad \varepsilon_1 = 1$$

$$T_2 = 500 \text{ K} \quad \varepsilon_2 = 1$$

FIGURE 8.36

Infinite parallel plates considered in exercise problem 8.35.

one surface is insulated. The third surface is at 490 K. The enclosure is evacuated and is 1 m wide on all the three sides. The emissivity of two sides of the enclosure is 0.8 (surface at 1200 K and the insulated one), while that of the third surface is 0.4. Determine the rate at which heat energy must be supplied to the heated side per unit length at 1200 K. What is the temperature of the insulated surface?

8.9 The space between two very long concentric tubes with diameter 25.4 mm and 50.8 mm is evacuated. The inner surface of the outer tube and the outer surface of the inner tube are diffuse and gray and have emissivities of 0.03 and 0.07,

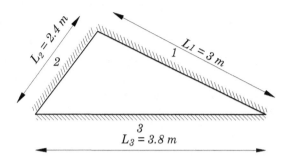

FIGURE 8.37

Two-dimensional, gray-diffuse evacuated enclosure considered in exercise problem 8.10.

respectively. The inner and outer tube surfaces are maintained at 260 K and 300 K, respectively.

a Determine all the view factors for this two surface enclosure.

b Determine the net radiation heat transfer between the two surfaces.

8.10 A two-dimensional, gray-diffuse evacuated enclosure (with no heat transfer to outside), as shown in Fig. 8.37, has all the surfaces maintained at uniform temperature. Relevant details are given below.

Surface 1: $T_1 = 1400$ K, $L_1 = 3.0$ m, $\varepsilon_1 = 0.50$.
Surface 2: $T_2 = 300$ K, $L_2 = 2.4$ m, $\varepsilon_2 = 0.95$.
Surface 3: $T_3 = 800$ K, $L_3 = 3.8$ m, $\varepsilon_3 = 0.15$.

a Calculate all the view factors for this three-surface enclosure.

b Determine the net radiation heat transfer from all the surfaces.

c Verify energy balance.

8.11 Consider two spheres with diameters 0.5 and 0.25 *m* that are placed concentrically as shown in Fig. 8.38. The surface emissivities of the outer and inner spheres are 0.4 and 0.7, respectively. The temperatures of the outer and inner spheres are maintained at 303 and 404 K, respectively. Answer the following questions:

a. Net heat transfer from the inner to the outer sphere.

b. Will there be any change in the heat transfer rate by altering the position of the inner sphere?

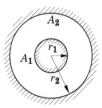

FIGURE 8.38

Radiation involving concentric spheres under consideration in problem 8.11.

8.12 A mixture of 40% of H_2O and 50% of CO_2 and 10% of NO_2(by volume) at a temperature of 2000 K and a total pressure of 1 atm is contained in a furnace in the shape of a cuboid of dimensions 1 m × 2 m × 1.5 m. The inside walls of the furnace are black, with the wall temperature being 1200 K.

    a. Determine the emissivity of the gas mixture.

    b. Determine the absorptivity of the gas mixture.

    c. Determine the radiation heat transfer to the walls of the furnace.

8.13 Repeat problem 8.12 for the case of a total pressure of 2 atm.

References

Balaji, C., 2014. Essentials of Radiation Heat Transfer. John Wiley and Sons, NJ, USA.

Hottel, Hoyt C., 1954. Radiant heat transmission. Heat Transmission. WH McAdams, New York.

Leckner, B., 1972. Spectral and total emissivity of water vapor and carbon dioxide. Combust, Flame 19 (1), 33–48.

Numerical heat transfer

9

9.1 Introduction

As engineers, we hope to obtain some quantities of interest, such as heat transfer or maximum temperature, in a physical situation. In the previous chapters, we saw that the ways to determine these quantities of engineering interest are to represent the physical problem as a system of differential equations, initial conditions, and boundary conditions, and then solve this system if possible.

We also saw in the previous chapters that solving these equations, even in relatively simple cases, can involve several analytical tricks. And each problem requires a trick of its own! Additionally, real problems require complex geometry, which, inconveniently for us, does not look like slabs or cylinders. Even more unfortunately, there could be nonlinearity, turbulence, etc. In short, the real world is very rarely like the analytically tractable, pen-and-paper world, and it is usually not possible to obtain a closed-form solution to heat transfer problems.

It is under these circumstances that we take recourse to numerical techniques. As you will see below, unlike analytical techniques, which require different tricks for different problems, numerical techniques use a broadly unified approach for almost every circumstance. Further, even though we will discuss only simple geometries in this chapter, the final section discusses how the same techniques can be used to handle real engineering problems.

9.2 Three broad approaches to numerical methods

Our purpose as heat transfer engineers is to obtain numbers for quantities of our interest and not necessarily solve complicated differential equations. In the whole of numerical heat transfer, there is, broadly speaking, a single overarching strategy that helps us obtain numbers of our interest without really obtaining analytical expressions for the differential equations. This strategy is to obtain numerical solutions only at discrete locations rather than as a continuous function of time and space like we did earlier. And how is this achieved? Through a trick called "discretization"— turning our continuous partial differential equations (PDEs) or ordinary differential equations (ODEs) into discrete, algebraic equations.

Heat Transfer Engineering. http://dx.doi.org/10.1016/B978-0-12-818503-2.00009-5

295

The idea behind discretization is to start with a convenient, continuous form of the equations and approximate the terms in these equations, assuming that information is available only at discrete points. The details of this will become clear shortly, but let us consider two simple possibilities for any governing equation. If the governing equation is in differential form and involves derivatives, we approximate the derivatives (which require continuous values) by slopes (which only require values at discrete points). On the other hand, if the governing equation is in the integral form (for example, the control volume formulation), then we approximate the integral by approximate areas under the curve.

Remarkably, this "one weird trick" lets us handle equations not only in heat transfer but also in fields as different as quantum mechanics and relativity. You might be interested to know that this single idea is the basis of everything from weather prediction to movie simulations of fire. That said, discretization is an art, and there are still multiple ways of discretizing the same physical situation. Three of the most popular approaches in heat transfer are:

1. Finite difference method (FDM)—The fundamental idea in FDM is to start from the differential equation and express each derivative as a slope (or, equivalently, a difference). FDM works very well when the geometry is simple and when our points of interest are arranged regularly in an ordered mesh.
2. Finite volume method (FVM)—The fundamental idea in FVM is to start from the integral conservation form of the equations. For example, even for an irregularly shaped finite volume, we can say that the fluxes coming in from the sides should be balanced. This is the control volume form of the equations. We split the domain into many, usually irregular, finite volumes. This form of the equation involves integrals, which are then approximated as areas. It is a very elegant method and can handle complex irregular geometries. For this and other reasons, FVM is the industry standard for fluid mechanics and heat transfer problems, and most commercial solvers use FVM.
3. Finite element method (FEM)—The fundamental idea in FEM is to start with a weighted integral of the governing equations. This approach, which seems unnatural for fluid mechanics and heat transfer problems, often arises naturally within what is called the "variational formulation" in solid mechanics, which is where FEM was used first historically. In heat transfer, FEM works well in problems where convection is not dominant (such as pure conduction problems) and is not the first (or second) choice for typical heat transfer problems. Nonetheless, it is mathematically very powerful, and when we have multiple, additional physical effects such as aeroelastic effects, electromagnetic forces, thermomechanical stresses, non-Newtonian fluids, etc., FEM discretizations are more stable, powerful, and flexible than either FDM or FVM.

In this book, we will be covering numerical methods on an introductory level using the humble FDM. However, even within this simple framework, you should be able to see the power of discretization in handling multiple equations within the same overall framework.

9.3 Equations and their classification

As we saw in the previous section, the basic idea behind numerical methods is to solve "discrete" versions of our continuous differential equations. Over the last century of numerically solving differential equations, engineers and numerical scientists have found out that, amazingly enough, sometimes the physics of the discrete equations is different from the physics of the continuous equations. Further, often it is found that if we ignore the underlying physics of the problem, the numerical approach can give absurd results. So, before we carry out a "discretization" exercise, it is important that the physics of the underlying equations is respected as much as possible. In this section, we will look at some ways to characterize some aspects of the basic physics of the equations we encounter typically in heat transfer.

9.3.1 Classification based on linearity and order

A differential equation is simply an equation that involves a dependent variable (such as velocity, temperature, etc.) and its derivatives with respect to some independent variables (spatial location and time in our case).

$$\frac{\partial^2 T}{\partial x^2} + \frac{\partial^2 T}{\partial y^2} = \alpha \frac{\partial T}{\partial t} \tag{9.1}$$

For example, Eq. (9.1) is a PDE with the temperature T as the dependent variable, and it is a function of x, y, and t. That is, $T = T(x, y, t)$. The PDE involves an equation with second derivatives of space and the first derivative of time.

Linearity—A *linear* PDE is one where the dependent variable and its derivatives only appear as linear functions. Eq. (9.1), for example, is a linear equation. On the other hand, if our equation looked like

$$\frac{\partial^2 T}{\partial x^2} + \frac{\partial^2 T}{\partial y^2} = \beta \left(\frac{\partial T}{\partial t} \right)^2 \tag{9.2}$$

It would be nonlinear because the $(\partial T / \partial t)^2$ term is not a linear function of the derivative. Similarly, any equation involving radiative terms that involve T^4 would be nonlinear. The Navier-Stokes momentum equations are nonlinear due to terms of the form $u \dfrac{\partial u}{\partial x}$.

Physically, linear PDEs have two important consequences. Firstly, small changes in causes only result in small effects. For example, perturbations in the boundary conditions, geometry, etc., only cause small changes in the temperature field in case the equation is linear. Such systems cannot become chaotic. Secondly, in case the linearity is homogeneous (that is, it is simply proportional and does not have an extra constant term), we can use the superposition principle. That is, the sum of two solutions is also a solution to the PDE. This allows us to sometimes decouple several physical effects in a problem.

Order of PDE—The order of any PDE is the order of the *highest derivative* within the PDE. For example, Eq. (9.1) is a second-order PDE due to the order of the spatial derivatives. In fluid mechanics and heat transfer, our PDEs are usually second order in space.

While we will not be able to discuss this in detail in this book, it is useful to note that the order of the derivative of spatial terms has important physical implications. A first-order derivative in space is usually indicative of an advection process, that is, the unmodified transport of a physical quantity from one location to another like a river carries a leaf or a cold front transports temperature. A second-order spatial derivative, on the other hand, is usually indicative of a diffusion process. That is, the term spreads the property in space. The spreading of temperature by conduction or the spreading of a dye in a stationary fluid are diffusion processes and are governed by second-order terms.

9.3.2 Classification based on information propagation

For second-order PDEs, an important classification approach is the one based on the direction of information propagation within the domain. For example, it is intuitive physically that in an advection process, the effects are felt more downstream than upstream. For example, if we add a pollutant in a flowing river, the effects of the pollutant would be felt more downstream of the flow than upstream of the flow. In contrast, the effects of diffusion are felt uniformly in space. Mathematically, such dependencies of information (or effect) propagation are said to be contingent on something called "characteristics" of the equation. We will be skipping the details of the mathematics and just discussing the final results, which help us classify equations of our interest.

Let T be a function of two variables, that is, $T = T(x_1, x_2)$. Let $T(x_1, x_2)$ satisfy a second-order PDE of the form

$$A(x_1,x_2)\frac{\partial^2 T}{\partial x_1^2} + B(x_1,x_2)\frac{\partial^2 T}{\partial x_1 \partial x_2} + C(x_1,x_2)\frac{\partial^2 T}{\partial x_2^2} + \\ D(x_1,x_2)\frac{\partial T}{\partial x_1} + E(x_1,x_2)\frac{\partial T}{\partial x_2} + F(x_1,x_2)T = G(x_1,x_2) \tag{9.3}$$

Then, we classify the equation as being elliptic, hyperbolic, or parabolic based on the term $B^2 - 4AC$ as follows:

1. Elliptic PDE

The equation is called elliptic if $B^2 - 4AC < 0$. An example is the Laplace equation

$$\frac{\partial^2 T}{\partial x^2} + \frac{\partial^2 T}{\partial y^2} = 0 \tag{9.4}$$

Which, if you recall, governs steady state heat conduction (without heat generation). Here, $B^2 - 4AC = 0^2 - 4 \times 1 \times 1 = -4 < 0$. The Laplace equation is, therefore, elliptic. Elliptic equations have the property that any changes on the boundary immediately affect the field inside. Therefore, in a steady state heat conduction problem, any change in temperature of a boundary point will affect every point within the domain. This is called the *domain of influence*. Conversely, due to the elliptic nature of the problem, the temperature of any point within the domain depends on *all* the points in the boundary. This is called the *domain of dependence*. It is important that this property be reflected in any numerical method we derive for the heat conduction problem.

2. Hyperbolic PDE
The equation is called hyperbolic if $B^2 - 4AC > 0$.

$$\frac{\partial^2 u}{\partial t^2} - c^2 \frac{\partial^2 u}{\partial x^2} = 0 \tag{9.5}$$

An example is the one-dimensional wave equation Eq. (9.5). Eq. (9.5) effectively is a combination of two advective waves moving at speed $\pm c$. Here, $B^2 - 4AC = 0^2 + 4 \times 1 \times c^2 = 4c^2 > 0$. The wave equation is, therefore, hyperbolic. Hyperbolic equations govern processes that have limits on the speed of information propagation. For instance, if you speak, it is not possible for anyone to hear you till the sound wave reaches them. Therefore, unlike elliptic equations, changes in one place do not affect every point in the domain equally. Retaining this portion of the physics makes discretizing hyperbolic equations slightly tricky. While we will not deal with hyperbolic equations in this text, they are an important topic in compressible flows and also convection-dominated incompressible flows.

3. Parabolic PDE
The equation is called parabolic if $B^2 - 4AC = 0$. An example is the so-called diffusion equation Eq. (9.6).

$$\frac{\partial^2 T}{\partial x^2} - \alpha \frac{\partial T}{\partial t} = 0 \tag{9.6}$$

Eq. (9.6) governs unsteady (one-dimensional) conduction. Here, $B^2 - 4AC = 0^2 - 4 \times 0 \times 1 = 0$. The diffusion equation is, therefore, parabolic. Parabolic equations have the property that any changes in the given time only affect properties in the future. That is, effects cannot propagate into the past. This, of course, seems obvious, and you might wonder why mathematics is required to come to such a trivial conclusion. There is, however, a reason why we need mathematics here. Unlike the diffusion equation above, there are cases where there are timelike variables that are not given in a convenient form as t. A famous example is the boundary layer equations we encountered earlier. Consider the following equation for the steady state, flat-plate boundary layer.

$$u\frac{\partial u}{\partial x} + v\frac{\partial u}{\partial y} - v\frac{\partial^2 u}{\partial y^2} = 0 \qquad (9.7)$$

Notice that in Eq. (9.7) there are two independent variables—x and y. If we now apply our classification equation (try this as an exercise), we will see that the equation is parabolic with x playing the role that t did in the diffusion equation.

The understanding that x is a timelike variable in a parabolic equation gives us both profound insight into the nature of boundary layers as well as computational advantages. This was, in fact, one of the major triumphs of Prandtl's boundary layer simplification. Very briefly, firstly, since x behaves like time, any perturbations we introduce in a boundary layer can only travel downstream and not travel upstream. This means a flow becomes turbulent only downstream of perturbations. Secondly, when computing solutions for the boundary layer equations, one can "march" in space along x. That is, we need not solve for the field in the whole field simultaneously. This was used to great advantage in flow solvers for boundary layers for a long time.

9.4 Basics of the finite difference method

Let us now look at how any derivative in a differential equation can be approximated if we have only discrete points available for the function. Let us first consider an informal example. Let us say you were traveling by car on a highway, and your speedometer is not working. If you wish to estimate your speed, you could do so if you have mileposts and a stopwatch. That is, you would estimate your speed u as $u = \dfrac{dx}{dt} \approx \dfrac{x(t+\Delta t) - x(t)}{\Delta t}$. This type of approximation is the heart of the finite difference method. You can immediately, intuitively see an important feature of the method—your approximation gets better and better as you reduce Δt. But can we make this idea of increasing accuracy with decreasing intervals a bit more precise? We can do so using the Taylor series.

9.4.1 Taylor series and finite difference formulae

Recall that the Taylor series tells us the relation between a function's values at two points separated by an interval:

$$f(x+h) = f(x) + h\frac{df}{dx}\bigg|_x + \frac{h^2}{2!}\frac{d^2 f}{dx^2}\bigg|_x + \frac{h^3}{3!}\frac{d^3 f}{dx^3}\bigg|_x + \dots . \frac{h^r}{r!}\frac{d^r f}{dx^r}\bigg|_x + \dots \qquad (9.8)$$

Eq. (9.8) is the basis for most finite difference discretizations. Let us see a few examples:

1. First derivative

$$T(x_i + \Delta x) = T(x_i) + \Delta x \left.\frac{dT}{dx}\right|_{x_i} + \frac{\Delta x^2}{2!} \left.\frac{d^2T}{dx^2}\right|_{x_i} + \dots \tag{9.9}$$

Rearranging Eq. (9.9), we get

$$\left.\frac{dT}{dx}\right|_{x_i} = \frac{T(x_i + \Delta x_i) - T(x_i)}{\Delta x_i} + \left.\frac{d^2T}{dx^2}\right|_{x_i} \frac{\Delta x_i}{2} + \dots \tag{9.10}$$

As the interval Δx gets small, we can ignore the higher order terms and write our approximation for the first-order derivative as

$$\left.\frac{dT}{dx}\right|_{x_i} \approx \frac{T_{i+1} - T_i}{x_{i+1} - x_i} + O(\Delta x_i) \tag{9.11}$$

The term $O(\Delta x_i)$ denotes the "order" of error of the first term we have ignored, that is, $\left.\frac{d^2T}{dx^2}\right|_{x_i} \frac{\Delta x_i}{2}$. As we reduce the interval Δx_i, the error will reduce proportionately. Since the power of Δx in the error is 1, we call this a first-order approximation. The above approximation is called the forward difference approximation as it requires $T(x_{i+1})$. But what if we are at the right boundary in a slab and do not have this information, as there is no more point to the right? In such a case, you can immediately guess that we can calculate the backward difference approximation by using the expansion for $T(x_i - \Delta x_i)$. Let us follow a similar process as above:

$$T(x_i - \Delta x_i) = T(x_i) - \left.\frac{dT}{dx}\right|_{x_i} \Delta x_i + \left.\frac{d^2T}{dx^2}\right|_{x_i} \frac{\Delta x_i^2}{2} - \dots \tag{9.12}$$

Rearranging Eq. (9.12)

$$\left.\frac{dT}{dx}\right|_{x_i} = \frac{T(x_i) - T(x_i - \Delta x_i)}{\Delta x_i} + \left.\frac{d^2T}{dx^2}\right|_{x_i} \frac{\Delta x_i}{2} + \dots \tag{9.13}$$

$$\left.\frac{dT}{dx}\right|_{x_i} \approx \frac{T_i - T_{i-1}}{x_i - x_{i-1}} + O(\Delta x_i) \tag{9.14}$$

This too is first-order accurate. So, both forward and backward differences give us first-order accuracy. You might have thought of combining the forward and backward difference estimates. Let us see what happens if we do that.

$$T(x_{i+1}) = T(x_i) + \left.\frac{dT}{dx}\right|_{x_i} \Delta x_i + \left.\frac{d^2T}{dx^2}\right|_{x_i} \frac{\Delta x_i^2}{2!} + \left.\frac{d^3T}{dx^3}\right|_{x_i} \frac{\Delta x_i^3}{3!} + \dots \tag{9.15}$$

$$T(x_{i-1}) = T(x_i) - \left.\frac{dT}{dx}\right|_{x_i} \Delta x_i + \left.\frac{d^2T}{dx^2}\right|_{x_i} \frac{\Delta x_i^2}{2!} - \left.\frac{d^3T}{dx^3}\right|_{x_i} \frac{\Delta x_i^3}{3!} + \dots \tag{9.16}$$

Subtracting Eq. (9.16) from Eq. (9.15), we obtain

$$T(x_{i+1}) - T(x_{i-1}) = 2\frac{dT}{dx}\bigg|_{x_i}\Delta x_i + 2\frac{d^3T}{dx^3}\bigg|_{x_i}\frac{\Delta x_i^3}{3!} + \ldots\ldots\ldots \tag{9.17}$$

$$\therefore \frac{dT}{dx}\bigg|_i = \frac{T(x_{i+1}) - T(x_{i-1})}{2\Delta x_i} - \frac{d^3T}{dx^3}\bigg|_i\frac{\Delta x_i^2}{3!} + \ldots\ldots\ldots \tag{9.18}$$

$$\frac{dT}{dx}\bigg|_i \approx \frac{T_{i+1} - T_{i-1}}{2\Delta x_i} + O(\Delta x_i^2) \tag{9.19}$$

This, for obvious reasons (Eq. (9.19), is called the central difference method since it uses points both to the left and the right. Notice that the error term now goes as Δx^2 instead of simply Δx. This is, therefore, a second-order accurate approximation. An important implication for this second-order accuracy is that each time we reduce the grid interval by a factor of 10, the error goes down by a factor of 100 for the central difference method instead of merely a factor of 10 for the first-order accurate method. In fact, it is possible to get even higher-order methods if we take more points in the stencil; for example, if we include five points in the difference approximation $x_i, x_{i\pm1}, x_{i\pm2}$, it is possible to obtain a fourth accurate approximation. For most practical purposes, however, simple first- and second-order approximations are sufficient, and we will stick with these in this book.

2. Second derivative

The governing equations in heat transfer usually have a second-order derivative in space. These can also be approximated using the Taylor series. Suppose we wish to approximate d^2T/dx^2 at a given point x_i. We can still use the same stencil x_i, x_{i+1}, x_{i-1} and do so as follows.

Adding Eq. (9.16) and Eq. (9.15), we get

$$T(x_{i+1}) + T(x_{i-1}) = 2T(x_i) + 2\frac{d^2T}{dx^2}\bigg|_i\frac{\Delta x_i^2}{2!} + \frac{d^4T}{dx^4}\bigg|_i\frac{\Delta x_i^4}{12} + \ldots\ldots\ldots \tag{9.20}$$

$$\therefore \frac{d^2T}{dx^2}\bigg|_i = \frac{T(x_{i+1}) - 2T(x_i) + T(x_{i-1})}{(\Delta x_i)^2} + O(\Delta x_i^2) \tag{9.21}$$

As we can see from the error term, this approximation is second-order accurate. There is one more way of obtaining the above expression for the second derivative.

Consider a node i, as shown in Fig. 9.1. The node east of this is $i + 1$ while that to the west is $i - 1$. We also imagine nodes $i + 1/2$ and $i - 1/2$ situated in the middle of the computational nodes.

Now,

$$\frac{dT}{dx}\bigg|_{i+1/2,j} \approx \frac{T_{i+1,j} - T_{i,j}}{\Delta x_i} \tag{9.22}$$

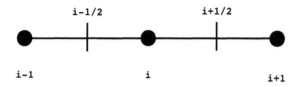

FIGURE 9.1

Finite difference stencil with midpoints.

$$\frac{dT}{dx}\bigg|_{i-1/2,j} \approx \frac{T_{i,j} - T_{i-1/2,j}}{\Delta x_i} \tag{9.23}$$

$$\therefore \frac{d^2T}{dx^2}\bigg|_{i,j} \approx \frac{\frac{dT}{dx}\big|_{i+1/2,j} - \frac{dT}{dx}\big|_{i-1/2,j}}{\Delta x_i} \tag{9.24}$$

$$= \frac{T_{i+1,j} - 2T_{i,j} + T_{i-1,j}}{(\Delta x_i)^2} \tag{9.25}$$

This is the same expression as before.

9.4.2 Overall process for the finite difference method

Even with the simple basics above, we are now ready to apply FDM to a range of problems. Regardless of the specific physical problem, there is an overall process that can be applied.

1. Mathematical modeling
Find a mathematical model for the physical problem in terms of a differential equation. If there is one independent variable, it is an ODE, or else, it is a PDE. All the previous chapters in this book showed how this could be done for various heat transfer problems.
2. Create a mesh or grid
Select a set of points within the physical domain where the solution is desired. In the case of unsteady problems, we need to "discretize" time as well into discrete time points. Remember from the Taylor series that the closer the mesh points are, the lower the interval, and consequently, the more accurate our approximation will be. There are many techniques to choose appropriate grids for problems. Within this book, we will be choosing a uniformly spaced grid and time points for convenience.
3. Discretize the governing equations
For every mesh point, apply the finite difference approximation to every derivative within the governing equation(s) and convert the equation into a difference equation for every mesh point. Also, convert all boundary conditions and initial conditions similarly into approximate equations.
If the governing equation is linear (like the Laplace equation), we will have a set of linear equations at the end of this step. In case the equation is nonlinear, we will have a set of algebraic equations.

4. Solve the set of equations

Now that we have a set of N equations where N is the total number of unknown variables, we can solve them simultaneously. Usually, we use an iterative method to solve these. The details for one such iterative method—Gauss-Seidel—are given in the next section.

5. Postprocess to find desired quantities

Often we require not only the field variables, such as velocity and temperature, but derived quantities, such as shear stress or heat transfer. The previous step only gives the field variables. We can then calculate our quantities of interest again using finite difference approximations. For example, the heat transfer can be calculated by approximating $q = -k\dfrac{\partial T}{\partial n}$ at the surface of interest.

In the following sections, we will apply the above process to various heat transfer problems to obtain numerical approximations.

9.5 Steady conduction

Let us consider some problems in steady state conduction to apply our understanding of the FDM. We start with the one-dimensional problem that we encountered earlier—heat transfer in a rectangular fin.

Example 9.1: *Consider a rectangular fin with a length of 10 cm, a thickness of 1 mm, and a thermal conductivity of 180 W/mK, which is insulated at its tip. The convection process is characterized by a fluid temperature at 25 °C and a heat transfer coefficient of 100 W/m² K. The base of the fin is at 100 °C. Using one-dimensional finite difference and grid pattern with five nodes indicated on the figure, solve the resulting nodal equations with the Gauss-Seidel method (detailed in the solution below), with an initial guess of 50 °C for all the five nodes and hence determine the five nodal temperatures. Perform eight iterations and verify whether the numerical solution respects the Neumann boundary condition at its tip.*
Solution:
1. Mathematical modeling: The mathematical model for the physical problem was developed in chapter 2 of this book. The govering equation is an ODE given below.

$$\frac{d^2T}{dx^2} - \frac{hP}{kA_c}(T - T_\infty) = 0$$

Comment: Recall from chapter 2 that the above equation itself has built several assumptions, such as uniform temperature across the cross-section, etc. We could have also chosen a more complex model that allows for temperature variation within the cross-section (what would the correct mathematical model be then?). Choosing the appropriate mathematical model is an important step in the solution process and requires physical insight and engineering common sense to determine the appropriate level of precision for the problem. The numerical method is only a substitute for solving the differential equation and not for determining the equation itself! Because of this, you will see that numerical packages always require you to use your physical understanding of the problem to choose among various governing equations. The package will only take care of the subsequent solution.

2. Create a mesh or grid: As shown in Fig. 9.2, the domain of the problem is discretized into six discrete points that constitute the mesh for this problem. An additional ghost point outside the computational domain has been created to impose the Neumann boundary condition.

Comment: The insulated tip results in the boundary condition $\frac{dT}{dx} = 0$ at the right end. This Neumann boundary condition can be handled in two different ways. The first one is to use just the 5 points given and use a backward difference scheme. So, we can apply $\frac{dT}{dx} \approx \frac{T_5 - T_4}{dx} = 0$, which results in $T_5 = T_4$ as the approximate boundary condition. However, this backward difference formula for the boundary condition is just first-order accurate, whereas we will be using second-order accurate formulae everywhere else in the domain. It is possible to improve the accuracy by cleverly introducing a "ghost" point outside of the domain. You can see below how this saves the day while retaining the simplicity of the method. Tricks such as this are a staple of the numerical community, and commercial packages have many such clever hacks thrown in but hidden from the user.

3. Discretize the governing equation and boundary conditions: The discretized form of the governing ODE and boundary conditions are as follows

- Governing equation.

$$\frac{T_{i+1} - 2T_i + T_{i-1}}{\Delta x^2} - \frac{hP}{kA_c}(T_i - T_\infty) = 0$$

$$T_i = \frac{T_{i+1} + T_{i-1} + m^2 \Delta x^2 T_\infty}{(2 + m^2 \Delta x^2)} \tag{9.26}$$

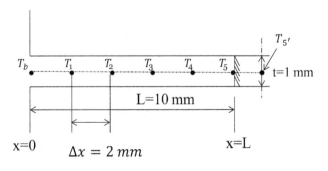

FIGURE 9.2

One-dimensional conduction in the fin for Example 9.1.

- Neumann boundary condition at the right end. At x = L, fin is insulated; that is,

$$\frac{dT}{dx} = 0 \Rightarrow \frac{T_{5'} - T_4}{2\Delta x} = 0 \Rightarrow T_{5'} = T_4$$

$$T_5 = \frac{T_{5'} + T_4 + m^2 \Delta x^2 T_\infty}{(2 + m^2 \Delta x^2)}$$

$$T_5 = \frac{2T_4 + m^2 \Delta x^2 T_\infty}{(2 + m^2 \Delta x^2)} (\because T_{5'} = T_4)$$

$$T_5 = \frac{2T_4 + 0.111}{2.004} \tag{9.27}$$

- Dirichlet boundary condition at the left end. At x = 0, $T_b = T_0$

$$T_1 = \frac{T_0 + T_2 + m^2 \Delta x^2 T_\infty}{(2 + m^2 \Delta x^2)}$$

$$T_1 = \frac{100.111 + T_2}{2.004} \tag{9.28}$$

- Discretized equations at the interior points.

$$T_2 = \frac{T_3 + T_1 + 0.111}{2.004} \tag{9.29}$$

$$T_3 = \frac{T_4 + T_2 + 0.111}{2.004} \tag{9.30}$$

$$T_4 = \frac{T_5 + T_3 + 0.111}{2.004} \tag{9.31}$$

Comment: Eqs. 9.26–9.31 are the combined set of governing equations and boundary conditions for the problem. Notice how both the ODE as well as the BCs have been combined into a set of equations. The equations are linear since the ODE, as well as the BCs, were linear. We have five equations and five unknowns, which can now be solved using multiple techniques. If we take smaller intervals, we will have a larger number of equations and unknowns.

4. Solve the set of equations: While linear systems of equations can be solved in multiple algorithms, all these fall under two major classes—iterative methods and direct methods. Direct methods such as Cramer's rule or Gauss elimination solve for the variables exactly without any guesswork. Iterative methods, on the other hand, start with a guess for the solution and iteratively improve the guess. Iterative methods are often preferred computationally, especially in numerical approximations, for their ease of implementation. We will employ one such iterative method—the

Gauss-Seidel algorithm—for the rest of the chapter. The algorithm is very intuitive and works as follows:

    a. Initialize: Take an initial guess for all the variables, for example, $T_1^0, T_2^0, T_3^0, T_4^0, T_5^0$

    b. Formulate equations: Rewrite each equation with one variable on the left and all the others on the right-hand side. That is, each variable is written in terms of the remaining variables. Eqs 9.26–9.31 are already written in this form.

    c. Update: Use the equations in the previous step to update each variable. For example, if all temperatures in the current example are initialized to 50, then this step will result in

$$T_1^1 = \frac{100.111 + T_2}{2.004} = \frac{100.111 + 50}{2.004} \approx 74.90$$

$$T_2^1 = \frac{T_3 + T_1 + 0.111}{2.004} = \frac{50 + 74.90 + 0.111}{2.004} \approx 62.38 \qquad (9.32)$$

Note that we always use the latest available value of the variables in the update. For example, in the update for T_2, we used the latest available value of T_1.

    d. Check stopping criterion: We stop our iterations if either we have reached a predecided number of iterations (which we have specified as eight in this example), or if our error between the previous step and the current step is small enough. The error value can be seen in the table below for the current example. You will notice that the error constantly decreases with the number of iterations.

On following the above process, the results of the first eight iterations are given in Table 9.1

5. Postprocess to find desired quantities: To verify if our numerical solution respects the Neumann boundary condition, we calculate the heat transfer at the tip. The heat transfer at the tip is given by $q'' = -k \frac{T_5 - T_4}{\Delta x}$, which equals 9 kW/m². We can see that the

Table 9.1 Progress of Gauss-Seidel iterations for Example 9.1.

Iteration no.	T_1,°C	T_2, °C	T_3, °C	T_4, °C	T_5, °C	$\sum_{i=1}^{4}(T_i^{j+1} - T_i^{j})^2$
0	50	50	50	50		-
1	74.90	62.38	56.13	53.02	52.97	829.18
2	81.08	68.53	60.71	56.78	56.72	125.09
3	84.15	72.34	64.48	60.54	60.47	66.41
4	86.05	75.17	67.77	64.05	63.98	47.12
5	87.46	77.52	70.70	67.26	67.18	36.63
6	88.64	79.57	73.32	70.17	70.08	29.28
7	89.65	81.38	75.67	72.79	72.70	23.64
8	90.57	83.01	77.80	75.16	75.06	19.15

$\therefore T_1 = 90.57\,°C, T_2 = 83.01\,°C, T_3 = 77.80\,°C, T_4 = 75.16\,°C, T_5 = 75.06\,°C$

numerical solution is inaccurate and yet to converge to the correct solution. Therefore, we need more numbers of iterations for the correct solution. The high value of the stopping criterion at the end of 8 iterations confirms this.

We now apply the same process for two-dimensional steady state heat transfer without any heat generation.

Example 9.2: *Consider steady two-dimensional heat transfer in a square metal plate, as shown in the figure below. The temperature at the boundaries is prescribed to be 300 K, all sides except at the top where the temperature is maintained at 400 K. The thermal conductivity of the body is k = 180 W/mK. There is no heat generation in the slab. Using the finite difference method with a mesh size of Δx = Δy = 0.1 m, determine the temperature at nodes 1, 2, 3, and 4. Perform iterations with the Gauss-Seidel method until* $\sum_{i=1}^{4}(T_i^{j+1} - T_i^{j})^2 \leq 1 \times 10^{-4}$, *where j is the iteration number, and find the heat flux at a point $P_3(0.1, 0.1)$. Start with an initial guess of 325 K for all the temperatures.*

Solution:

1. **Mathematical modeling:** The governing equation for two-dimensional steady state heat conduction without heat generation is

$$\frac{\partial^2 T}{\partial x^2} + \frac{\partial^2 T}{\partial y^2} = 0$$

2. **Create a mesh or grid:** The two-dimensional grid is discretized into 4 × 4 uniformly spaced discrete points as shown in Fig. 9.3.

3. **Discretize the governing equations:** Using the central difference method, we discretize the governing equation as follows.

$$\frac{T_{i+1,j} - 2T_{i,j} + T_{i-1,j}}{(\Delta x)^2} + \frac{T_{i,j+1} - 2T_{i,j} + T_{i,j-1}}{(\Delta y)^2} = 0$$

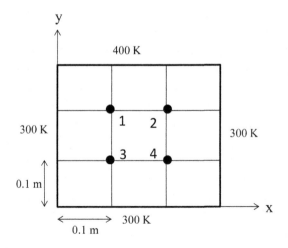

FIGURE 9.3

Two-dimensional conduction without heat generation for Example 9.2.

Table 9.2 Progress of Gauss-Seidel iterations for Example 9.2.

Iteration no.	T_1, K	T_2, K	T_3, K	T_4, K	$\sum_{i=1}^{4}(T_i^{j+1}-T_i^j)^2$
1	325	325	325	325	-
2	337.5	340.63	315.63	314.06	607.91
3	339.06	338.28	313.28	312.89	14.801
4	337.89	337.7	312.7	312.6	2.146
5	337.60	337.55	312.55	312.52	0.1341
6	337.52	337.51	312.51	312.51	0.0084
7	337.51	337.50	312.50	312.50	0.0005
8	337.50	337.50	312.50	312.50	3×10^{-3}

$\therefore T_1 = 337.5K, T_2 = 337.5K, T_3 = 312.5K, T_4 = 312.5K$

In this problem, $\Delta x = \Delta y$

$$\therefore T_{i,j} = \frac{T_{i+1,j} + T_{i-1,j} + T_{i,j+1} + T_{i,j-1}}{4}$$

$$T_1 = \frac{T_2 + T_3 + 300 + 400}{4}$$

$$T_2 = \frac{T_1 + T_4 + 300 + 400}{4}$$

$$T_3 = \frac{T_1 + T_4 + 300 + 300}{4}$$

$$T_4 = \frac{T_2 + T_3 + 300 + 300}{4}$$

4. Solve the set of equations: Using the Gauss-Seidel method of iteration, we get Table 9.2.

5. Postprocess to find desired quantities: The heat flux vector at P_3 is given by

$$q''_{P_3} = -k\left(\frac{T_4 - 300}{2\Delta x}\vec{i} + \frac{T_1 - 300}{2\Delta y}\vec{j}\right) = -11.25kW\vec{i} - 33.75kW\vec{j}.$$ Therefore the heat flux at this point is 35.576 kW/m²K.

We now consider two-dimensional, steady state conduction with heat generation in the next example.

Example 9.3: *Consider steady two-dimensional heat transfer in a rectangular metal plate with constant thermophysical properties, as shown in Fig. 9.4. The temperature at the boundaries are as prescribed in the figure. The thermal conductivity of the*

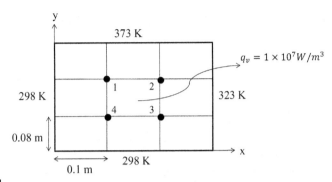

FIGURE 9.4

Two-dimensional conduction with heat generation for Example 9.3.

body is k = 150 W/m K. Assume that heat is generated in the plate uniformly at a rate of $q_v = 1 \times 10^7$ W/m³, as shown in the figure. Write the finite difference formulation for the given two-dimensional heat conduction problem with internal heat generation. Using the finite difference method with a mesh size of $\Delta x = 0.1$ and $\Delta y = 0.08$ m, determine the temperature at nodes 1, 2, 3, and 4. Perform iterations with the Gauss-Seidel method until $\sum_{i=1}^{4} (T_i^{j+1} - T_i^{j})^2 \le 1 \times 10^{-4}$, where j is the iteration number. Start with an initial guess of 310 K for all the temperatures.

Solution:

1. **Mathematical modeling:** The governing equation for two–dimensional, steady state heat conduction with heat generation is

$$\frac{\partial^2 T}{\partial x^2} + \frac{\partial^2 T}{\partial y^2} + \frac{q_v}{k} = 0$$

2. **Create a mesh or grid:** The two-dimensional grid is discretized into 4 × 4 uniformly spaced discrete points as shown in Fig. 9.4.

3. **Discretize the governing equations:** Using the central difference method, we have

$$\frac{T_{i+1,j} - 2T_{i,j} + T_{i-1,j}}{(\Delta x)^2} + \frac{T_{i,j+1} - 2T_{i,j} + T_{i,j-1}}{(\Delta y)^2} + \frac{q_v}{k} = 0$$

$$\frac{\Delta x^2(T_{i,j+1} + T_{i,j-1}) + \Delta y^2(T_{i+1,j} + T_{i-1,j}) - 2T_{ij}(\Delta x^2 + \Delta y^2)}{\Delta x^2 \Delta y^2} + \frac{q_v}{k} = 0$$

$$T_{ij} = \frac{\Delta x^2(T_{i,j+1} + T_{i,j-1}) + \Delta y^2(T_{i+1,j} + T_{i-1,j}) + \frac{q_v}{k}\Delta x^2 \Delta y^2}{2(\Delta x^2 + \Delta y^2)}$$

$$T_1 = \frac{\Delta x^2(373 + T_3) + \Delta y^2(T_2 + 298) + \frac{10^7}{150}\Delta x^2 \Delta y^2}{2(\Delta x^2 + \Delta y^2)}$$

$$T_2 = \frac{\Delta x^2 (373 + T_4) + \Delta y^2 (T_1 + 323) + \dfrac{10^7}{150} \Delta x^2 \Delta y^2}{2(\Delta x^2 + \Delta y^2)}$$

$$T_3 = \frac{\Delta x^2 (298 + T_1) + \Delta y^2 (T_4 + 298) + \dfrac{10^7}{150} \Delta x^2 \Delta y^2}{2(\Delta x^2 + \Delta y^2)}$$

$$T_4 = \frac{\Delta x^2 (298 + T_2) + \Delta y^2 (T_3 + 323) + \dfrac{10^7}{150} \Delta x^2 \Delta y^2}{2(\Delta x^2 + \Delta y^2)}$$

After inserting the values of Δx and Δy, we have

$$T_1 = 0.3049 T_3 + 0.1951 T_2 + 301.95 \tag{9.33}$$

$$T_2 = 0.3049 T_4 + 0.1951 T_1 + 306.82 \tag{9.34}$$

$$T_3 = 0.3049 T_1 + 0.1951 T_4 + 279.08 \tag{9.35}$$

$$T_4 = 0.3049 T_2 + 0.1951 T_3 + 283.96 \tag{9.36}$$

4. Solve the set of equations: Using the Gauss-Seidel method of iteration, we get Table 9.3.

5. Postprocess to find desired quantities: As we do not want to find any derived quantity (like heat transfer rate), we can stop here and need not do any further postprocessing.

Notice the progression in the solved examples. We went from one-dimensional steady state to two-dimensional steady state without heat generation and finally added

Table 9.3 Progress of Gauss-Seidel iterations for Example 9.3.

Iteration no.	T_1, K	T_2, K	T_3, K	T_4, K	$\sum_{i=1}^{4} (T_i^{j+1} - T_i^j)^2$
0	310	310	310	310	-
1	456.95	490.48	478.88	526.94	129756.42
2	543.65	573.55	547.65	565.68	20646.37
3	580.82	592.62	566.54	575.18	2192.23
4	590.30	597.36	571.28	577.55	140.51
5	592.67	598.55	572.46	578.14	8.78
6	593.27	598.84	572.76	578.29	0.5496
7	593.42	598.92	572.84	578.33	0.0343
8	593.46	598.94	572.86	578.34	0.0021
9	593.46	598.94	572.86	578.34	0.0001

$\therefore T_1 = 593.46 K, T_2 = 598.94 K, T_3 = 572.86 K, T_4 = 578.34 K$

heat generation as well. If you try to solve these problems analytically, you will notice that each of these problems requires vastly different analytical methods and tricks. Contrast this with the numerical process. Notice that, throughout, our finite difference process hardly changed, and we were able to easily adapt it to the problem even as the equations changed. Indeed, it is this easy adaptability of the numerical approximation approach that makes it such a powerful tool for practical problems.

9.6 Unsteady conduction

The previous section looked at steady state conduction cases, usually governed by $\nabla^2 T = 0$. However, if we are interested in either the transient (i.e., *how* we got to the steady state), or a generally unsteady problem, we need to add one more wrinkle—time dependence—into the problem and see how to handle it numerically. Note that the steady state conduction problem is the steady state of a general, unsteady equation of the form

$$\nabla^2 T = \alpha \frac{\partial T}{\partial t} \tag{9.37}$$

We will start with an extremely simple case and then proceed to a more complex example. The example has been a little bit open-ended deliberately to encourage exploration and discussion and foster some intuition about discretization and its effect on the numerical solution.

Example 9.4: *Consider a slab of thickness L = 1.6 cm with walls maintained at temperatures T_L = 25 and T_R = 100, on the left and right respectively. The initial temperature in the slab is uniformly T_{int} = 25 everywhere. The slab's diffusivity is $\alpha = 1 \times 10^5$ cm²/s. Answer the following questions:*
1. Write the governing equation, boundary conditions, and initial condition for this problem.
2. Discretize the slab with uniformly spaced points. Write approximate expressions for the governing equation. Comment on the order of accuracy of your approximation.
3. Are there alternate versions of the discretization that you can think of? Comment on the order of accuracy of these alternates.
4. Rearrange your discretization to write an expression for how the temperature of any location evolves in time. Does any nondimensional parameter naturally arise from this? What does it remind you of?
5. Calculate the temperatures in the center of the slab at $Fo = 0.125, 0.25, 0.5, 1, 2$ using an explicit scheme. Including the boundary points, try with 3,5 points in the computational domain. What are the effects of changing Δt in your discretizations? What can you conclude from this?
6. Repeat the above for the implicit scheme.

Solution:

1. The governing equation for the temperature $T(x,t)$ is the unsteady conduction equation and is given by

$$\frac{\partial T}{\partial t} = \alpha \frac{\partial^2 T}{\partial x^2} \quad x \in [0,L], t \in [0,\infty) \tag{9.38}$$

The boundary conditions are Dirichlet and are given by

$$T(0,t) = T_L \quad t > 0 \tag{9.39}$$

$$T(x,t) = T_R \quad t > 0 \tag{9.40}$$

The initial condition is given by

$$T(x,0) = T_L \quad x \in (0,L) \tag{9.41}$$

2. If there are just three points in the domain, there is just one unknown temperature, T_{mid}. Let us say the three temperatures are $T_1 = T_L$, $T_2 = T_{mid}$, and $T_3 = T_R$. Note that we have to discretize in time too. This is very much like the two-dimensional grid we had used purely in space earlier. The only difference is that since the direction y is a spatial direction, the customary space-time notation is to use the index n as a superscript for stepping in time and i for stepping in space. That is, T_i^n denotes $T(x_i,t^n)$, where the superscript n denotes the index in time.

Fig. 9.5 shows a typical space-time stencil for unsteady problems. We can now discretize each of the derivatives in the governing equation as follows.

$$\frac{\partial^2 T}{\partial x^2} \approx \frac{T_{i+1}^n - 2T_i^n + T_{i-1}^n}{(\Delta x)^2} \tag{9.42}$$

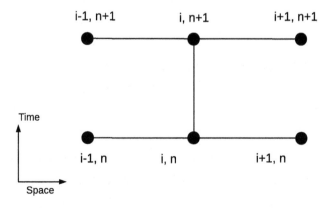

FIGURE 9.5

Space-time stencil for a one-dimensional, unsteady problem.

$$\frac{\partial T}{\partial t} \approx \frac{T_i^{n+1} - T_i^n}{\Delta t} \tag{9.43}$$

The order of accuracy of the two derivatives are $O(\Delta x^2)$ and $O(\Delta t)$ respectively. The overall scheme is therefore given by

$$\frac{T_i^{n+1} - T_i^n}{\Delta t} = \alpha \frac{T_{i+1}^n - 2T_i^n + T_{i-1}^n}{(\Delta x)^2} \tag{9.44}$$

and is $O(\Delta x^2, \Delta t)$ in accuracy.

Comment: You might ask, "Why did we choose the forward difference operator for the time derivative? Why not the central or the backward difference?" This would be an astute question. The current choice that we have made is called FTCS (forward time central space). The other two would also be legitimate choices and lead to CTCS (center time center space) and BTCS (backward time center space) methods. It turns out CTCS, for various reasons, is not a good choice despite seemingly being $O(\Delta x^2, \Delta t^2)$. We will, however, examine BTCS shortly.

3. Some alternate discretizations are as follows:

$$\frac{T_i^n - T_i^{n-1}}{\Delta t} = \alpha \frac{T_{i+1}^n - 2T_i^n + T_{i-1}^n}{(\Delta x)^2} \tag{9.45}$$

This, of course, is the BTCS discussed just previosly and is also $O(\Delta x^2, \Delta t)$. There exists an alternate discretization with a nonobvious trick that enables $O(\Delta x^2, \Delta t^2)$ without resorting to CTCS. This is called the Crank-Nicolson scheme and is somewhat like an average of the FTCS and BTCS. The scheme is as follows:

$$\frac{T_i^n - T_i^{n-1}}{\Delta t} = \frac{\alpha}{2} \left(\frac{T_{i+1}^n - 2T_i^n + T_{i-1}^n}{\Delta x^2} + \frac{T_{i+1}^{n-1} - 2T_i^{n-1} + T_{i-1}^{n-1}}{\Delta x^2} \right) \tag{9.46}$$

Comment: We will not prove the $O(\Delta x^2, \Delta t^2)$ accuracy of Crank-Nicolson here. You can try it as an advanced exercise yourself by writing out a Taylor series expansion of both sides in two variables.

4. For finding out how the temperature at any node will evolve in time, we need to find out how values at the future time step (t^{n+1}) depend on the values at the current time step (t^n). So, we will rearrange terms so that terms containing different time levels appear on different sides. We will consider two discretizations here—FTCS and BTCS.

For FTCS, rearranging Eq. (9.44), we obtain

$$T_i^{n+1} = T_i^n + \frac{\alpha \Delta t}{\Delta x^2} \left(T_{i+1}^n - 2T_i^n + T_{i-1}^n \right) \tag{9.47}$$

Notice the appearance of the term $\dfrac{\alpha \Delta t}{\Delta x^2}$. This is a nondimensional term. It is called the diffusion number, and we will label it with D from now on. It should remind you of the Fourier number $\dfrac{\alpha t}{L^2}$. In terms of D, the equation is

$$T_i^{n+1} = (1-2D)T_i^n + D\left(T_{i+1}^n + T_{i-1}^n\right) \tag{9.48}$$

Comment: Notice how naturally the diffusion number appears out of the discretization. If $\Delta x = L$, we would, in fact, directly obtain the Fourier number. It is remarkable how many times numerical discretizations throw up physical parameters naturally.

Consider BTCS now. For convenience, we will replace the (future, present) time levels $(n-1,n)$ in Eq. (9.45) with $(n,n + 1)$. Rearranging Eq. (9.45) and substituting $D = \dfrac{\alpha \Delta t}{\Delta x^2}$, we obtain

$$-D\left(T_{i+1}^{n+1} + T_{i-1}^{n+1}\right) + (1+2D)T_i^{n+1} = T_i^n \tag{9.49}$$

Comment: Notice that, unlike FTCS, where it was possible to calculate T_i^{n+1} for each i independent of the values at other nodes, in BTCS we obtain T_i^{n+1} as a part of a system of equations that have to be solved simultaneously for all nodes. Similarly, Crank-Nicolson also requires solution of a system of equations. Since the expression for the updated temperature is available explictly for each node in FTCS, it is commonly called the *explicit scheme*, whereas BTCS and Crank-Nicolson are called *implicit schemes*. Clearly, implicit schemes require more computational effort. Is there any reason to use them then? We will see shortly that it does make sense to use them in several situations.

5. Consider just three points in the domain. Two of them are known and fixed due to Dirichlet boundary conditions. So, $T_1^n = T_L = 100$ and $T_3^n = T_R = 25$. The only unknown temperature is T_2^n. Since there are only two intervals, $\Delta x = L/2 = 0.8$ cm $= 0.008$ m. The initial condition is that at $t = 0$ $T(x = L/2, 0) = T_{init} = 25\,°C$. We now evolve this in time using the evolution Eq. (9.48). This becomes, for our case,

$$T_2^{n+1} = (1-2D)T_2^n + D(100+25) \tag{9.50}$$

The evolution of the temperature depends on the diffusion number D, which, in turn, depends on our choice of Δt. We want our first solution at $Fo = \dfrac{\alpha t}{L^2} = 0.125$. This corresponds to a time of $t = \dfrac{0.125L^2}{\alpha} = 3.12 \times 10^{-6}\,\text{s}$.

Let us choose our time step Δt such that we obtain the $Fo = 0.125$ solution in one time step. This corresponds to $D = \dfrac{\alpha \Delta t}{\Delta x^2} = 10 \times 3.12 \times 10^{-6}/0.008^2 = 0.5$. For $D = 0.5$, Eq. (9.50) gives

$$T_2^1 = (1-2\times0.5)T_2^0 + 0.5 \times (100+25) \tag{9.51}$$

That is, $T_2^1 = 62.5\,°C$, the average temperature of the ends at $Fo = 0.125$. You might suspect that something is not quite accurate here. After all, the mid-plane temperature is supposed to be the average temperature only at steady state and not at early times such as $Fo = 0.125$. The reason we obtain this inaccurate temperature is because of the finite Δx and Δt that we are using. So, with numerical methods one always has to be careful about a certain amount of numerical error, as we have been seeing so far in this chapter.

Now, we might ask, if we have already reached steady state at $Fo = 0.125$, what happens beyond this time? Let us check. If we take one more time step with the same Δt from here, we obtain

$$T_2^2 = (1 - 2 \times 0.5)T_2^1 + 0.5 \times (100 + 25) \tag{9.52}$$

That is, $T_2^2 = 62.5\,°C$, and the center-point stays at the steady state temperature, which is physically sound.

A natural question to ask now is to see what happens as Δt from the one we chose here. Let us take two extreme limits to understand what happens as $\Delta t \to 0$ and $\Delta t \to \infty$.

In the lower limit of time steps, as $\Delta t \to 0$, $D \to 0$ we obtain from Eq. (9.48), $T_i^{n+1} = T_i^n$. That is, the value at every point will be stuck at the initial value itself. Extrapolating from here, for very low time steps ($D \ll 0.5$) we will obtain very slow convergence to steady state.

In the upper extreme, as $\Delta t \to \infty$, we notice that $D \to \infty$. We obtain from Eq. (9.48) that $T_2^1 = T_2^0 + D(100 + 25 - 2 \times 25)$. So, as $D \to \infty$, $T_2^1 \to \infty$, which is obviously absurd. Reaching such unphysically large temperatures is called *numerical instability*, and one of the most common ways in which it occurs in numerical methods is by choosing large time steps. For FTCS on the one-dimensional conduction equation, one can prove that any choice of $D > 0.5$ will ultimately result in unphysical temperatures. So, FTCS for one-dimensional conduction works only for $0 < D \leq 0.5$. In general, it is important to remember that **explicit schemes always work only under certain ranges of time steps**. The determination of these ranges is typically covered in full-scale numerical methods courses, and we will avoid that here.

In Table 9.4 there are the results of numerical experiments for various values of Δt that are within the above stability range. You will notice that the solution gets more

Table 9.4 Values of central temperature using three points with explicit scheme for Example 9.4.

		D			
Fo	**t**	**0.5**	**0.25**	**0.1**	**T_{exact}**
0.125	3.2e-06	62.5	57.812	52.670	48.596
0.25	6.4e-06	62.5	61.328	59.279	58.451
0.5	1.28e-05	62.5	62.427	62.154	62.157
1.0	2.56e-05	62.5	62.5	62.496	62.498
2.0	5.12e-05	62.5	62.5	62.5	62.5

Table 9.5 Values of central temperature using five points with explicit scheme for Example 9.4.

Fo	t	D 0.5	D 0.25	D 0.1	T_{exact}
0.125	3.2e-06	53.125	51.614	49.759	48.596
0.25	6.4e-06	60.156	59.433	58.690	58.451
0.5	1.28e-05	62.354	62.257	62.159	62.157
1.0	2.56e-05	62.499	62.498	62.497	62.498
2.0	5.12e-05	62.5	62.5	62.5	62.5

accurate as Δt reduces. This is as predicted by the Taylor series analysis, which says that the error is $O(\Delta t, \Delta x^2)$.

What happens if we use FTCS but with five points in the domain now? Notice that our Δx is now 0.004, and our error, therefore, should reduce. The scheme stays the same as before (see Eq. (9.48)). The difference now is that T_2^n, T_3^n, T_4^n have to evolve at each time step. The results for the center temperature are given in Table 9.5. As expected, you can see that the prediction is better as Δt and Δx decrease.

6. Let us consider the same experiments as above, but with BTCS. We start with three points in the domain. Recall that, as before, $T_1^n = T_L = 100\,°C$ and $T_3^n = T_R = 25\,°C$. The only unknown temperature is T_2^n. Since there are only two intervals, $\Delta x = L/2 = 0.8$ cm $= 0.008$ m. The initial condition is that at $t = 0$ $T(x = L/2, 0) = T_{init} = 25\,°C$. From Eq. (9.49), we obtain for the evolution of T_2

$$T_i^{n+1} = \frac{T_i^n}{1+2D} + \frac{D}{1+2D}(T_L + T_R) \tag{9.53}$$

Let us consider the two extreme cases, $\Delta t \to 0$ and $\Delta t \to \infty$ now. In the lower extreme, $\Delta t \to 0$, we obtain that $T_i^{n+1} = T_i^n$. That is, there is no evolution at all, since we are stuck at the same time. This is the same as it was in the explicit case. So, as we take smaller time steps, evolution will become slower.

What is interesting, however, is the other extreme case $\Delta t \to \infty$. In this case, $D \to \infty$. Taking limits, we obtain $\frac{T_i^n}{1+2D} \to 0$ and $\frac{D}{1+2D} \to \frac{1}{2}$. So we see that as $\Delta t \to \infty$, $T_i^{n+1} \to \frac{T_L + T_R}{2}$.

The above result is very useful for two reasons. Firstly, it is completely physical that the correct steady state temperature is reached as time tends to ∞. Secondly, we no longer seem to have the time step restrictions from which the explicit scheme suffered. That is, we may take as large a time step as we please without becoming unstable. Indeed, this is a major advantage of implicit schemes. *Implicit schemes are usually stable for practically any time step.*

In the Table 9.6, we run some numerical experiments with the implicit BTCS and just three points in space.

Table 9.6 Values of central temperature using three points with implicit scheme for Example 9.4.

Fo	t	D			
		0.5	1.0	2.0	T_{exact}
0.125	3.2e-06	62.5	62.5	62.5	48.596
0.25	6.4e-06	62.5	62.5	62.5	58.451
0.5	1.28e-05	62.5	62.5	62.5	62.157
1.0	2.56e-05	62.5	62.5	62.5	62.498
2.0	5.12e-05	62.5	62.5	62.5	62.5

Let us now increase the number of points to five within the domain. Notice now that we have T_2^n, T_3^n, T_4^n as unknown points within the domain. At each time step, we see from Eq (9.49) that we have to solve a simultaneous system of equations for T_2, T_3, T_4. The equation is:

$$\begin{bmatrix} 1+2D & -D & 0 \\ -D & 1+2D & -D \\ 0 & -D & 1+2D \end{bmatrix} \begin{Bmatrix} T_2 \\ T_3 \\ T_4 \end{Bmatrix}^{(n)} = \begin{Bmatrix} T_2 \\ T_3 \\ T_4 \end{Bmatrix}^{(n-1)} + D \begin{Bmatrix} T_1 \\ 0 \\ T_5 \end{Bmatrix}^{(n-1)}$$

For the first time step, this looks like

$$\begin{bmatrix} 1+2D & -D & 0 \\ -D & 1+2D & -D \\ 0 & -D & 1+2D \end{bmatrix} \begin{Bmatrix} T_2 \\ T_3 \\ T_4 \end{Bmatrix} = \begin{Bmatrix} 25 \\ 25 \\ 25 \end{Bmatrix} + D \begin{Bmatrix} 25 \\ 0 \\ 100 \end{Bmatrix} \qquad (9.54)$$

For various values of D, the solutions are as follows (Table 9.7):

Notice that we are able to use much larger time steps and also obtain good accuracy with the implicit scheme. This is particularly advantageous when we wish to obtain solutions in problems with very small, inherent timescales.

Table 9.7 Values of central temperature using five points with implicit scheme for Example 9.4.

Fo	t	D			
		0.5	1.0	2.0	T_{exact}
0.125	3.2e-06	50.023	51.239	53.028	48.596
0.25	6.4e-06	58.017	57.991	58.096	58.451
0.5	1.28e-05	61.926	62.786	61.563	62.157
1.0	2.56e-05	62.491	62.482	62.458	62.498
2.0	5.12e-05	62.5	62.5	62.5	62.5

Summary of unsteady conduction

We have covered a lot of ground rapidly, so we will summarize the essence of numerical methods for unsteady conduction here.

1. The unsteady conduction equation is parabolic in nature. Therefore, it is useful to take one-sided differences in time and central differences in space.

2. It is possible to use either explicit (e.g., FTCS) or implicit (e.g., BTCS or Crank-Nicolson) methods for unsteady conduction.

3. For a single given time step, explicit methods are faster than implicit methods.

4. Explicit schemes are, however, restricted in terms of the size of the time step that can be taken. Typically, the time restriction is of the form $\Delta t \le D_{cri} \dfrac{\Delta x^2}{\alpha}$. For one-dimensional conduction $D_{cri} = 0.5$. Beyond this time step, explicit schemes are unstable and lead to "blow up."

5. Implicit schemes are usually highly stable and, in cases such as BTCS for conduction, they can be unconditionally stable. That is, one can take as large a time step as one desires while not blowing up. Though implicit schemes require solving a system of equations, they are often preferred for unsteady conduction problems, as the explicit time restriction can be prohibitively restrictive.

9.7 Introduction to methods for convection

While convection is an extremely important part of heat transfer, we will be only briefly talking about the numerical aspects here, as it requires a full-fledged discussion to be applied practically.

By means of introduction, we will note here that the discretization of convective terms such as $u\dfrac{\partial T}{\partial x}$ that occur within the energy equation requires special care. The reason is that the physics of these terms is different.

Recall that conduction is a diffusive process and that this diffusion does not distinguish one direction from the other. This is why our spatial derivatives in conduction are symmetric and central. However, convection *has* a preferred direction. This is precisely in the direction of local flow. Not obeying this aspect of physics typically causes instability in the numerical method (it is remarkable to see how often violating physics causes numerical problems).

So, convective terms are usually *upwinded*. That is, we take one-directional derivatives in the direction of local convection. For instance, if the flow is in the positive x direction, that is, $u > 0$, then $u\dfrac{\partial T}{\partial x}$ would be discretized as

$$u\frac{\partial T}{\partial x} \approx u_i \frac{T_i - T_{i-1}}{x_i - x_{i-1}} \tag{9.55}$$

On the other hand, if the flow is in the opposite direction, then $u < 0$, and we should write the derivative in the other way. We would obtain, therefore,

$$u\frac{\partial T}{\partial x} \approx u_i \frac{T_{i+1} - T_i}{x_{i+1} - x_i} \tag{9.56}$$

In other ways, discretization of the convective equation stays similar to everything else we have seen so far.

9.8 Practical considerations in engineering problems

As mentioned in the introduction to this chapter, numerical methods are now an indispensable part of industrial design, research, and development. Many industries use commercial software packages for heat transfer and flow predictions. The essentials even of these packages remain similar to what we have discussed here. However, in the interests of "making things as simple as possible but not any simpler," let us see how practical problems differ from the toy problems seen in this chapter. We will use the same framework we discussed earlier in Section 9.2 to highlight the additional considerations required in practice.

1. Mathematical modeling: Deciding on an appropriate mathematical model for a given industrial problem is often a matter of balancing detail with the computational expense. For example, if we are modeling a heat exchanger, to what physical accuracy do we model rust? More commonly, we need to select the appropriate level of turbulence modeling required for turbulent flows. Chemical reactions, phase change, mass transfer, etc., are additional complications. All these require detailed computations by means of providing additional PDEs or by using empirical correlations to model terms. In addition, modeling interface and boundary conditions are also, usually, not straightforward. In theory, it is always possible to compute everything (for example, by simulating every single molecule!), but in practice, a good engineer shines at this step by choosing a model that is expected to provide adequate accuracy in a reasonable amount of time. Software packages regularly ask the user to specify the required mathematical model for a problem within a prespecified number of choices. Some packages also have the option of providing new PDEs to model novel physical effects.

2. Creating a mesh: Unlike our toy problems that conveniently lived in a "rectangle world," real geometries are complex. Imagine, for instance, the interior of an engine! These do not lend themselves easily to the kind of simple, uniform mesh we have used so far. Usually, such complex domains require both

- Nonuniform meshes—grid spacing is not equal everywhere in space.
- Non-Cartesian meshes—we use shapes other than rectangles and cuboids to fill the space; for example, triangles and tetrahedrons fill complex domains more easily.

There are dedicated pieces of software for this step alone, and it is an important portion of industrial solutions. Obtaining a "good mesh" can often be the difference between an excellent and a useless solution.

3. Discretize the governing equation: The discretization we choose should be consistent with the mesh. If the mesh is non-Cartesian, we use finite volume or finite element methods instead of finite difference.

Further, as we discussed earlier, discretization must take into account the physics of the problem. Convection-dominated problems must have some form of *upwinding*. There are many other tricks of the trade that are applied at this step that enhance either speed or accuracy, or both. Choosing the appropriate discretization scheme is also given as a choice to the user in many software packages for this reason.

4. Solve the set of equations: There are a variety of methods that are used to solve the resultant set of linear or nonlinear equations. Often there is a balance between speed and accuracy that must be maintained. Some popular general choices (which we have not discussed here) include the multigrid method and the conjugate gradient method. These tend to work for large classes of problems with a high degree of speed and accuracy.

5. Postprocessing: All the effort above is all in the service of this step—obtaining insight from our computation. Computation is not all about numbers. It is about inferring something from the data.

Something that beginners in numerical methods often forget is to have a clear physical picture of what the final plots are expected to look like. Without this preliminary picture, it is easy to accept anything that the computer throws up.

It is also important to choose variables to display that will enable us to determine if anything has gone wrong in the previous steps. Choosing which primary variables, such as temperature and velocity, or derived variables, such as stress, vorticity, etc., to plot, monitor, and display is a matter of insight and engineering sense.

As you can see, practical problems follow the same overall framework as our problems here but differ in subtle (but important) ways. We hope that this chapter has whet your appetite enough to pursue a more detailed study of numerical methods.

Problems

9.1 Revisit problem 2.4 of chapter 2. Do the following.

    a. Write down the governing equation and the boundary conditions for this problem. (Hint: use symmetry at the center.)

    b. Discretize the governing equation using the finite difference formulation at any point i inside the wire.

    c. Take five points between the center and the outer rim of the wire. Write the discrete formulation for these points.

    d. Write the appropriate finite difference formulation of the boundary conditions.

    e. Solve the above system of equations using the Gauss-Siedel method.

    f. Write a computer program using 50 points between the center and the outer rim.

    g. Compare your results with the analytical solution.

9.2 Revisit problem 2.2 of Chapter 2. Do the following.
  a. Write the governing equation and the boundary conditions.
  b. Discretize the governing equation using the finite-difference formulation.
  c. What kind of system of equations you will get after discretization, Linear or Non-linear?

9.3 Revisit problem 2.10 of chapter 2. Do the following.
  a. Write down the governing equation for the fin temperature distribution, along with the boundary conditions.
  b. Discretize the governing equation using finite difference method.
  c. What will be the boundary condition at the tip of the fin?(Hint: use energy balance at the tip.) Write down the finite difference formulation for this boundary condition.
  d. Take five internal points along the length of the fin and write the finite difference equations for these points.
  e. Solve the above system of equation using the Gauss-Siedel method.
  f. Compare the solution with the exact solution.
  g. Write a computer program for this problem. Experiment with the internal number of points.

9.4 Consider a cylindrical pin fin with a length of 5 cm, a diameter of 12 mm, and a thermal conductivity of 15 W/mK that is insulated at its tip. The convection process is characterized by a fluid temperature at 30 °C and a heat transfer coefficient of 25 W/m²K. The base of the fin is at 100 °C. Do the following.
  a. Write down the governing equation along with the boundary conditions.
  b. Write down the finite difference formulation at any internal point i along the fin length.
  c. Take five internal points along the length of the fin and write down the finite difference formulation for these points.
  d. Write down the appropriate finite difference formulation for the tip of the fin.
  e. Solve the above system of equations using the Gauss-Seidel method.
  f. Write a computer program and experiment on the number of points. Comment on the results.

9.5 A square slab of dimensions 10 cm × 10 cm is very deep in the direction perpendicular to the plane of the paper. The slab is made of a material with thermal conductivity of $k = 15$ W/mK. Steady state prevails in the slab, there is no heat generation, and all the properties are assumed to be constant. The boundary conditions are given in the accompanying Fig. 9.6. Do the following.
  a. Write the governing equation for the temperature distribution inside the plate along with the boundary conditions.
  b. Discretize the above equation for any internal point (i,j) in the plate.
  c. Take a 2 × 2 grid with equal Δx, Δy inside the plate. Write down the finite difference equations for these points.
  d. Solve the above set of equations using the Gauss-Siedel method for five iterations.

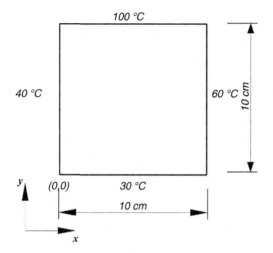

FIGURE 9.6

Schematic Representation of problem 9.5.

    e. Write a computer program using a 10 × 10 grid. Decide upon an appropriate stopping criteria for the iterations.

    f. Compare the temperature distribution with the analytical series solution.

    g. Experiment with the number of grid points. Comment on the results.

9.6 Consider steady two-dimensional heat transfer in a square metal. The temperature at the boundaries is prescribed to be 300 K on all sides except the top where the temperature is maintained at 400 K. The thermal conductivity of the body is $k = 180$ W/mK. There is heat generation of 10^7W/m^3 in the slab. Using the finite difference method with a mesh size of $\Delta x = \Delta y = 0.1$ m, determine the temperatures at nodes 1, 2, 3, and 4. Perform iterations with the Gauss-Seidel method till $\mathbf{max(T_{ij}^{(n+1)} - T_{ij}^{(n)})} < 10^{-4}$, where the superscript n is the iteration number. Take the initial guess for the temperature at all nodes as 375 K.

9.7 The time constant of a K-type (Chromel-Alumel) thermocouple of diameter 0.70 mm has been determined to be 1 s, and the temperature indicated by the thermocouple at 0.5 s is 57.54 °C. The specific heat and density of Chromel-Alumel are 420 J/kgK and 8600 kg/m^{3}respectively, and the temperature of the medium is 100 °C. Using the lumped heat capacity model, do the following.

    a. Write down the governing equation for this problem.

    b. Use a first-order, finite difference approximation for the time derivative at any time t.

    c. By taking $\Delta t = 0.05$ s, calculate the temperature until t = 1s.

    d. Using the same Δt, march backward in time and find the initial temperature of the thermocouple.

    e. Plot the temperature with respect to time and compare it with the analytical solution.

9.8 Consider a slab of thickness $L = 1$ m with walls maintained at temperatures $T_L = T_R = 0\,°C$. The initial temperature distribution in the slab is given by $T(x) = 4x - 4^2(x)$. The slab's diffusivity is $\alpha = 0.1$ m^2 s^{-1}. The slab is now allowed to cool for the next 1 second in a quiescent environment. Based on the information given, answer the following questions.

a. Write the governing law, boundary conditions, and initial condition for this problem. Specify the assumptions you made.

b. To solve this problem with FDM, we discretize the domain into five uniformly spaced points along x and t axes, respectively, that is, $\Delta x = \Delta t = 0.25$. Using FTCS, write a computer program to calculate the temperature distribution of the slab after 1 second.

c. Recall that truncation errors in the approximation of derivatives depend upon the step sizes. It seems logical to reduce the step sizes for Δx and Δt to get a more accurate solution. Therefore run the previous computer program with reduced step size of 0.1, that is, $\Delta x = \Delta t = 0.1$. Do you get some counterintuitive results? If yes, explain why.

d. Repeat previous problems with the Crank-Nicolson method. Do you observe any nonphysical results with the reduced grid size? Explain why or why not.

Machine learning in heat transfer

10

10.1 Introduction

This chapter introduces how to apply machine learning to heat transfer. It is practically impossible for one not to have heard of machine learning. The buzz and hype surrounding machine learning seem to have reached such a fever pitch that it is being connected (perhaps forcibly) to everything from grocery shopping to preventing pandemics! Given the hype, it is natural to ask, is the application of a "trendy" field like machine learning to a classical subject such as heat transfer forced or natural? It turns out that this application is quite natural and, in fact, has been explored quite vigorously by a section of the heat transfer community over the last 15 years.

This is usually treated as an advanced research topic. Still, due to its increasing practical importance, we have included this chapter to give readers a very brief overview of the ways in which machine learning has and can be used in heat transfer. We expect this topic to grow in importance, especially in engineering practice. The purpose of this brief chapter is to serve as an appetizer for future explorations of engineering heat transfer.

10.2 Physics versus data methods

We have implicitly used two approaches in our study and modeling of heat transfer thus far. The first approach is to use algebraic or differential equations to describe the problem at hand. These differential equations are derived, in turn, from conservation laws, and we call them *physics-based models* for this reason. Examples of this are the ordinary differential equations (ODEs) governing fin heat transfer or partial differential equations (PDEs), such as the Navier-Stokes equations, and the energy equation governing general flow and heat transfer.

A second modeling approach involves correlations and is the one we used heavily in our convection studies. These models were driven primarily by empirical, experimental studies and depended on past data. We call these *data-based models*. Correlations such as Churchill-Chu, Dittus-Boelter, and others belong to this class of models. In this sense, the reader may have already seen the tremendous importance that data-driven models have in the practical engineering of heat transfer. Let us see further how these data models may be formally designed, and what other uses such data-driven models may have.

Heat Transfer Engineering. http://dx.doi.org/10.1016/B978-0-12-818503-2.00010-1

10.2.1 **Physics and data in heat transfer**

Recall from the earlier chapters that there are three main purposes for the study of engineering heat transfer. (See Bejan (1993) for a fuller discussion on this). These are:

1. The increase of heat transfer from or to a system.
2. The decrease of heat transfer from or to a system.
3. The control of temperature in a system.

All of these require us to model the system in order to understand the relationship between the parameters of the system and the quantity we wish to control. Once we understand this relationship, we can proceed to find out how to design systems, create devices, and develop ideas that control heat transfer; this is the essence of the entire "enterprise" of engineering heat transfer.

However, there is an important wrinkle in the above process. In engineering practice, more often than not, it is very difficult to obtain the above relationship between the parameters and variables analytically. For instance, let us say the heat transfer in a particular system such as a heat exchanger is given by:

$$Q = f\left(x_1, x_2, ..., x_N\right) \tag{10.1}$$

Here, Q is the heat transfer, which we wish to maximize, and x_i are the parameters that affect it. These parameters include density, viscosity, and other properties of cold and hot fluids; time at which we are calculating heat transfer; and geometric parameters, such as size of the pipes, number of tubes, arrangement of tubes, roughness, emissivity (if radiation is not negligible), laminar or turbulent flow, etc. As one can see, the number of parameters in a practical problem can become explosively large. Under such circumstances, solving the governing equations analytically to obtain the above relationship is impossible, even in theory.

An alternate possibility is to obtain the heat transfer for a given system configuration computationally by solving the governing equations. We saw this approach in the chapter on numerical methods (Chapter 9), and it is, in fact, standard engineering practice to use computations to study complex engineering systems. However, even for a single system configuration, this approach can often be extremely expensive in terms of the time and computational resources required for computation. When it comes to the design and optimization of thermal systems, we require multiple computations to evaluate different system configurations to design a single system. The computational cost for this can be prohibitively substantial.

This deadlock can be broken if we can somehow exploit our past experience of design, experiments, and computations on similar systems. This way, we would not have to recalculate every configuration from scratch using fundamental conservation laws. Statistical and computational techniques that allow us to do this are abstracted into the single umbrella of data-based models. The key in such data-based models is to collect and learn from our past computations, experiments, and designs. The central point is to note that the data is not divorced from the physics. On the contrary, the data embeds all of the physics, often much more than a model can.

A very early example of such data-based models is the charts and correlations we have copiously used in our book so far that are an essential part of engineering practice. For example, the Dittus-Boelter correlation is a "learned" correlation based on massive amounts of observations made on flow and heat transfer in pipes. Notice the advantage Dittus-Boelter has over a purely physics-based model. While it is approximate, we can still quickly calculate the heat transfer without doing detailed simulations. The difference in computational time between a full, physics-based computational solution and Dittus-Boelter can be as much as years vs. seconds. Consequently, we can quickly design systems for optimal heat transfer by using the data-based relationship instead of painstakingly computing the flow for every configuration. This relationship can also work as a "surrogate model" for solving inverse heat transfer problems, such as finding material properties, given the heat flux or temperature distribution, which has been mentioned earlier in the book.

So, data-based models can be exploited to:

1. Speed up computations of complex configurations.
2. Speed up the design cycles of thermal systems.
3. Speed up solutions of inverse problems in heat transfer.

While we can see the use of data-based models, Dittus-Boelter and similar correlations exist only for specific configurations. If we wish to build a data-based model for a new configuration, we need to have a process or algorithm that can come up with a correlation between the variables and the parameters. In the context of engineering heat transfer, machine learning gives us such algorithms. An essential algorithm in this family is the artificial neural network (ANN) family. However, for context, before we look at ANNs, let us take a brief look at machine learning.

10.2.2 Artificial intelligence and machine learning

Machine learning or data-based methods are often conflated using the more general term of artificial intelligence. For clarity, it is useful to distinguish between these two terms.

Artificial intelligence: This is a subfield of computer science. In the historical sense of the word, even traditional pieces of software such as commercial computational fluid dynamics solvers may be said to use artificial intelligence. In a more modern sense, artificial intelligence refers to techniques that try to automate portions of human cognition, decision, or action. In our opinion, this term is unnecessarily mystified. As of the writing of this book, artificial intelligence is simply an umbrella term for a family of formal algorithms of varying complexity and efficacy that help in automation.

Machine learning: Machine learning is a subset of artificial intelligence and refers to algorithms that improve with more data. An example of a (currently nonexistent) machine learning system for heat transfer would be a commercial flow solver that automatically improves accuracy or speed as we make it compute new flows. Similarly, imagine that we have a software version of the Dittus-Boelter correlation that can be continuously fed inputs with new experimental data on pipe flows with heat transfer and keeps improving its predictions. This would be a machine learning system, and you can imagine how useful it would be. While nothing of this sort currently exists, we will see shortly that the process is well understood theoretically and easily within the realm of possibility in the coming years. We cover the simple basics of this idea in the following sections.

10.2.3 Common algorithms in machine learning

As mentioned previously, machine learning refers to a collection of algorithms that can draw better inferences as one increases the amount of data. There is a confusingly large set of algorithms within this. Still, one can broadly classify these algorithms into three rough categories depending on the type of learning.

Supervised learning: This refers to a "labeled data set" where data is specified as input/output pairs. "Label" here refers to the explicit specification of data in the output for any given input. For example, if one collects pipe heat transfer data and provides Re, Pr (the inputs) as well as Nu_D (the desired output), the algorithm will try and find an optimal function relating the input to the output. In modern machine learning, speech recognition and face recognition are usually treated as supervised learning tasks.

Unsupervised learning: This refers to an "unlabeled data set" in which we provide data and expect the machine to learn interrelationships or possible groupings without being given any prior information. Here, "unlabeled" refers to the fact that data is simply provided as inputs without giving any example output data. A common use for this is the unsupervised, automatic separation of data into clusters with similar behavior. For example, if we provide the unlabeled heat transfer data for a system under both turbulent and laminar conditions without being *a priori*—which flow is which—the unsupervised algorithm could potentially identify the existence of two different clusters of physics. This can be very useful when we suspect different physical regimes within a problem, but the data is too large for us to draw inferences manually. In commercial machine learning, the same principle is used for customer clustering, etc.

Reinforcement learning: When the machine has to learn an optimal sequence of steps with only a single end "reward" to go by, it is called reinforcement learning. This is potentially useful for control algorithms where the end "reward" could reduce drag, reduce turbulence, etc. For example, recent studies, such as Rabault et al. (2019), show that it is possible to determine optimal strategies for active flow control and reduce drag using reinforcement learning. Methods such as this could potentially help us in situations where traditional control strategies have difficulty due to complex physics. These could be applications such as creating shape morphing fins or wings that can adapt their shape to control drag and heat transfer. In modern machine learning, the same principles are used to create impressively strong chess algorithms and game players.

Of all the above learning approaches, supervised techniques are the most developed and utilized. Within supervised learning techniques, artificial neural networks

(ANNs) account for a large part of the modern machine learning revolution. Even within heat transfer, ANN-based techniques are the most commonly applied ones. For this reason, within this book, we will be focusing only on neural networks and their applications in heat transfer.

10.3 **Neural networks for heat transfer**

We saw above that, in engineering applications, we need to learn the relationships between the parameters that influence a problem and the output quantity such as heat transfer. Mathematically, the relationship to be discovered is expressed as a function, as in Eq. (10.1). In cases where we cannot find the exact function analytically (which is more the norm rather than the exception in heat transfer), we can approximate it using some data modeling technique. As mentioned in the previous section, such tasks fall under the category of supervised learning problems. Artificial neural networks (ANNs) are the most flexible among supervised learning techniques. In this section, we will see how, given sufficient data, ANNs can be used as a powerful and universal function approximation method.

A note on terminology: Some readers may have also seen the commonly used terms "deep learning" or "deep neural networks (DNNs)." While, technically, they are a subset of machine learning and a superset of neural networks, for all intents and purposes, heat transfer engineers may use these terms interchangeably with ANNs.

10.3.1 **The learning paradigm**

Suppose we are trying to determine an expression for the temperature at a given point in a long fin as a function of various physical parameters. Through a careful non-dimensionalization of the governing equation and the boundary conditions, we can come to the conclusion that this would be an unknown function of the form $\theta = f(\theta_b, m, x)$, where $\theta = T - T_\infty$ is the temperature (difference) at the location x, $\theta_b = T_b - T_\infty$ is the base temperature (difference), and $m = \sqrt{hP/kA}$ is a parameter. Recall that we determined the exact function f relating these quantities purely through physics in Chapter 2 of this book. But, if we have little to no knowledge of physics, or if the physical model is too expensive to compute, we could collect lots of prior data and try to fit an approximate model to the data instead of relying on physics.

The above would be an example of a supervised learning task. The general version of this is to find an approximate function relating the parameters of a problem to the output, purely from prior data. Machine learning has many possible algorithms to solve such general supervised learning tasks. Most of these algorithms, including neural networks, follow a general "learning process."

Mathematically, the process simply finds a way to optimally relate inputs and outputs given past data pairs. For example, for the fin problem, our output is the temperature θ at a given location and the inputs are the base temperature θ_b, location x, and fin parameter $\sqrt{hP/kA}$. Mathematically, we denote the output variable by y and the input variables by x. Note that both x and y can have multiple components and can be thought of as "vectors." In our fin problem, for instance, $x = (\theta_b, x, m)$ while $y = \theta$ is a scalar.

We can now formally define the learning task as follows: ***Given sufficient example*** (x, y) ***pairs, find a function*** $y = f(x)$ ***that approximates this dataset as well as possible.***

The above task is known in the statistical literature as a **regression** task. You might have heard of linear regression. Even if not, you would certainly be familiar with drawing "trend" or visual best fit lines for some given (x, y) pairs on a two-dimensional graph paper. Analogously, we may view our general learning task as "learning" some best fit curve between higher-dimensional inputs and outputs.

The learning process
We can now outline the learning process as including the following steps:

1. Mathematical formulation
2. Selecting the hypothesis function
3. Data collection
4. Learning optimum parameters
   a. Initializing parameters
   b. Forward prediction pass
   c. Feedback for finding the optimum parameters

Let us now look at these steps in detail.
Step 1: Mathematical formulation—In this step, we decide the appropriate inputs **x** and outputs **y** for the problem. This is obvious in some cases, such as in our fin example above and slightly difficult in other cases, such as cases where we need to specify geometrical shape as an input. This step often requires engineering judgment and is an important place where we can incorporate our knowledge of physics, by appropriate selection and non-dimensionalization of the inputs and outputs.
Step 2: Selecting the hypothesis function—Though we are trying to find the "best" function that fits all the past data, we cannot search in the infinite space of *all* possible functions. To limit the search, we prespecify the *form* of the function. For example, if we have a function of one variable $y = f(x)$, we could prespecify the form as being linear. This would mean that the function is fixed to be in the form $y = w_0 + w_1 x$. In this expression, the form of the function is fixed but the *parameters* w_0 and w_1 are unknown and variable.

Notice now that every function can be thought of having two conceptual parts—a fixed form and some variable parameters. Consider another example. Say y is a function of a vector input with two components, that is, $y = f(x) = f(x_1, x_2)$, and we hypothesize that it can be approximated as $y \approx w_0 + w_1 x_1 + w_2 x_2 + w_3 x_1^2 + w_4 x_2^2 + w_5 x_1 x_2$. In this case, the form of the function is fixed to be a quadratic in two variables and the parameters are $w_0, w_1, w_2, w_3, w_4, w_5$.

The functional form we choose to approximate the data is called the **hypothesis function.** The fundamental trick in learning is to *fix the form of the hypothesis function and learn the best possible parameters that fit the data.*

Note that, in machine learning, an important part—the form of the function—is specified by us. The learning algorithm only learns the parameters, not the form. As we will see, neural networks may also be thought of as hypothesis functions for which the form is called the "architecture," and we have to learn the parameters called weights.

In general, we are free to choose any hypothesis function whatsoever—linear, quadratic, neural network, or some other function we wish to create. Regardless of this choice, the learning process that follows from here will work.

We will denote the hypothesis function relation by the notation

$$\hat{y} = h(x; w) \tag{10.2}$$

Notice the following points implicit in the notation here; $\hat{y}$ denotes the approximate model output and should be distinguished from the actual data that exists in reality (see the next step). The hypothesis function depends both on the form and the parameters. The form is denoted by the part before the semicolon in $h(x; w)$ and the dependence on the parameters by the w following the semicolon.

Step 3: Data collection—It is useful in any learning task to collect as much relevant past data as possible. We need to collect past data as (x; y) pairs. The output collected here y is called the **ground truth,** in order to distinguish it from the model or approximate output, which will be produced by our hypothesis function.

It is usual in this step to split the data into **training data** (typically 70 % of the data) and **testing data** (the remaining 30 %). The training data is used to learn the best parameters, and the testing data is used to evaluate how well our model performs.

We can think of it this way – If a professor gave some example heat transfer problems to "train" her students to learn, she could not use the same problems to test you in the exam. Your performance on the previously seen problems would not be a fair evaluation of your true knowledge. The professor would have to keep some unseen test problems in order to truly evaluate your performance. Similarly, we keep the training and testing datasets separate for learning and evaluation, respectively.

Step 4: Learn optimum parameters—Recall that, in Step 2, we already fixed the form of the hypothesis function that we wish to learn. All that remains now is to determine the "best" parameters for the hypothesis function that fit the training data collected in Step 3. In rare cases, such as linear hypothesis functions, it is possible to directly obtain these optimum parameters using analytical techniques. Usually, however, we have to follow a few iterative substeps in order to converge to the optimum parameters numerically.

1. **Initialize parameters**—We first initialize the values for the parameters in this substep. It is understood and intuitive that better initializations will lead us to the optimal parameters faster. However, determining good initial parameters is an

FIGURE 10.1

Schematic of the forward prediction pass. The unknown function we are trying to approximate is represented by a *black box*.

open research problem. Sometimes, our engineering knowledge of the problem may give some bounds, etc., on the parameters to be used. If not, we will assume some suitably bounded random initial values for the parameters.

2. **Forward prediction pass**—Now that the initial guessed values of the parameters w are available, we can find the predictions made by our model hypothesis function for every input data point. For example, if our hypothesis function was $\hat{y} = h(x) = w_0 + w_1 x$ and we gave initial guess for $w = (w_0, w_1)$ as $(w_0 = 0.25, w_1 = 1)$, then for an input data point with $x = 0.2$, we would obtain the model prediction as $\hat{y} = 0.25 + 1 \times 0.2 = 0.45$. We can similarly find our model predictions $\hat{y}$ for every training data x. Fig. 10.1 shows the schematic of the forward prediction pass described above.

3. **Feedback for optimum parameters**—The prediction $\hat{y}$ from the previous step will not, in general, match the ground truth y. For example, in the example given above, our prediction for $x = 0.1$ is $\hat{y} = 0.45$, but the ground truth could have been $y = 2$. In general, since our initial guess for the parameters was random, the model prediction $\hat{y}$ will be vastly different from the ground truth y for most, if not all, points. There is, therefore, a prediction error. We can intuitively recognize that "feedback" has to go to the parameters and change them to reduce this error.

For this, we first need to define the error $J(y, \hat{y})$, which measures the difference between the prediction and the ground truth. This error is also called the **cost function** in machine learning literature. A very common error measure is the **mean squared error,** defined as

$$J = \frac{1}{m} \sum_{i=1}^{m} (y_i - \hat{y}_i)^2 \tag{10.3}$$

where m is the number of training points available in the dataset.

Having defined the error, we can now attempt to keep changing the parameters until we minimize the error. Therefore, our original learning problem can be posed as the following optimization problem: "For what values of the parameters w is the error J between the model prediction $\hat{y} = h(x; w)$ and the ground truth y the minimum?"

This optimization can be performed analytically in some rare cases (such as linear hypothesis functions), but is usually achieved computationally by means of numerical optimization algorithms. The simplest of these numerical algorithms is called **gradient descent**, and the parameter update step is given by

$$w^{new} = w^{old} - \alpha \frac{\partial J}{\partial w} \tag{10.4}$$

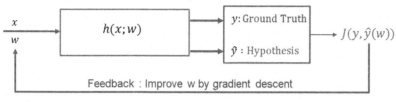

FIGURE 10.2

Schematic of the learning process.

This is a user-specified parameter called the **learning rate**, and it affects the speed at which the parameters converge to the minimum. The above iteration for w is continued till we reach a prespecified convergence limit (that is, our w^{old} and w^{new} are very close).

One should not be intimidated by this step, as we do not really need gradient descent explicitly in the rest of this chapter. We have only provided it so that you can be familiar with this commonly used term. It suffices to understand here that there are canned algorithms (such as gradient descent) that will perform the optimization once the expression for the error J and the parameters w are given. The interested reader can refer to any text on optimization such as Balaji (2019), for further details on the gradient descent method and its derivation.

A schematic of the iterative learning process (Step 4) is provided in Fig. 10.2. It is worth emphasizing that the same learning process can be used for various learning tasks and hypothesis functions.

10.3.2 **Linear regression**

As mentioned above, neural networks can be seen as sophisticated regression techniques. Before seeing how neural networks work, we will describe the simplest possible regression technique—linear regression—and see how it can be seen as a learning algorithm under certain circumstances.

Linear regression is, in essence, a regression algorithm with a linear model as the hypothesis function. While we have included this model in the current chapter as a stepping stone for neural networks, it is also independently useful in heat transfer. For instance, there are many situations in conduction where we encounter a linear model. Recall that for one-dimensional, steady-state conduction across a slab with constant properties and no heat generation, the temperature profile is linear in the axial direction. Other linear relations also occur in a "hidden" form within heat transfer. For instance, correlations involving the Nusselt number are power laws which will look linear when logarithms are taken.

As mentioned above, for single variable inputs and outputs x, y the linear hypothesis model is

$$\hat{y}_i = w_0 + w_1 x \tag{10.5}$$

Let us say that for a single variable problem we collect m data points (x_i, y_i) for $i = 1, 2...m$, and have to find the "best" line that passes through them. Based on the discussion in the previous section, this line would have to minimize the mean squared error or cost function given by

$$J = \frac{1}{m} \sum_{i=1}^{m} (y_i - \hat{y}_i)^2$$

$$= \frac{1}{m} \sum_{i=1}^{m} (y_i - w_0 - w_1 x_i)^2$$

(10.6)

At this point, our optimal parameters for the hypothesis are those that minimize the above sum. These optimal parameters can be calculated in one of two ways:

1. **Direct optimization**—This involves solving for the parameters directly by imposing the following conditions at the minimum point

$$\frac{\partial J}{\partial w_0} = 0$$

$$\frac{\partial J}{\partial w_1} = 0$$

(10.7)

Applying the condition (10.7) to the cost function given by (10.6) results in the following analytical expressions for optimal parameters (the proof is left as an exercise for the reader; see Problem 10.1):

$$w_1 = \frac{m \sum x_i y_i - \left(\sum x_i \right) \left(\sum y_i \right)}{m \sum x_i^2 - \left(\sum x_i \right)^2}$$

$$w_0 = \frac{\sum y_i - w_1 \sum x_i}{m}$$

(10.8)

where all summations $\sum$ should be read as $\sum_{i=1}^{m}$.

2. **Learning or iterative optimization**—The above method of direct optimization has two important disadvantages. Firstly, it works only for simple, linear, or near-linear hypothesis functions. Secondly, every time we add a new data point, we would have to recalculate w_0, w_1 from scratch using Eq. (10.8). There is no way for us to utilize our previously calculated parameter values and improve them. The attentive reader would recognize the above as precisely the learning problem!

The solution to this learning problem was given before in Step 4 of the learning process. This solution is to initialize the parameters and iteratively keep refining them as new data comes in. We can use gradient descent or any of the multiple optimization algorithms. Within this text, we will be assuming that some effective algorithm (direct or iterative) has been used to obtain the optimized parameters and will be skipping the details. The interested reader is advised to look at specialized machine learning texts such as Goodfellow et al. (2016) or optimization texts such as Balaji (2019) for further details on optimization algorithms.

Example 10.1: *A long mild steel slab of thickness 10 cm having a thermal conductivity of $k = 44.5W/mK$ can be considered to be insulated on the top and bottom sides. There is no heat generation in the slab, and thermophysical properties can be assumed to be constant. The steady-state temperature recorded by the thermocouples at five selected locations inside the slab is given below. Determine the steady-state heat flux in the slab using appropriate assumptions. Use the direct method for finding the optimal parameters. The slab is $1\,m^2$ in the area in the direction normal to the heat transfer (Table 10.1) Fig. 10.3).*

Table 10.1 Steady-state temperature distribution recorded by thermocouples for Example 10.1.

S. no.	x (in cm) from left end	T (in °C)
1	1	34.51
2	3	34.16
3	5	32.50
4	7	32.24
5	9	30.41

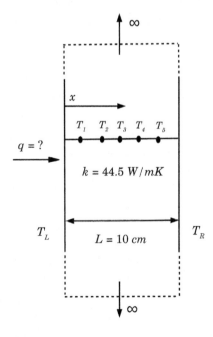

FIGURE 10.3

Schematic diagram for Example 10.1.

Solution:

At steady state, the flux in this problem is constant across the slab. In particular, $q = -k\dfrac{dT}{dx}$. The question, therefore, boils down to finding $\dfrac{dT}{dx}$.

The chief difficulty here is that we have only discrete values of the temperature. Further, as we will see below, the measured temperatures do not fall exactly on a straight line as expected, theoretically, in case of the temperature in a slab. This is because of "experimental errors" including noise in the thermocouple measurements. Therefore, we resort to finding the "best" line that fits this data. This now falls within the framework of the learning process we listed above.

Let us now follow the template given in the learning process in order to solve this problem.

Step 1: Formulation—The input and output variables for this problem are obvious. We choose x as the input and $y = T$ as the output variable.

Step 2: Hypothesis—Before imposing a hypothesis function, we plot the given data points to see if we can observe a trend. Fig. 10.4 shows the plot of how our "output," the temperature, varies with the "input" x, the location. We can immediately observe that the temperature variation has a decreasing trend that is roughly linear. Of course, we also know this from our knowledge of physics of the problem. A good hypothesis function, therefore, would be the linear function

$$\hat{y} = h(x;w) = w_0 + w_1 x \tag{10.9}$$

Step 3: Data collection—The data is already collected in this problem and given the size of the data (just 5 points), it does not make sense to split it further into training and testing sets.

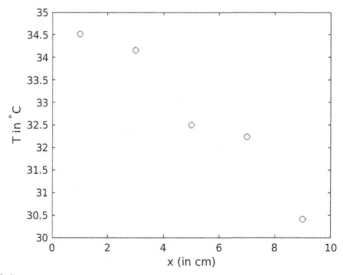

FIGURE 10.4

Location versus temperature in example 10.1.

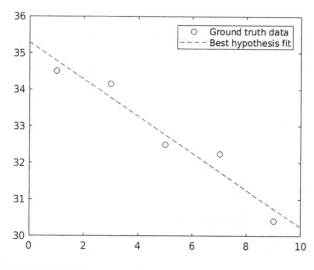

FIGURE 10.5

Location versus temperature in example 10.1.

Step 4: Optimal parameters—The hypothesis function is linear and hence the parameters can be computed using the expressions given in Eq. (10.8). Calculating these (see Exercise 10.2) results in

$$w_0 = 35.29 \quad w_1 = -0.5060 \tag{10.10}$$

Since $\hat{y} = w_0 + w_1 x$, the proposed model is

$$T(x) \approx 35.29 - 0.506x \tag{10.11}$$

A plot of the above best fit along with the original thermocouple data is shown in Fig. 10.5. One can notice that even though the model fit does not pass through *any* of the data points, it still fits the data quite well.

Now, $\dfrac{dT}{dx} = w_1 = -0.506 \ ^{\circ}C/cm$. Therefore, the heat flux is given by

$$q = -k\frac{dT}{dx} = 44.5 \times 0.506/10^{-2} = 2251.7W/m^2 \tag{10.12}$$

10.3.3 Neural networks

In the above example, we used a linear hypothesis function, as our knowledge of the physics of the problem warranted it. But what about highly nonlinear phenomena involving multiple variables? For a long time, this was solved either by some high order polynomial or power-law regression. These methods work well when there is only a small amount of data or when the nonlinearity is not severe. However, in modern times, with "big" data available, there is scope to apply more complex and highly nonlinear hypothesis functions. One such class of hypothesis function that is very powerful in the nonlinearities it can express is the artificial neural network.

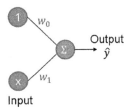

FIGURE 10.6

Pictorial representation of the linear model.

The earliest artificial neural networks were biologically inspired. McCulloch and Pitts in 1943 were among the first to try and abstract the processes in a biological neuron to their mathematical and computational equivalents. In this early view of the human mind, the brain was simply a computational device that discerns and learns patterns through unit computational processes taking place in a simple computational unit—the neuron. Whether this view is accurate or not is a subject of much debate, but the outcome of the work of several brilliant minds has been the refinement of neural networks as a simple and powerful technique for approximating highly nonlinear relations. In this book, we avoid the biological analogy (and the ensuing controversy) and simply view the neural network as a powerful class of hypothesis functions.

To understand how neural networks function, consider a pictorial representation (Fig. 10.6) of the linear hypothesis function we discussed previously.

Notice the following features of the figure:

- Each circle gives as output a scalar variable. We call these circles **neurons**. These are the fundamental building blocks of neural networks.
- The computation takes place sequentially from left to right. We call the first layer of neurons the **input layer** and the final layer as the **output layer.**
- The neurons in the input layer are multiplied by the parameters represented by the connecting lines. These parameters w are called **weights**. Biologically, they are supposed to represent the strength of the connections between neurons. Mathematically, it is clear that the higher this weight between two neurons, the stronger the connection between these two "concepts" or variables.
- The final neuron has both inputs and outputs. And the output is given by a **weighted sum** of the input: $\hat{y} = w_0 \times 1 + w_1 \times x$.

We will call the above type of neuron in a linear model a linear neuron. The linear neuron gives as output simply the weighted sum of the inputs $\hat{y} = w_0 + w_1 x$. For convenience, we use the notation that $x_0 = 1$ and $x_1 = x$. The constant "neuron" with $x_0 = 1$ is called the **bias unit**. We can now write the output as

$$\hat{y} = w_0 x_0 + w_1 x_1$$
$$= \sum_{i=0}^{n-1} w_i x_i \tag{10.13}$$

where n is the number of neuron connections coming in. That is,

$$\text{Output of Linear Neuron} = \sum \text{Weight} \times (\text{Input to Neuron}) \tag{10.14}$$

The neural network works on a very similar principle as the linear model with just two small modifications:

1. Nonlinearity
2. Hidden layers

Let us look at these modifications in detail.

Nonlinearity—If we have only linear operations in the neuron, we can only represent linear functions. In order to represent nonlinear functions, we just need to make a small change to the linear neuron output given by Eq. (10.15).

$$\text{Output of Linear Neuron} = g\left(\sum \text{Weights} \times (\text{Inputs to Neuron})\right) \tag{10.15}$$

where g is some nonlinear function. This function g is known as the nonlinearity or the **activation function,** because biologically it was supposed to represent the fact that a neuron is activated nonlinearly based on the sum of its incoming signals. We represent the output of this nonlinear neuron as $\hat{a}$. This output can be seen as the combination of a linear and nonlinear step as follows:

$$\textbf{Linear Step } z = \sum_{i=0}^{n-1} w_i x_i \tag{10.16}$$

Nonlinear step $\hat{a} = g(z)$

In summary, we can write the output of the neuron as $\hat{a} = g\left(\sum_{i=1}^{n-1} w_i x_i\right)$. Note that this is the most general expression for any neuron in any neural network, no matter how complicated. A very common choice for the nonlinear activation function is the so-called **sigmoid function** and is calculated as

$$\sigma(z) = \frac{1}{1 + \exp(-z)} \tag{10.17}$$

where σ is the symbol representing the sigmoid. Fig. 10.7 shows the variation of the sigmoid with z.

It can be seen both mathematically and graphically that the sigmoid nonlinearity varies between 0 and 1. The output nearly vanishes for highly negative inputs and asymptotes to 1 for highly positive inputs. This is supposed to represent the biological "waking up" or activation of the neuron. Small or negative signals are insufficient to activate it, while large signals make the neuron fire. Regardless of the biological origins, the sigmoid is a commonly used nonlinearity, and, for the rest of the chapter, it will be assumed that whenever we use g, we mean the sigmoid nonlinearity σ. Let us now consider an example of the computation inside an artificial neuron.

Example 10.2: *Fig. 10.8 shows a single neuron in a neural network. The inputs to this neuron are shown in the figure. Assuming that all connections have a weight of 1, and that the nonlinearity is a sigmoid, determine $\hat{a}$, the output of the neuron.*

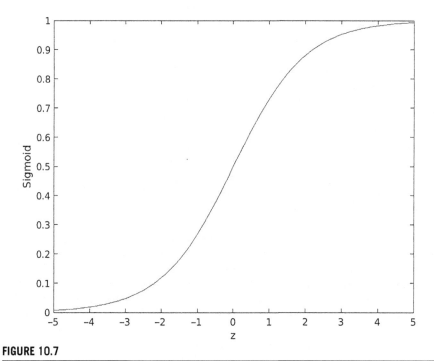

FIGURE 10.7

Output of the Sigmoid nonlinearity.

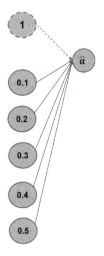

FIGURE 10.8

Figure for Example 10.2.

Solution: All the weights are given to be $w_i = 1$. The output is given by $\hat{a} = g\left(\sum_{i=1}^{n-1} w_i x_i\right)$. For clarity, we calculate this in two steps.

Linear step—For convenience, we will use our earlier notation that $x_0 = 1$.

$$z = \sum_{i=0}^{5} w_i x_i$$
$$= w_0 x_0 + w_1 x_1 + w_2 x_2 + w_3 x_3 + w_4 x_4 + w_5 x_5 \qquad (10.18)$$
$$= 1 + 0.1 + 0.2 + 0.3 + 0.4 + 0.5$$
$$= 2.5$$

Nonlinear step

$$\hat{a} = g(z)$$
$$= \frac{1}{1 + \exp(-2.5)} \qquad (10.19)$$
$$= \mathbf{0.9241}$$

As you see, the process is not complicated at all! It is just a nonlinear function applied over a linear model. Nonetheless, the above process is essentially what takes place in every neuron of every neural network, no matter how complex. The only difference would be the nonlinear function being used. In fact, the sigmoid is good enough for most practical purposes.

Hidden layers—The second difference between the linear model shown in Fig. 10.6 and a full neural network is the existence of intermediate layers called "hidden" layers. In the linear model, there were only the input and output layers. However, it can be shown that there are a large number of functions that cannot be approximated by this structure.

The solution is to add intermediate calculation layers in the middle with an arbitrary number of neurons. Fig. 10.9 shows a schematic of such a neural network. The

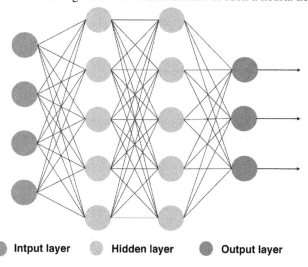

Intput layer   **Hidden layer**   **Output layer**

FIGURE 10.9

A typical deep neural network.

figure may look complicated, but note that the calculation of each neuron's output is *exactly* the same process as discussed in Example 10.2. Therefore, the calculation of any forward pass through the network is not complicated and can be easily programmed.

Note:

- There are two hidden layers in this network. Each hidden layer has 5 neurons.
- Each connecting line represents a weight. All the weights are parameters of the neural network hypothesis function.
- It is conventional in neural network diagrams not to show the "bias" unit (the unit with a constant neuron with a value of 1). This is the reason why the bias unit in Fig. 10.8 was shown with a dotted line. The lines emerging from the "bias" units are also similarly hidden. We must assume that these exist in every layer.
- The input layer has 4 neurons, meaning that there are 4 components in the input vector x. Similarly, there are 3 components in the output vector y. If we include the bias units, then are 5 neurons in the input layer and 4 in the output layer.
- Not including the bias weights, there are a total of $4 \times 5 + 5 \times 5 + 5 \times 3 = 60$ adjustable weights in this network. (How many weights are there including the bias weights?). If we did not have the hidden layer, we would have had only 12 weights. This is one of the purposes of the hidden layer—the more the number of adjustable weights, the more the number of functions we can approximate well.
- The forward pass through this network simply involves systematically calculating the output of each neuron from left to right and progressing through the network.
- Just like it was possible to write the linear model as $\hat{y} = w_0 + w_1 x$, it is possible to write an analytical expression for the neural network shown above. This analytical expression would look like $\hat{y} = NN(\mathbf{x}; \mathbf{w})$. Here, as discussed above, x and y have 4 and 3 components respectively and there are 60 components in w. The analytical expression for this network would obviously be very complicated! Nonetheless, it is possible to write it, and it is very insightful to think of the picture of a neural network as simply a graphical representation of a complicated mathematical function. This idea is used very effectively in an approach called physics informed neural networks by Raissi et al. (2019) for solving inverse problems.
- Since there are multiple weights, it makes sense to decide on some uniform notation denoting a given weight. For this, notice that there is a single line or weight connecting them for two neurons in adjacent layers. We will use the notation $w_{ij}^{(k)}$ to denote the weight of the ith neuron in the kth layer joining with the jth neuron in the next layer.

Given that the neural network is just two simple modifications away from a linear model, what gives it power, and why is it so widely used? The secret behind this is a powerful mathematical theorem called the universal approximation theorem.

Universal approximation theorem—Given sufficient data and neurons, a neural network with even a single hidden layer can approximate any function to any desired accuracy.

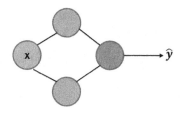

FIGURE 10.10

Figure for Example 10.3.

This theorem is the key to understanding why neural networks are possible. Regardless of whatever heat transfer problem we are approximating, whether laminar or turbulent, as long as we can collect sufficient data, and as long as we are willing to keep adding neurons (and, therefore, adjustable parameters), we can always get a precise "surrogate" neural network model. There are, however, several caveats to this, which we discuss below in Section 10.4.

Let us look at an example with an extremely simple neural network to understand how a typical neural network works.

Example 10.3: *Fig. 10.10 shows a simple neural network. For the given data and weights, do a forward pass through the network and calculate the corresponding output $\hat{y}$ (assuming that the nonlinearity is a sigmoid). The bias units are not shown but must be assumed to be present.*

$$x = 5.0$$
$$w_{01}^{(1)} = 1.0 \quad w_{02}^{(1)} = 0.8 \quad w_{11}^{(1)} = 0.7 \quad w_{12}^{(1)} = 0.9$$
$$w_{01}^{(2)} = 1.0 \quad w_{11}^{(2)} = 0.7 \quad w_{21}^{(2)} = 0.9$$

Solution:

For clarity, we draw the network diagram with the bias units and weight explicitly shown in Fig. 10.11. Note that we have split the figure into two to show the weights clearly. Some weights and connections are missing in each figure for viewing clarity.

Both figures omit some portions of the network for clarity of labeling weights.

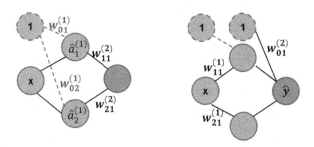

FIGURE 10.11

Figure for Example 10.3, with the weights and biases shown.

Let us now sequentially calculate the outputs of the neurons one by one. In the first (and only) hidden layer,

$$
\begin{aligned}
\hat{a}_1^{(1)} &= g\left(w_{01}^{(1)}x_0 + w_{11}^{(1)}x_1\right) \\
&= g\left(1 \times 1 + 0.7 \times 0.5\right) \\
&= g\left(1.35\right) \\
&= 0.7941
\end{aligned}
\tag{10.20}
$$

Similarly,

$$
\begin{aligned}
\hat{a}_2^{(1)} &= g\left(w_{02}^{(1)}x_0 + w_{12}^{(1)}x_1\right) \\
&= g\left(0.8 \times 1 + 0.9 \times 0.5\right) \\
&= g\left(1.25\right) \\
&= 0.7773
\end{aligned}
\tag{10.21}
$$

In the output layer,

$$
\begin{aligned}
\hat{y} &= g\left(w_{01}^{(2)}x_0 + w_{11}^{(2)}\hat{a}_1^{(1)} + w_{21}^{(2)}\hat{a}_2^{(1)}\right) \\
&= g\left(1 \times 1 + 0.7 \times 0.7941 + 0.9 \times 0.7773\right) \\
&= g\left(2.255\right) \\
&= 0.9051
\end{aligned}
\tag{10.22}
$$

The calculation in the above exercise is the essence of a forward pass through any neural network. At the end of this exercise, we hope you understand that forward pass through a network, even if tedious, is extremely straightforward.

We have now discussed everything about neural networks except for the most important thing—"How do we find the optimal weights for a given dataset?" The answer to this was already given implicitly in Step 4 of the learning process given above. To reiterate this, we first initialize the weights to random values. Then, with the given weights, we do a forward pass for every data point and obtain a model prediction $\hat{y}$. The error J is then calculated, and finally, we correct the weight through some optimization technique.

An important portion of the weight optimization above is that it usually requires the gradient $\frac{\partial J}{\partial w}$. This gradient calculation is usually the most computationally expensive part of the neural network process and was, in fact, one of the reasons why neural networks were impractical for a long time—finding optimal weights was and is time-consuming. The discovery of an efficient algorithm called **backpropagation** for gradient calculation and its implementation on modern computational architectures is part of the reason for the modern machine learning resurrection. Backpropagation is, in essence, the chain rule of partial differentiation applied to neural networks. Due to its technical nature, we are skipping it in this chapter. The interested reader can refer to Goodfellow et al. (2016) for further details.

Let us now summarize our understanding of neural networks through the next example.

Example 10.4: *We are trying to obtain a model for the temperature at any given point in a long fin purely using prior data. Sketch the process required to achieve this using neural networks. Comment on whether a linear model could work here.*
Solution:

Note: Since the real problem requires a lot of data, we have outlined the theoretical process below without actual numerical data.

We follow the usual learning process here.

Step 1: As was discussed in Section 10.3.1, the mathematical formulation of the long fin involves defining $x = (\theta_b, x, m)$ and $y = \theta$. Note that this requires us to already understand the physical parameters of the problem sufficiently to know that $\theta = T - T_\infty$ is a useful reduction of parameters in the problem.

Step 2: We could naively select a linear model here. In this case, we would have $\theta = w_0 + w_1\theta_b + w_2 x + w_3 m$. It should be clear from the physics of the problem that this is a bad model because, for a long fin, θ cannot keep increasing or decreasing unboundedly with x, which would happen in the linear model we have provided above. Similar problems will occur with any polynomial model. This is why the use of neural networks is particularly useful here. It allows us to fit a model that we can confidently use to approximate any nonlinear relation.

However, this leads to a question. How many layers and neurons should we use? Each network "architecture" is effectively a different hypothesis function. For engineering problems, it is always best to build intuition about the size of the network required by first working with a single hidden layer. It is a good idea to play with the number of neurons in this layer to see the number of neurons beyond which you will not gain any performance improvement. In practice, for the fin problem, it turns out that anywhere between 3 and 6 neurons in the hidden layer work really well.

Fig. 10.12 shows the schematic for the architecture in this problem. Note that while the neurons in the hidden layer will have a nonlinearity (say, a sigmoid), the

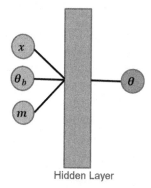

Hidden Layer

FIGURE 10.12

Schematic of ANN for Example 10.4, with one hidden layer.

output neuron cannot have a sigmoid as nonlinearity. Ideally, it is better to leave this neuron as a purely linear neuron (see Problem 10.5).

Step 3: Data collection simply involves collecting available data of long fins at varying values of θ_b, x, m and measuring the corresponding θ. However, for the purposes of testing how neural networks work, we could generate synthetic data computationally by simply calculating $\theta = \theta_b e^{-mx}$. This would require us to specify ranges of θ_b, x, m for which we will create the synthetic data. Even experimental data would have some range. For instance, we could choose to create 1000 such "synthetic" data points. These should now be split into training and testing sets separately, and only the training set should be utilized for finding the optimum parameters.

Step 4: The architecture of the network (number of neurons and layers) we chose in Step 2 fixes the form of the hypothesis function. We can now find the optimum parameters by optimizing over the training dataset. For neural networks, it is impossible to perform direct optimization due to their nonlinearity. We, therefore, use some iterative algorithms. Typically, a nonlinear optimization called Levenberg-Marquardt (see Balaji [2019] for details) is a good choice for these problems.

At the end of Step 4, we now have the complete surrogate model

$$\theta = \text{NN}(\theta_b, x, m; w) \tag{10.23}$$

This surrogate model can now be used wherever the analytical expression was. For example, we could use it in the design, optimization of fins, or for inverse problems in fins. While this is a simple problem, you can use the same process for any complicated problem.

10.4 Practical considerations in engineering problems

If the process of fitting neural networks is as simple as outlined in the previous section, why is the use of neural networks not universal across engineering applications of heat transfer? There are several reasons why this is so. We list some of the important ones below. These also serve as warning signs when an engineer attempts to apply the theory above to their practice. The issues are

- A neural network is best thought of as an interpolating, regression model. Usually, its applicability is limited to the range of data it has seen. It cannot extrapolate to discover new physics. In short, it is mathematics, not magic!
- In practice, it takes time and effort to pose the problem correctly. Even the specification of inputs and outputs in a problem is often nontrivial in engineering situations. Think about heat exchanger design, for instance. The appropriate inputs and outputs for this design are nontrivial. What one chooses to measure and designate as input and output is as important as how one fits a network to it. This requires **domain expertise** and deep knowledge of the field at hand.
- Collecting and curating data is an extremely important part of the problem. There might be a variety of past data of varying reliability, completeness, and

precision. Once again, think of collecting past data about heat exchanger design. Combining heterogeneous data statistically and offering it in a digestible form to the algorithm is a very time consuming and difficult part of the overall process. In fact, sometimes, at the end of data cleaning, there is not much data left to train sophisticated neural networks.

- Deciding on the appropriate architecture (neurons, layers, nonlinearity, etc.) of the network is not always easy. In practice, there is a lot of iteration required in coming up with an accurate, cheap, and trainable network. Determining the appropriate architecture, in many cases, is a time-consuming task.
- Finding the optimal parameters for a network is, in practice, an extremely hard task in industrial problems. Networks can be stubborn in not converging to optimal parameters. Optimizing neural networks is, therefore, a very rich and active open research problem. It is perhaps the hardest technical part of the neural network framework.
- Even if one gets past the above steps, the neural network is not interpretable. That is, we do not know what physical significance (if any) the neurons in the hidden layer have. If the network suggests an outlandish or nonintuitive design, it is very difficult to trust it without having a "feel" for the problem. Similarly, in case of faulty predictions, it is difficult for the engineer to know where to fix the surrogate model. Due to these reasons, in critical applications engineers may prefer a simpler, interpretable model to a more accurate but opaque model, such as a neural network.

10.5 Applications in heat transfer

1. Consider a heat exchanger that is being used in a factory setting.. From this exchanger, we obtain continuous data and can train the effectiveness of the exchanger as a function of the pertinent parameters through a neural network. This network can then be used to predict performance at off-design conditions or pave the way for possible optimization. The key advantage is that the network can be adaptively trained, meaning that the network will be dynamically updated with new data as and when they are collected and weights and biases too will keep evolving.

2. In continuation of (1) it is known that the heat transfer rate Q and pressure drop Δp are in fundamental conflict in a heat exchanger. With past data it is possible to develop one or two networks (i.e., either one for Q and Δp together or two for each of these separately) to characterize the quantities of interest against the controlling variables and then try to optimize combining the two criteria. The network can be built with past data from an existing heat exchanger and, as explained before, can be dynamically updated, which will intrinsically take care of issues like fouling.

3. At a much simpler level, almost all of the charts we introduced and used in this text (e.g., Heisler charts for transient heat transfer, fin efficiency charts, and heat

exchanger effectiveness charts) can be generated with a trained neural network once we have access to the original data. The advantages of a network will be convenience to use and easy programmability to look for, say, alternate fin designs followed by optimization.

4. In many other areas of heat transfer, where we solve an inverse problem—that is, obtaining the cause (e.g., thermal conductivity or thermal diffusivity) from the effects (e.g., temperature distribution) without explicitly mentioning about them—we can develop a neural network with temperatures as inputs and causes as output, which then is a straight solution to the inverse problem. Even so, large training data sets are required and this approach is easier said than done for involved problems such as those in convective heat transfer.

5. Advanced applications like determination of emissivity of special coatings used in spacecraft can be solved using measurements, a surrogate model, and an algorithm to give samples of trial emissivity values, which minimizes the difference between measurements and simulations of, for example, the temperature time history of the specimen with the coating whose emissivity is to be determined.

6. Almost all of the correlations that have been developed in convective heat transfer can be reworked using neural networks. The advantage is that experimental or numerical results obtained by several investigators can be combined with a view to come out with an ANN (artificial neural network)-based robust correlation. In fact many of the highly popular correlations such as those of Churchill and coworkers are based on this strategy, albeit without using a network.

7. Finally, there is a fascinating new subfield that has come to be known as physics-informed neural networks (PINNs), where we actually solve engineering problems, including ones in heat transfer by incorporating a network and its derivatives in the original differential equation with a view to obtain the best set of weights to solve the problem. While data will be required for doing this, once the network is trained it will return results in a fraction of the time taken by a conventional solver. However, it is still early days as yet for PINNs, and they have a long way to go.

10.6 Summary

Neural networks are a subset of machine learning techniques for supervised learning. In heat transfer, they can be used to serve as approximate models for both forward and inverse problems. Neural networks can be seen as a regression technique using a nonlinear hypothesis function and, given sufficient data, can serve as universal approximators. Despite their flexibility, there are some lacunae while applying neural networks to practical problems. Nonetheless, machine learning, in general, and neural networks, in particular, hold tremendous potential for heat transfer applications. Recent methods such as those by Raissi et al. (2019) have shown tremendous promise in having novel methods for inverse heat transfer using a combination of physics and data. We think that this direction is particularly interesting and that, in the future, there will be tighter integration of data and physics in heat transfer engineering.

Problems

10.1 Complete the proof given in Section 10.3.2. That is, show that for m univariate data points using the linear model $\hat{y} = w_0 + w_1 x$, the optimal parameters w_0 and w_1 for minimizing the mean squared error are given by:

$$w_1 = \frac{m \sum x_i y_i - \left(\sum x_i \right) \left(\sum y_i \right)}{m \sum x_i^2 - \left(\sum x_i \right)^2}$$

$$w_0 = \frac{\sum y_i - w_1 \sum x_i}{m}$$

where all summations $\sum$ should be read as $\sum_{i=1}^{m}$.

10.2 Determine the optimum parameters for a quadratic fit to the data above. That is, assuming a model of the form $\hat{y} = w_0 + w_1 x + w_2 x^2$, find the expressions for the optimum parameters w_0, w_1, and w_2 for minimizing the mean squared error.

10.3 The coefficient of thermal expansion, α, of steel is given at discrete values of the temperature in Table 10.2. We wish to fit a model to the data in order to find thermal expansions at intermediate temperatures not available in the table. Find the expressions and parameters for

    a.   The best fit linear model

    b.   The best fit model for the given data. Which of these models fits the data better? (Hint: Use the results of Problems 10.1 and 10.2).

10.4 Consider a variant of Example 10.1 where the temperatures at 5 different locations are the same as given in the example. We are also given that the flux at $x = 0$ in the slab is $2250 \, W/m^2$. However, the thermal conductivity of the material is unknown. From the data given, estimate the thermal conductivity. This is an example of an inverse problem.

Table 10.2 Variation of coefficient of thermal expansion of steel with temperature.

Temperature, T (°F)	Coefficient of thermal expansion, α (in/in °F)
80	6.47×10^{-6}
40	6.24×10^{-6}
0	6.00×10^{-6}
−40	5.72×10^{-6}
−80	5.43×10^{-6}

10.5 Consider a very simple neural network with the following features. There is neuron only in the input, hidden, and output layers. Assuming no bias units and a sigmoid activation function at both layers, answer the following.

    a. Write down the analytical expression for the hypothesis function this network represents.

    b. How many weights are there in the problem?

    c. Assuming a least squared cost function, find expressions for $\dfrac{\partial J}{\partial w_i}$ of all the weights.

10.6 Consider the fin problem in Example 10.4. Answer the following questions.

    a. Why is it important that the output neuron not have a sigmoid nonlinearity? (Hint: Think about the range of the sigmoid function vs. the range of the data we want the output neuron to represent.)

    b. Assuming a single hidden layer with 6 neurons, how many adjustable weights does the full network have? How many bias weights (simply called biases generally) does the network have?

    c. Assuming the same architecture as the previous part, and that all weights and biases have been initialized to 1, what is the output of the network for an input data having $\theta_b = 70$, $m = 4$, and $x = 0.1$?

References

Bejan, A., 1993. Heat Transfer. John Wiley and sons, New York.

Balaji, C., 2019. Thermal System Design and Optimization. Ane Books Pvt, New Delhi.

Goodfellow, I., Bengio, Y., Courville, A., 2016. Deep Learning. MIT press, Cambridge, MA.

McCulloch, W., Pitts, W., 1943. A logical calculus of ideas immanent in nervous activity. Bull. Math. Biophys. 5 (4), 115–133.

Raissi, M., Perdikaris, P., Karniadakis, G.E., 2019. Physics-informed neural networks: A deep learning framework for solving forward and inverse problems involving nonlinear partial differential equations. J. Comput. Phys. 378, 686–707.

Rabault, J., Kuchta, M., Jensen, A., Réglade, U., Cerardi, N., 2019. Artificial neural networks trained through deep reinforcement learning discover control strategies for active flow control. J. Fluid Mech. 865, 281–302.

Boiling and condensation

11

11.1 Introduction

Conduction heat transfer and single-phase convective heat transfer were presented in the previous chapters. Two-phase heat transfer, which involves conversion of liquid into vapor, referred to as boiling, or conversion of vapor into liquid, referred to as condensation, takes place in equipment such as boilers and condensers in thermal power plants, evaporators and condensers in refrigeration systems, water-cooled nuclear reactors, major equipment in process industry, and modern heat sinks for thermal management of electronics. Of particular importance in the design of the equipment are the rates of heat transfer or heat transfer coefficients and the associated pressure losses. In boiling and condensation, the coupling between the fluid dynamic process and the heat transfer process is stronger than what exists in single-phase flows. In a two-phase flow with phase change, there is a continuous variation in the fraction and distribution of each phase and hence the flow pattern, which influences the local heat transfer processes. Therefore, the flow at any axial location in the tube can never be fully developed thermally or hydrodynamically, unlike a single-phase flow. The flow also involves transient properties and deviations from thermodynamic equilibrium. In this chapter, the commonly observed regimes of pool boiling and flow patterns in flow boiling are discussed. Prior to the presentation of the correlations used to predict the heat transfer coefficients and the critical heat flux, the chapter presents a brief discussion on the wall superheat required for nucleation from a heating surface. This is followed by a discussion on the film condensation that occurs on flat plates and horizontal tubes and a presentation of the respective heat transfer coefficient correlations. The chapter ends with an introduction to the prediction of pressure drop in two-phase flows with phase change.

11.2 Boiling

Boiling can occur with the heating surface immersed in a pool of an initially stagnant liquid, referred to as pool boiling, or with the liquid forced over the heating surface by some external means such as a pump, which is referred to as flow boiling or forced convective boiling. As long as the heating surface temperature is greater than the saturation temperature of the liquid, there is a possibility of boiling, whether the bulk liquid is at saturation temperature (saturated boiling) or below the saturation temperature (subcooled boiling). Heat transferred from the heating surface to the liquid is given by the expression

$$q = h\left(T_w - T_{sat}\right) = h\Delta T_{sup} \tag{11.1}$$

where $\Delta T_{sup} = \left(T_w - T_{sat}\right)$ is the wall superheat.

Single-phase convection is influenced by fluid density, thermal conductivity, viscosity, specific heat, geometric parameter, and velocity in case of forced convection and the difference between the temperatures of the heating surface and the fluid and volumetric coefficient of thermal expansion in case of natural convection. Boiling, on other hand, is influenced by additional properties that include latent heat of vaporization, saturation temperature, vapor density, surface tension, and heating surface characteristics. Because of the large number of influencing variables and different mechanisms of heat transfer involved depending on the flow patterns and the wall superheat, the boiling process is complex, and it is difficult to obtain an equation (or a correlation) for the heat transfer coefficient as a function of pertinent parameters regardless of whether our approach is theoretical or experimental.

11.3 Pool boiling

Pool boiling, especially nucleate pool boiling, due to its high heat transfer coefficients, finds applications in chemical and petrochemical industries, refrigeration and desalination of sea water (for example, flooded evaporators), metallurgical industry (for example, quenching process), and cooling of electronic components (for example, vapor chamber heat sink). The typical temperature profile in the saturated pool boiling is shown in Fig. 11.1. There is a large temperature gradient close to the heating surface, with the temperature varying from the wall temperature, T_w, at the heating surface to nearly the saturation temperature, T_{sat}, across the thin thermal boundary

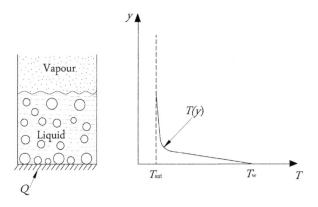

FIGURE 11.1

Temperature profile in the liquid during pool boiling.

layer. The liquid is superheated in this region. There are different regimes of pool boiling, and the heat transfer mechanism in each of the regimes differs greatly.

11.3.1 Pool boiling curve

A typical pool boiling curve for water boiling at atmospheric pressure is shown in Fig. 11.2. When the heat flux or wall superheat is low, no vapor bubbles emanate from the heating surface and the heat transport occurs by natural convection currents in the region AB. This stage is shown in Fig. 11.3A. The evaporation that occurs at the free surface is called "silent evaporation" or "silent boiling." In laminar natural convection, the heat transfer coefficient (h) varies as $\Delta T_{sup}^{1/4}$ or $q^{1/5}$, and in turbulent natural convection, it varies as $\Delta T_{sup}^{1/3}$ or $q^{1/4}$. As the heat flux or the wall superheat increases, vapor bubbles form on the heating surface when a certain wall superheat is attained (B). This is called the onset of nucleate boiling (ONB). Because of the formation of vapor bubbles, there is an increase in the heat transfer coefficient that leads to a sudden decrease in the wall temperature (B'). With further increase in the heat flux, more bubbles are formed on the heating surface and the heat transport occurs by both vapor bubbles and natural convection in the areas not influenced by the bubbles. This results in an increase in the heat transfer coefficient and hence the slope of the boiling curve (B''). Fig. 11.3B shows typical nucleate boiling. A further increase in the heat flux or the wall temperature results in the increased nucleations on the heating surface, and the nucleation site density will be so high that there will not be any area available for natural convection, and the heat transfer coefficient increases drastically (C). In nucleate boiling, h typically varies as ΔT_{sup}^{3} or $q^{3/4}$. At still higher heat fluxes, the vapor bubble generation frequency and the nucleation site density will be so high that the increased vapor generation rate results in the formation of

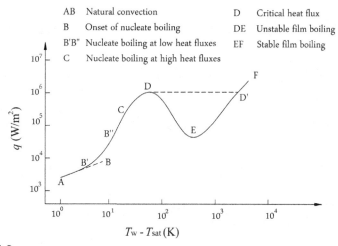

FIGURE 11.2

Pool boiling curve for water at atmospheric pressure.

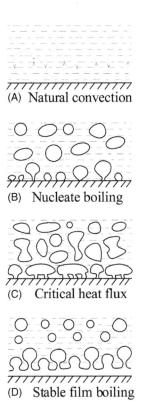

(A) Natural convection

(B) Nucleate boiling

(C) Critical heat flux

(D) Stable film boiling

FIGURE 11.3

Different stages in the pool boiling curve.

a vapor film, which tends to prevent the surrounding liquid from coming in contact with the heating surface (D). This causes a sharp reduction in the heat transfer coefficient, which leads to the wall temperature shooting up (DD') in the case of constant heat flux of the heating surface. If the heating surface is maintained at constant wall temperature, there is a reduction in the wall heat flux. This point is commonly known as the critical heat flux, and terms such as "boiling crisis" and "burnout" are also used, although it need not necessarily result in the actual burnout that depends on the metallurgical properties. Sometimes, this is also referred to as the departure from nucleate boiling (DNB), a term that is more appropriate from the point of view of physics and is illustrated in Fig. 11.3C. With further increase in the wall temperature (for the case of wall temperature–controlled heating surface), there will be a reduction in the heat flux, and the region DE is known as the unstable (or metastable) film boiling, characterized by the rapid formation and collapse of the vapor film, and at point E the vapor film becomes stable. If the wall temperature is increased further, there will be an increase in the heat flux (EF), caused by the contributions from both convection (conduction through the vapor film) and radiation, and the film remains stable, which is shown in Fig. 11.3D. The point E, where the heat flux is minimum, is known as the Leidenfrost point, named after J. G. Leidenfrost, who first described the effect of insulating vapor film formed between the heating surface and the liquid. It may be noted that resistance heating (Joule heating) and nuclear reactions (in nuclear reactors) result in constant wall heat flux, while constant wall temperature can be maintained or controlled by the phase change process (for example, the condensation of vapor at different pressures). The temperature of the combustion gases used in a boiler can be controlled by varying the excess air (or air-fuel ratio).

11.3.2 Nucleation

Consider a spherical vapor bubble of radius r in a liquid. The vapor pressure $\left(p_g\right)$ will be higher than the liquid pressure p_f as given by

$$p_g - p_f = \frac{2\sigma}{r} \tag{11.2}$$

where σ is the surface tension.

The increased vapor pressure results in a higher number of vapor molecules striking the interface and being absorbed by the interface compared to that when the interface was planar. In order to maintain phase equilibrium, there must be a corresponding increase in the number of liquid molecules emitted through the interface. Hence, the liquid adjacent to the vapor bubble is superheated with respect to the liquid pressure. Let T_g be the temperature of the superheated liquid. To determine the liquid superheat, the Clausius-Clapeyron equation can be used.

$$\frac{dp}{dT} = \frac{h_{fg}}{v_{fg}T} \tag{11.3}$$

From the ideal gas law

$$v_g = \frac{RT}{p} \tag{11.4}$$

Using $v_{fg} \approx v_g$ and substituting Eq. (11.4) in Eq. (11.3)

$$\frac{dp}{p} = \frac{h_{fg}}{RT^2}dT \tag{11.5}$$

Using $p = p_f$ and $T = T_{sat}$ and substituting Eq. (11.2) in Eq. (11.5)

$$T_g - T_{sat} = \Delta T_{sat} = \frac{RT_{sat}^2}{h_{fg}}\frac{2\sigma}{p_f\,r} \tag{11.6}$$

It can be seen from the Eq. (11.6) that the liquid superheat requirement is higher for a smaller size of equilibrium vapor nucleus (r).

It may be noted that Eq. (11.6) is obtained without considering the curvature effect on the relationship between liquid temperature and vapor pressure. Curvature of the interface slightly decreases the vapor pressure $\left(p_g\right)$ inside the vapor nucleus compared with that for a planar interface, for the same liquid temperature. The equation of equilibrium vapor nucleus derived considering this effect eventually simplifies to Eq. (11.6) for $\left(2\sigma / p_f r\right) << 1$ and $T_g \sim T_{sat}$. For details, see Collier and Thome (1994).

Homogeneous nucleation refers to the process of vapor formation in a superheated (metastable) liquid, as, for example, in the microwave heating of a liquid. But the most common one is heterogeneous nucleation, which refers to the process of vapor formation from pre-existing nuclei such as non-condensable gas bubbles suspended in the liquid, and vapor or gas-filled cavities and scratches on the container wall. These sites where the pre-existing nuclei are present are called nucleation sites.

Eq. (11.6) gives the liquid superheat required for a vapor nucleus in a liquid of uniform temperature field. For nucleation from the heating surface, the temperature gradient away from the heating surface needs to be taken into account. Fig. 11.4 shows a hemispherical vapor nucleus of radius r_c, sitting at the mouth of a conical cavity (nucleation site). The liquid temperature profile shown in the figure is assumed to be linear through the thermal boundary layer of thickness δ, which is approximately equal to $\left(k_f / h\right)$. According to Hsu (1962), the criterion for nucleation from this cavity is that the temperature of the liquid close to the top of the bubble should be greater than that necessary for the equilibrium of the vapor nucleus (Eq. 11.6). Nucleation occurs if the liquid temperature line intersects the equilibrium vapor nucleus curve. As the heat flux increases, at a certain heat flux, the liquid temperature profile makes a tangent to the equilibrium vapor nucleus curve, which results in the first nucleation site to be activated, and the corresponding wall temperature is $T_{w,ONB}$, as shown in Fig. 11.4.

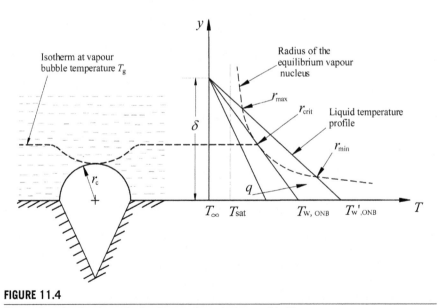

FIGURE 11.4

Onset of nucleation from the heating surface during pool boiling.

If the distortion of the isotherm at vapour bubble temperature is neglected, then the isotherm is located at a distance of r_c from the heating surface. With this simplification, and if the liquid pool is saturated $(T_\infty = T_{sat})$, the following analysis results in a relationship between r_c and the thermal boundary layer thickness δ. The liquid temperature profile is

$$\frac{T - T_{sat}}{T_w - T_{sat}} = \frac{\delta - y}{\delta} \tag{11.7}$$

At the point of tangency,

$$y = r_c \text{ and } \frac{dT}{dy} = \frac{dT}{dr}\bigg|_{r=r_c} \tag{11.8}$$

Using $T_g = T$ and substituting Eq. (11.6) and Eq. (11.7) in Eq. (11.8)

$$\frac{T_w - T_{sat}}{-\delta} = \frac{-RT_{sat}^2 \, 2\sigma}{h_{fg} P_f r_c^2} \tag{11.9}$$

$$\frac{T_w - T_{sat}}{-\delta} = \frac{RT_{sat}^2 \, 2\sigma}{h_{fg} P_f r_c}\left(-\frac{1}{r_c}\right) \tag{11.10}$$

Using $T_g = T$ and substituting Eq. (11.6) in Eq. (11.10)

$$\frac{T - T_{sat}}{T_w - T_{sat}} = \frac{r_c}{\delta} \tag{11.11}$$

Using $y = r_c$ and substituting Eq. (11.7) in Eq. (11.11),

$$\frac{\delta - r_c}{\delta} = \frac{r_c}{\delta}$$

which results in

$$r_c = \frac{\delta}{2} \tag{11.12}$$

Substituting Eq. (11.12) in Eq. (11.11)

$$T - T_{sat} = \frac{1}{2}(T_w - T_{sat}) \tag{11.13}$$

Note that here $T = T_g$, the temperature of the superheated liquid. Eq. (11.13) indicates that the liquid superheat close to the top of the bubble is half the wall superheat. In practice, a cavity size of $r_c = \delta/2$ (Eq. 11.12) may not result in nucleation, as cavities of larger size are not normally active sites (they are flooded and do not trap the vapor required as an embryo for nucleation). If there are no active nucleation sites of size r_c, the wall temperature must be increased further to initiate nucleation, as shown in Fig. 11.4, though the equilibrium vapor nucleus curve (Eq. 11.6) shown in Fig. 11.4 suggests that rougher surfaces (that contain cavities of larger size) readily nucleate, that is, they need lower wall superheat for nucleation. Even so, this need not necessarily be the case, as they may not be able to trap the vapor required for nucleation. Example 11.1 presents some calculations on the wall superheat required for nucleation for water at atmospheric pressure. The thermal boundary layer thickness δ can be calculated using single phase natural convection correlations such as the one below for turbulent natural convection on a horizontal surface (Fishenden and Saunders, 1950).

$$\frac{hD}{k_f} = 0.14\left[\left(\frac{g\beta\Delta T D^3 \rho_f^2}{\mu_f^2}\right)Pr_f\right]^{1/3} \tag{11.14}$$

where β and Pr are the volumetric coefficient of thermal expansion and the liquid Prandtl number, respectively.

Surface wettability is another important factor that influences nucleation. For solid-liquid systems with low surface wettability, superheat required for nucleation at the surface is low. For a perfectly non-wetting surface (contact angle measured through the liquid is 180°), no superheat is required for vapor bubble nucleation at the surface. For details on the influence of surface wettability, cavity shape and cavity mouth radius on the possibility of nucleation (vapor being trapped in the cavity), the wall superheat required for nucleation, and the stability of nucleation sites, the interested reader can refer to advanced texts such as Collier and Thome (1994) and Tong and Tang (1997).

Example 11.1: *Estimate the wall superheat required for nucleation from a horizontal surface immersed in water at atmospheric pressure, with the liquid pool at the saturation temperature, if cavities (nucleation sites) of all sizes are active, and if cavities (nucleation sites) of sizes 3 μm or smaller only are active.*

Solution:
Fluid properties corresponding to 100 °C:

$$\rho_f = 958\,\text{kg/m}^3,\ \mu_f = 283\times10^{-6}\ \text{Ns/m}^2,\ Pr_f = 1.75,$$
$$k_f = 0.681\ \text{W/mK},\ \beta = 7.51\times10^{-4}\ \text{K}^{-1}, R = 462\ \text{J/kgK}$$

Using Eq. (11.14)

$$\frac{hD}{k_f} = 0.14\left[\left(\frac{g\beta\Delta T_{sup} D^3 \rho_f^2}{\mu_f^2}\right)Pr_f\right]^{1/3}$$

$$\frac{h}{k_f} = 0.14\left[\left(\frac{g\beta\Delta T_{sup} \rho_f^2}{\mu_f^2}\right)Pr_f\right]^{1/3}$$

Since $\dfrac{h}{k_f} = \dfrac{1}{\delta}$, where δ is the thermal boundary layer thickness,

$$\frac{1}{\delta} = 0.14 \left[\frac{9.81 \times 7.51 \times 10^{-4} \times 958^2 \times 1.75}{(283 \times 10^{-6})^2} \right]^{1/3} \Delta T_{sup}^{1/3}$$

$$\delta = \frac{1.35 \times 10^{-3}}{\Delta T_{sup}^{1/3}}$$

The liquid superheat required for equilibrium vapor nucleus of size r_c (cavity radius) is given by Eq. (11.6).

$$\Delta T_{sat} = \frac{RT_{sat}^2 \, 2\sigma}{h_{fg} p_f r_c}$$

$$= \frac{462 \times 373^2 \times 2 \times 0.05878}{2257 \times 10^3 \times 1.013 \times 10^5 \times r_c}$$

$$= \frac{3.305 \times 10^{-5}}{r_c}$$

When cavities (nucleation sites) of all sizes are active, the minimum wall superheat required for nucleation results in $\left(\Delta T_{sup} \right)_{ONB} = 2\Delta T_{sat}$, and $r_c = \dfrac{\delta}{2}$.
Therefore,

$$\left(\Delta T_{sup} \right)_{ONB}^{1/3} = \frac{1.35 \times 10^{-3}}{2r_c}$$

$$\left(\Delta T_{sup} \right)_{ONB}^{2/3} = \frac{4 \times 3.305 \times 10^{-5}}{1.35 \times 10^{-3}}$$

$$\left(\Delta T_{sup} \right)_{ONB} = 0.03 \, \text{K}$$

$$\delta = 0.00434 \, m = 4.34 \, \text{mm}$$

$$r_c = \delta / 2 = 2.17 \, \text{mm}$$

When only cavities (nucleation sites) of sizes 3 µm or smaller are active, from Fig. 11.4

$$\left(\Delta T_{sup} \right)_{ONB} \approx \Delta T_{sat}$$

$$= \frac{3.305 \times 10^{-5}}{3 \times 10^{-6}}$$

$$= 11.0 \, \text{K}$$

$$\delta = \frac{1.35 \times 10^{-3}}{\Delta T_{sup}^{1/3}}$$

$$= \frac{1.35 \times 10^{-3}}{11^{1/3}}$$

$$= 6.07 \times 10^{-4} \, \text{m}$$

Note:

$$r_c = 0.03 \times 10^{-4} \, m \ll \delta = 6.07 \times 10^{-4} \, m$$

From this example, it is clear that if nucleation sites of all sizes are active, even very small superheat can trigger nucleation. Hence, boiling can be engineered by microstructured surfaces, coatings, and fabricated nucleation sites.

11.3.3 Nucleate boiling

Nucleate boiling is a complex phenomenon. High heat transfer coefficients associated with nucleate boiling can be attributed to different heat transfer mechanisms, the degree of which may vary. Vapor bubbles grow from active nucleation sites once the surface attains the wall superheat required for nucleation. Initially, the bubble growth is driven by the excess vapor pressure not balanced by the surface tension forces, and the inertia of the surrounding liquid controls the process. This stage is very short (a few ms after the inception). In the later stage, the bubble growth is governed by the rate of heat conducted from the superheated liquid to the liquid-vapor interface. There is evaporation of a thin "microlayer" underneath the bubble, leading to high transient heat transfer coefficients. The bubble departs from the surface once the buoyancy force overcomes the inertial force or surface tension force. The bubble departure causes disruption of the thermal boundary layer, because of which there is transient conduction during the waiting period. Once the thermal boundary layer is reformed and the wall superheat requirement for nucleation is attained, the bubble grows rapidly, pushing the superheated liquid away. Other mechanisms that contribute to the heat transfer are drift currents behind the rising bubble that cause a suction effect on the boundary layer and convection caused by the surface tension gradient resulting from the temperature variation in the superheated boundary layer. The latter is called the Marangoni effect. Natural convection always occurs in the areas uninfluenced by the bubbles. The bubbles departure diameters and frequencies influence the heat transfer, and for water at atmospheric pressure, these are in the range 1–3 mm and 20–50 s^{-1}, respectively. For a detailed study, the reader can refer to Collier and Thome (1994), Stephan (1992), and Tong and Tang (1997).

The most widely used Rohsenow (1952) correlation for nucleate pool boiling is of the form

$$Nu_b = C_1 Re_b^x Pr_f^y \tag{11.15}$$

The characteristic length L_b is based on the balance between surface tension force and buoyancy force,

$$L_b = \left[\frac{\sigma}{g(\rho_f - \rho_g)} \right]^{\frac{1}{2}} \tag{11.16}$$

The velocity used is the superficial liquid velocity towards the surface,

$$u_f = \frac{q}{h_{fg} \rho_f} \tag{11.17}$$

The boiling Reynolds number Re_b is given by

$$Re_b = \frac{\rho_f u_f L_b}{\mu_f} = \frac{\rho_f \left(\dfrac{q}{h_{fg}\rho_f}\right) \left[\dfrac{\sigma}{g(\rho_f - \rho_g)}\right]^{\frac{1}{2}}}{\mu_f} \tag{11.18}$$

The boiling Nusselt number Nu_b is given by

$$Nu_b = \frac{hL_b}{k_f} = \frac{q\left[\dfrac{\sigma}{g(\rho_f - \rho_g)}\right]^{\frac{1}{2}}}{\Delta T_{sup} k_f} \tag{11.19}$$

The Rohsenow correlation contains an arbitrary constant C_{sf} to account for the influence of liquid-surface combination on the nucleation properties.

The Rohsenow (1952) correlation is

$$\frac{c_{pf}\Delta T_{sup}}{h_{fg}} = C_{sf}\left[\frac{q}{h_{fg}\mu_f}\left(\frac{\sigma}{g(\rho_f - \rho_g)}\right)^{\frac{1}{2}}\right]^n \left(\frac{\mu_f c_{pf}}{k_f}\right)^{m+1} \tag{11.20}$$

Eq. (11.20) can be rearranged as

$$q = \mu_f h_{fg}\left(\frac{g(\rho_f - \rho_g)}{\sigma}\right)^{\frac{1}{2}}\left(\frac{c_{pf}\Delta T_{sup}}{h_{fg} C_{sf} Pr_f^{m+1}}\right)^{\frac{1}{n}} \tag{11.21}$$

where $n = 0.33$ and $m = 0.7$. Later, Rohsenow recommended that the value of m be changed to zero for water alone. All fluid properties correspond to the saturated state. The values of the constant C_{sf} for some liquid-surface combinations are given in Table 11.1.

Table 11.1 Values of C_{sf} for the Rohsenow (1952) correlation for various liquid-surface combinations.

Liquid-surface combination	C_{sf}
Water - polished copper	0.0128
n-Pentane - polished copper	0.0154
Carbon tetrachloride - polished copper	0.007
Water - lapped copper	0.0147
Water - scored copper	0.0068
Water - ground and polished stainless steel	0.008
Water - Teflon pitted stainless steel	0.0058
Water - chemically etched stainless steel	0.0133
Water - mechanically polished stainless steel	0.0132

11.3.4 Critical heat flux

Zuber (1958) derived an expression for the critical heat flux in pool boiling considering the Taylor and Helmholtz instabilities as

$$q_{crit} = q_{max} = K h_{fg} \rho_g^{1/2} \left[\sigma g (\rho_f - \rho_g) \right]^{1/4} \tag{11.22}$$

where K is a constant that lies between 0.13 and 0.16 based on the experimental data. The recommended value of K is 0.149. Typical q_{crit} values are in the order of 10^6 W/m^2 for water at near atmospheric pressure.

Example 11.2: *Estimate the wall superheat at the onset of departure from nucleate boiling (DNB) or critical heat flux for saturated pool boiling of water on a horizontal surface at atmospheric pressure. Assume the constant C_{sf} to be 0.013 in the Rohsenow correlation.*

Solution:

The properties of saturated water at atmospheric pressure are

$$T_{sat} = 100 \text{ °C}, \rho_f = 958 \text{ kg/m}^3, \rho_g = 0.598 \text{ kg/m}^3, h_{fg} = 2257 \text{ kJ/kg},$$
$$\sigma = 0.05878 \text{ N/m}, \mu_f = 283.1 \times 10^{-6} \text{ Ns/m}^2, c_{pf} = 4.218 \text{ kJ/kg K},$$
$$Pr_f = 1.75, C_{sf} = 0.013$$

Using Eq. (11.22),

$$q_{crit} = 0.149 \, h_{fg} \rho_g^{1/2} \left[\sigma g (\rho_f - \rho_g) \right]^{1/4}$$
$$= 0.149 \times 2257 \times 10^3 \times (0.598)^{1/2} \left[0.05878 \times 9.81 \times (958 - 0.598) \right]^{1/4}$$
$$= 1.26 \times 10^6 \text{ W/m}^2$$

According to the Rohsenow correlation (this is very widely used and is the workhorse in pool boiling calculations for obtaining the first cut estimates in an engineering problem), Eq. (11.20),

$$\Delta T_{sup} = \left(\frac{q}{\mu_f h_{fg}} \right)^n \left(\frac{\sigma}{g(\rho_f - \rho_g)} \right)^{n/2} \frac{h_{fg} C_{sf} Pr_f^{m+1}}{c_{pf}}$$

$$= \left(\frac{1.26 \times 10^6}{283.1 \times 10^{-6} \times 2257 \times 10^3} \right)^n \times \left(\frac{0.05878}{9.81 \times (958 - 0.598)} \right)^{n/2} \times$$
$$\frac{2257 \times 10^3 \times 0.013 \times (1.75)^{m+1}}{4.218 \times 10^3}$$

$$= (1971.96)^n \times \left(6.258 \times 10^{-6} \right)^{n/2} \times 6.955 \times (1.75)^{m+1}$$
$$= 4.933^n \times 6.955 \times 1.75^{m+1}$$

The original Rohsenow correlation had values of $n = 0.33$ and $m = 0.7$.

$$\Delta T_{sup} = 4.933^{0.33} \times 6.955 \times 1.75^{1.7}$$
$$= 30.49 \text{ K}$$

For $n = 0.33$ and $m = 0$ (recommended by Rohsenow, later for only water),

$$\Delta T_{sup} = 4.933^{0.33} \times 6.955 \times 1.75$$
$$= 20.60 \text{ K}$$

If one looks at Examples 11.1 and 11.2, it is evident that ΔT_{sup} for critical heat flux (or DNB) is higher than that is required for nucleation, as it should be.

11.3.5 Film boiling

Berenson (1960) derived an expression for the minimum heat flux (Leidenfrost point E in Fig. 11.2):

$$q_{min} = 0.09 \rho_g h'_{fg} \left[\frac{g(\rho_f - \rho_g)}{\rho_f + \rho_g} \right]^{\frac{1}{2}} \left[\frac{\sigma}{g(\rho_f - \rho_g)} \right]^{\frac{1}{4}} \qquad (11.23)$$

where,

$$h'_{fg} = h_{fg} + 0.68 \, c_{pg} \Delta T_{sup} \qquad (11.24)$$

Heat transfer in stable film boiling occurs by convection (conduction) through the vapor film and also by radiation across the film.

Berenson (1960) proposed the following expression for the convective heat transfer coefficient for film boiling on flat surfaces:

$$h_c = 0.425 \left(\frac{k_g^3 h'_{fg} \rho_g g (\rho_f - \rho_g)}{\mu_g (T_w - T_{sat}) \left(\frac{\sigma}{g(\rho_f - \rho_g)} \right)^{1/2}} \right)^{\frac{1}{4}} \qquad (11.25)$$

Bromley (1950) proposed the following expression for film boiling on horizontal cylinders:

$$h_c = 0.62 \left(\frac{k_g^3 h'_{fg} \rho_g g (\rho_f - \rho_g)}{\mu_g (T_w - T_{sat}) D} \right)^{\frac{1}{4}} \qquad (11.26)$$

For film boiling on spheres, Eq. (11.26) is used with the constant 0.62 replaced by 0.67.

Note that in Eq. (11.25) and Eq. (11.26), the vapor properties are evaluated at pressure P_f and temperature $(T_w + T_{sat})/2$ and liquid properties at T_{sat}.

Radiation heat transfer across the vapor film can be calculated using the following expression:

$$h_r = \frac{1}{\dfrac{1}{\varepsilon_w} + \dfrac{1}{\alpha_f} - 1} \frac{\sigma\left(T_w^4 - T_{sat}^4\right)}{\left(T_w - T_{sat}\right)} \tag{11.27}$$

where ε_w is the emissivity of solid (wall), and α_f is the absorptivity of liquid, which is equal to the emissivity of the liquid.

Bromley (1950) proposed the following expression for the combined effects of convection and radiation:

$$h = h_c \left(\frac{h_c}{h}\right)^{\frac{1}{3}} + h_r \tag{11.28}$$

For $h_r < h_c$, Bromley (1950) suggested the following approximation:

$$h = h_c + 0.75 h_r \tag{11.29}$$

11.4 Flow boiling

Flow boiling (or forced convection boiling) occurs when a liquid is forced through a tube or over a surface that is maintained at a temperature greater than the saturation temperature of the liquid. The applications of flow boiling include steam generators for thermal and nuclear power plants. The fluid dynamics and heat transfer mechanisms of flow boiling are more complex than in pool boiling, as the forced flow influences the formation of vapor bubbles, their separation from the heating surface, and the two-phase flow patterns along the tube.

11.4.1 Flow boiling regimes

Fig. 11.5 shows the variation of the wall and fluid temperatures, flow patterns, and heat transfer regions for the flow boiling in a vertical uniformly heated tube (constant wall heat flux). Heat transfer to the subcooled liquid takes place by single-phase convection alone until the wall temperature is higher than the saturation temperature, at which point the boiling commences while the liquid may still be subcooled. There is a decrease in the wall temperature due to the increased heat transfer coefficient as a result of boiling. The size of the vapor bubbles formed increases as the bulk liquid attains saturation temperature. At lower qualities, there are small bubbles dispersed in the liquid (bubbly flow), and the heat transfer mechanism is mainly nucleate boiling. The local heat transfer coefficient mainly depends on the heat flux and is not affected by the quality in nucleate boiling. As the quality (or the void fraction) increases along the flow, bubbles coalesce and form larger bubbles separated by liquid slugs, and the flow pattern is termed as a slug flow. With further increase in the vapor quality along the flow, a vapor core is formed, which is irregular, termed as a churn-annular flow.

Flow boiling regimes

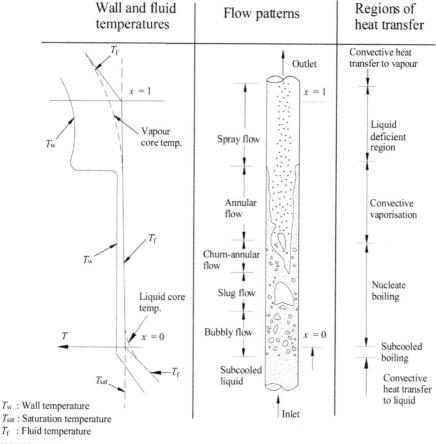

| Wall and fluid temperatures | Flow patterns | Regions of heat transfer |

T_w : Wall temperature
T_{sat} : Saturation temperature
T_f : Fluid temperature

FIGURE 11.5

Wall and fluid temperatures, flow patterns, and heat transfer regions for flow boiling in a vertical heated tube.

A continuous vapor core is then formed, surrounded by the liquid, and is termed as an annular flow. Because of the higher velocities caused by the increased vapor fraction, the nucleation from the heating surface is suppressed. This is due to the thinner thermal boundary layer resulting from the increased velocities. The heat transfer is mainly due to the convective evaporation at the liquid-vapor interface. The vapor core contains entrained liquid droplets formed as a result of the interaction between the liquid and vapor phases. The thickness of the liquid film decreases due to evaporation, and the velocity increases due to the increased quality, both of which result in an increase in the heat transfer coefficient (decrease in the wall temperature) along the flow. At some point, the liquid film completely dries out due to evaporation, and the vapor comes in contact with the wall, which results in a sudden decrease in the heat transfer coefficient (increase in the wall temperature). This point is called "dry-out", which is very important from the viewpoint of the design and operating conditions. Further downstream, the evaporation of the entrained liquid droplets increases the velocity, and hence there is a slight increase in the heat transfer coefficient. The vapor flow with entrained droplets is called a spray flow or mist flow, with the wall being liquid deficient. Eventually the tube is completely filled with vapor, at which point a single-phase convection starts. The quality (x) shown in the figure indicates thermodynamic quality obtained from energy balance. The actual quality can be different from the thermodynamic quality, mainly in the subcooled boiling and spray flow conditions.

In actual applications, one may not see all the flow patterns described here, as the flow patterns depend on the inlet condition, mass flux, heat flux, tube diameter, and tube length. For the case of the wall temperature controlled heating, the local heat flux depends on the local two-phase heat transfer coefficient. Fig. 11.6 shows the influence of heat flux on the heat transfer coefficient variation along the flow, for a constant mass flux. It can be seen that with the increase in heat flux (q_2), the heat transfer coefficient increases for the same vapor quality, but the dryout occurs early (at a lower quality). With further increase in the heat flux (q_3 and q_4), there occurs no

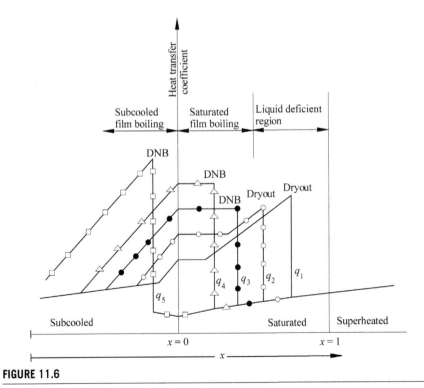

FIGURE 11.6

Heat transfer coefficient variation with quality for constant mass flux and heat flux with $q_1 < q_2 < q_3 < q_4 < q_5$.

transition to annular flow, and there is DNB in the saturated nucleate boiling, which causes a sudden reduction in the heat transfer coefficient. The DNB occurs even in the subcooled boiling at higher heat flux (q_s), as shown in the figure. Note that the DNB indicates film boiling similar to that in pool boiling and "dryout" indicates the dryout of liquid film. Normally the wall temperature increase caused by DNB is much higher than that by dryout. Note that q (DNB-subcooled) > q (DNB-saturated) > q (dryout).

It must be noted that boiling (both pool boiling and flow boiling) heat transfer is highly sensitive to the heating surface characteristics. Boiling characteristics change with surface ageing, especially for water (Collier and Thome, 1994; Jayaramu et al., 2019), and hence it is very important that the experimental runs be repeated until the repeatable data is obtained, which can be used for the heat transfer analysis, comparison with the literature data, or development of correlations.

11.4.2 The Chen correlation

By far the most successful and widely used two-phase correlation for flow boiling of water and organic fluids in conventional size tubes is the one proposed by Chen (1963). The correlation predicts the data for both the saturated nucleate boiling region and the convective vaporization region. It assumes that both nucleate boiling and convection occur in some proportion over the entire range of correlation, and their contributions are additive.

$$h_{TP} = h_{NcB} + h_c \tag{11.30}$$

where h_{TP}, h_{NcB}, and h_c are the local two-phase heat transfer coefficient, the nucleate boiling contribution, and the convection contribution, respectively; h_c can be evaluated by a Dittus-Boelter type equation that was presented in Chapter 5.

$$\frac{h_c D}{k_{TP}} = 0.023 \, Re_{TP}^{0.8} \, Pr_{TP}^{0.4} \tag{11.31}$$

Chen (1963) argued that heat is essentially transferred from wall to a liquid film in both dispersed and annular flows, so it is appropriate to use the liquid thermal conductivity in Eq. (11.31). Furthermore, the magnitudes of the liquid and vapor Prandtl numbers are approximately the same, and the two-phase value should not be very different from this value $\left(Pr_f \text{ or } Pr_g \right)$.

Eq. (11.31) can be written as

$$h_c = 0.023 \left[\frac{G(1-x)D}{\mu_f} \right]^{0.8} \left(\frac{\mu c_p}{k} \right)_f^{0.4} \left(\frac{k_f}{D} \right)(F) \tag{11.32}$$

where,

$$F = \left[\frac{Re_{TP}}{Re_f} \right]^{0.8} = \left[\frac{Re_{TP}}{\left(\frac{G(1-x)D}{\mu_f} \right)} \right]^{0.8} \tag{11.33}$$

The parameter F is expressed as a function of the Lockhart-Martinelli parameter, X_{tt}.

$$X_{tt} = \left(\frac{\rho_g}{\rho_f}\right)^{0.5}\left(\frac{\mu_f}{\mu_g}\right)^{0.1}\left(\frac{1-x}{x}\right)^{0.9} \tag{11.34}$$

$$F = 1 \text{ for } \frac{1}{X_{tt}} \le 0.1$$

$$F = 2.35\left(\frac{1}{X_{tt}} + 0.213\right)^{0.736} \text{ for } \frac{1}{X_{tt}} > 0.1 \tag{11.35}$$

The quantity X_{tt}^2 is the ratio of the frictional pressure gradient with liquid phase alone in the channel to that with vapor (gas) phase alone in the channel, with both phases assumed to be turbulent. Based on the experimental data, the parameter X_{tt} is related to the parameter F as in Eq. (11.35). With increase in the quality x, the parameter X_{tt} decreases, which causes an increase in the parameter F. The parameter F can be understood as an "enhancement factor" that enhances the forced convective evaporation with increase in the quality x. As discussed earlier in the *Flow boiling regimes* section, increasing the quality increases the velocity, which, together with the thinning of liquid film (caused by evaporation), increases the heat transfer coefficient in the annular regime.

Figure 11.4 shows that the liquid superheat is not constant across the thermal boundary layer. The mean superheat of the liquid (ΔT_e) in which the vapor bubbles grow is lower than the superheat at the wall (ΔT_{sup}). Since the thermal boundary layer is thick in pool boiling, the difference between ΔT_e and ΔT_{sup} is small. However, the thinner thermal boundary layer and the steeper temperature gradient in the forced convection result in a significant difference between ΔT_e and ΔT_{sup}.

Chen uses the Forster and Zuber (1955) analysis with ΔT_e and the corresponding vapor pressure difference (Δp_e) to write the equation for h_{NcB} as

$$h_{NcB} = 0.00122\left[\frac{k_f^{0.79}c_{pf}^{0.45}\rho_f^{0.49}}{\sigma^{0.5}\mu_f^{0.29}h_{fg}^{0.24}\rho_g^{0.24}}\right](\Delta T_e)^{0.24}(\Delta p_e)^{0.75} \tag{11.36}$$

A suppression factor, S, is then defined as the ratio of ΔT_e to ΔT_{sup}

$$S = \left[\frac{\Delta T_e}{\Delta T_{sup}}\right]^{0.99} \tag{11.37}$$

$$S = \left[\frac{\Delta T_e}{\Delta T_{sup}}\right]^{0.24}\left[\frac{\Delta T_e}{\Delta T_{sup}}\right]^{0.75}$$

$$S = \left[\frac{\Delta T_e}{\Delta T_{sup}}\right]^{0.24}\left[\frac{\Delta p_e}{\Delta p_{sup}}\right]^{0.75} \tag{11.38}$$

Eq. (11.38) is based on the Clasius-Clapeyron equation,

$$\frac{\Delta p_e}{\Delta T_e} = \frac{\Delta p_{sup}}{\Delta T_{sup}} = \frac{h_{fg}}{v_{fg} T_{sat}}$$

(11.39)

Substituting Eq. (11.38) in Eq. (11.36),

$$h_{NcB} = 0.00122 \left[\frac{k_f^{0.79} c_{pf}^{0.45} \rho_f^{0.49}}{\sigma^{0.5} \mu_f^{0.29} h_{fg}^{0.24} \rho_g^{0.24}} \right] (\Delta T_{sup})^{0.24} (\Delta p_{sup})^{0.75} (S)$$

(11.40)

S approaches unity at low flow rates and zero at high flow rates. Based on the experimental data, S can be expressed as

$$S = \frac{1}{1 + 2.53 \times 10^{-6} Re_{TP}^{1.17}}$$

(11.41)

The experimental data used for the development of the Chen correlation considered the two-phase flow in vertical tubes, and the fluids were water (0.55 to 34.8 bar, upward flow and downward flow), benzene, cyclohexane, heptane, methanol, and pentane, all at 1.03 bar with upward flow. The average deviation between the predicted (calculated) and measured data from the experimental cases was ±12%. Though the Chen correlation finds extensive applications for different fluids and operating conditions, it cannot be used for liquid metals or for horizontal tubes due to the asymmetrical nature of flow patterns, such as the stratified flows in horizontal tubes. Many variations of the Chen correlation were proposed later to extend its applicability to a wider range of fluids and operating conditions, and even for different channel geometries (e.g., offset strip fins). In fact, the Chen correlation's basic philosophy of the nucleate boiling suppression factor S and the liquid-phase convection enhancement factor F has been used even in the development of some of the correlations (called Chen-type correlations) for flow boiling in mini/micro-channels (Jayaramu et al., 2019).

11.4.3 Critical heat flux in flow boiling

The occurrence of the critical heat flux (CHF) and the nature of the CHF (dryout and DNB) in flow boiling were discussed earlier. This section presents the estimation of CHF for flow boiling in vertical uniformly heated round tubes. Flow is in the upward direction.

For uniformly heated tubes, the onset of the CHF occurs first at the tube (or channel) exit, and hence overheating first occurs at the tube exit, in the absence of flow and thermal instabilities.

Boiling cannot occur if the wall temperature is less than the saturation temperature. Therefore, the lower limiting value of the critical heat flux corresponds to the condition of the tube exit wall temperature being equal to the saturation temperature. The fluid temperature at the tube exit can be determined using the inlet condition, the mass flux, and the heat flux. The tube exit wall temperature can be determined using the heat flux, the single phase heat transfer coefficient (h_{f0}), and the fluid temperature at the exit. For a tube of diameter D, length z, mass flux G and the inlet subcooling $(\Delta T_{sub})_i$, the lower limiting value of the critical heat flux is

$$q_{crit,min} = \frac{(\Delta T_{sub})_i}{\frac{4z}{Gc_{pf}D} + \frac{1}{h_{fo}}} \qquad (11.42)$$

The upper limiting value of the critical heat flux corresponds to the condition of the exit quality being equal to $1(x(z)=1)$,

$$q_{crit,max} = \frac{GDh_{fg}}{4z}\left[1 + \frac{c_{pf}(\Delta T_{sub})_i}{h_{fg}}\right] \qquad (11.43)$$

Based on the extensive experimental data on the CHF for water in vertical uniformly heated round tubes, the following correlation was proposed by Bowring (1972):

$$q_{crit} = \frac{A' + DG\bar{c}_{pf}\frac{(\Delta T_{sub})_i}{4}}{C' + z} \qquad (11.44)$$

where,

$$A' = 2.317\left[\frac{DGh_{fg}}{4}\right]\frac{F_1}{1.0 + 0.0143F_2D^{1/2}G} \qquad (11.45)$$

$$C' = \frac{0.077F_3DG}{1.0 + 0.347F_4(G/1356)^n} \qquad (11.46)$$

D, G, and h_{fg} are in m, kg/m²s, and J/kg, respectively. The exponent n is given by

$$n = 2.0 - 0.00725p \qquad (11.47)$$

with the operating pressure p in bar. The pressure functions F_1, F_2, F_3, and F_4 are given in Table 11.2.

The empirical correlation was obtained based on the experiments with water in the following parameter ranges:

p: 2–190 bar
D: 0.002–0.045 m
z: 0.15–3.7 m
G: 136–18600 kg/m²s

The RMS error is 7%, and 95% of the test data lie within ±14%. The empirical correlation does not differentiate between the types of CHF - dryout and DNB (film boiling). Experimental data on the variation of CHF with different parameters is available in Collier and Thome (1994) and Stephan (1992).

Table 11.2 Functions F_1, F_2, F_3, and F_4 for the Bowring critical heat flux correlation, for only water (Bowring, 1972).

Pressure (bar)	F_1	F_2	F_3	F_4
1	0.478	1.782	0.4	0.0004
5	0.478	1.019	0.4	0.0053
10	0.478	0.662	0.4	0.0166
15	0.478	0.514	0.4	0.0324
20	0.478	0.441	0.4	0.0521
25	0.48	0.403	0.401	0.0753
30	0.488	0.39	0.405	0.1029
35	0.519	0.406	0.422	0.138
40	0.59	0.462	0.462	0.1885
45	0.707	0.564	0.538	0.2663
50	0.848	0.698	0.647	0.3812
60	1.403	0.934	0.89	0.7084
68.9	1.000	1.000	1.000	1.000
70	0.984	0.995	1.003	1.03
80	0.853	0.948	1.033	1.322
90	0.743	0.903	1.06	1.647
100	0.651	0.859	1085	2.005
110	0.572	0.816	1.108	2.396
120	0.504	0.775	1.129	2.819
130	0.446	0.736	1.149	3.274
140	0.395	0.698	1.168	3.76
150	0.35	0.662	1.186	4.227
160	0.311	0.628	1.203	4.825
170	0.277	0.595	1.219	5.404
180	0.247	0.564	1.234	6.013
190	0.22	0.534	1.249	6.651
200	0.197	0.506	1.263	7.32

11.4.4 A brief overview of flow boiling in micro-channels

Rapid miniaturization of electronic devices poses challenges to the thermal management of electronics. Data center cooling and cooling of rocket motors and laser diodes involve high heat fluxes. As micro-channels have high heat transfer surface area to fluid flow volume ratio, and flow boiling heat transfer coefficients are high, heat sinks based on the flow boiling in micro-channels is one of the promising techniques to efficiently dissipate high heat fluxes. Evaporating vapor bubbles in micro-channels are confined by the channel cross-sectional dimensions (Gedupudi et al., 2011), and hence the flow patterns during flow boiling in micro-channels can be different from those in the conventional size tubes. The confined vapour bubble growth leads to pressure fluctuations, which can result in transient flow reversal if upstream compressibility is present (Gedupudi et al., 2011; Jain et al., 2019). Another characteristic is the possibility of occurrence of intermittent dryout caused by the cyclic passage of liquid slugs and long confined bubbles (Jain et al., 2019). Enhancing the two-phase heat transfer coefficient and the critical heat flux and reducing the flow instabilities and pressure drop are the research challenges in this area (Kandlikar et al., 2006; Karayiannis and Mahmoud, 2017).

11.5 Condensation

The condensation process is the reverse of the boiling process and involves the change of a vapor phase to a liquid phase. Just as liquid superheat is required to induce the nucleation of bubbles in boiling, vapor subcooling is required to induce the nucleation of droplets in condensation. If the condensate forms a continuous film, then it is called film-wise condensation (or simply, film condensation), which occurs on wetted surfaces. In drop-wise condensation (or simply, drop condensation), the vapor condenses into small liquid droplets of different sizes that coalesce and fall down the cooled surface, and it occurs on non-wetted surfaces. Drop condensation, as seen from the above definition, offers less thermal resistance, and this results in heat transfer coefficients as much as 5 to 10 times the values in film condensation. Though drop condensation would be preferred to film condensation, it is difficult to achieve or maintain drop condensation. Much of the recent research focuses on promoting drop condensation using microstructured surfaces, which is beyond the scope of this textbook. Additionally, there is a lack of reliable theories of drop condensation.

11.6 Film condensation on a vertical plate

Nusselt (1916) made the following simplifying assumptions to solve the film condensation problem analytically, which is often considered a classic treatise on the subject of condensation:

1. The condensate flow in the film is laminar.
2. The plate temperature, T_w, is uniform and less than the vapor saturation temperature, T_{sat}.
3. The fluid properties are constant.
4. Negligible shear stress at the liquid-vapor interface.
5. Momentum changes in the film are negligible.
6. The heat transfer across the film is by conduction only, and the condensate temperature profile is linear.

Fig. 11.7 shows the typical growth of condensate film as well as velocity and temperature profiles. The analysis is made per unit width of the plate in the direction

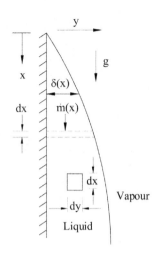 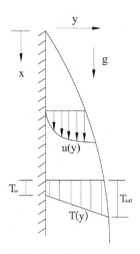

Growth of condensate film   Velocity and temperature profiles

FIGURE 11.7

Film condensation on a vertical plate.

perpendicular to the plane of the paper. Consider a force balance on a differential element of the film at any axial location, x, shown in Fig. 11.7.

$$\rho_f g dx dy + \tau(y + dy) dx + p(x) dy = \tau(y) dx + p(x + dx) dy \tag{11.48}$$

$$\frac{\partial \tau}{\partial y} = -\rho_f g + \frac{\partial p}{\partial x} \tag{11.49}$$

Using the boundary layer type approximation $\left(\frac{\partial p}{\partial y} = 0 \right)$,

$$\frac{dp}{dx} = \rho_g g \tag{11.50}$$

Therefore, Eq. (11.49) changes to

$$\frac{\partial \tau}{\partial y} = -\left(\rho_f - \rho_g \right) g \tag{11.51}$$

According to Newton's law of viscosity,

$$\tau = \mu_f \frac{\partial u}{\partial y} \tag{11.52}$$

Substituting Eq. (11.52) in Eq. (11.51),

$$\frac{\partial^2 u}{\partial y^2} = \frac{-g}{\mu_f} \left(\rho_f - \rho_g \right) \tag{11.53}$$

Integrating Eq. (11.53) twice and applying the boundary conditions $u = 0$ at $y = 0$, and $\dfrac{\partial u}{\partial y} = 0$ at $y = \delta$, the velocity profile becomes

$$u(y) = \frac{(\rho_f - \rho_g) g \delta^2}{\mu_f} \left[\frac{y}{\delta} - \frac{y^2}{2\delta^2} \right] \tag{11.54}$$

The condensate mass flow rate at any location x is given by

$$\dot{m}(x) = \int_0^{\delta(x)} \rho_f u(y) dy = \frac{\rho_f (\rho_f - \rho_g) g \delta^3}{3\mu_f} \tag{11.55}$$

$$d\dot{m} = \frac{\rho_f (\rho_f - \rho_g) g \delta^2}{\mu_f} d\delta \tag{11.56}$$

From the heat transferred by conduction,

$$d\dot{m} = \frac{k_f (T_{sat} - T_w)}{\delta h_{fg}} dx \tag{11.57}$$

From Eq. (11.56) and Eq. (11.57),

$$\delta^3 d\delta = \frac{k_f (T_{sat} - T_w) \mu_f}{\rho_f (\rho_f - \rho_g) g h_{fg}} dx \tag{11.58}$$

On integrating Eq. (11.58) with the boundary condition $\delta = 0$ at $x = 0$,

$$\delta(x) = \left[\frac{4 k_f (T_{sat} - T_w) \mu_f}{\rho_f (\rho_f - \rho_g) g h_{fg}} x \right]^{1/4} \tag{11.59}$$

To obtain the local heat transfer coefficient,

$$h_x dx (T_{sat} - T_w) = \frac{k_f (T_{sat} - T_w)}{\delta(x)} \tag{11.60}$$

$$h_x = \frac{k_f}{\delta(x)} \tag{11.61}$$

Substituting Eq. (11.61) in Eq. (11.59),

$$h_x = \left[\frac{\rho_f (\rho_f - \rho_g) g h_{fg} k_f^3}{4\mu_f (T_{sat} - T_w) x} \right]^{1/4} \tag{11.62}$$

The average heat transfer coefficient is

$$\bar{h}_L = \frac{1}{L}\int_0^L h_x\, dx = 0.943\left[\frac{\rho_f\left(\rho_f - \rho_g\right)g h_{fg} k_f^3}{\mu_f\left(T_{sat} - T_w\right)L}\right]^{1/4} \tag{11.63}$$

All liquid properties in Eq. (11.62) and Eq. (11.63) are best evaluated at the mean film temperature $\left(T_{sat} + T_w\right)/2$, and h_{fg} and ρ_g at T_{sat}. Although Eq. (11.63) has been derived for a vertical flat plate, the expression can also be used for film condensation on the inside or outside surfaces of a vertical tube provided the radius is large compared with the film thickness at the bottom of the tube.

For an inclined flat plate that makes an angle θ with the horizontal, it can be readily seen that Eq. (11.63) modifies to

$$\bar{h}_L = 0.943\left[\frac{\rho_f\left(\rho_f - \rho_g\right)g\sin(\theta) h_{fg} k_f^3}{\mu_f\left(T_{sat} - T_w\right)L}\right]^{1/4} \tag{11.64}$$

The average heat transfer coefficient can also be expressed as a function of the condensate flow rate at the bottom of the plate.

From energy balance,

$$\bar{h}_L L\left(T_{sat} - T_w\right) = \dot{m}_L h_{fg}$$
$$L\left(T_{sat} - T_w\right) = \frac{\dot{m}_L h_{fg}}{\bar{h}_L} \tag{11.65}$$

Substituting Eq. (11.65) in Eq. (11.63),

$$\bar{h}_L = 0.925\left[\frac{\rho_f\left(\rho_f - \rho_g\right)g k_f^3}{\mu_f \dot{m}_L}\right]^{1/3} \tag{11.66}$$

The film Reynolds number at the bottom of the plate is

$$Re_\delta = \frac{\rho_f \bar{u}(L)D_h}{\mu_f} = \frac{\rho_f \bar{u}(L)4\delta}{\mu_f} = \frac{4\dot{m}_L}{\mu_f} = \frac{4\rho_f\left(\rho_f - \rho_g\right)g\delta^3}{3\mu_f^2} \tag{11.67}$$

Substituting Eq. (11.67) in Eq. (11.66) and assuming $\left(\rho_f - \rho_g\right) \approx \rho_f$,

$$\frac{\bar{h}_L}{k_f}\left[\frac{\mu_f^2}{\rho_f^2 g}\right]^{1/3} = 1.47 Re_\delta^{-1/3} \tag{11.68}$$

Experimental data suggests that the wave-free laminar regime exists at the bottom of the plate for $Re_\delta \le 30$.

It may be noted that Re_δ in Eq. (11.68) is unknown. Re_δ can be expressed as

$$Re_\delta = \frac{4\bar{h}_L \left(\dfrac{\mu_f^2}{\rho_f^2 g} \right)^{1/3}}{k_f} \phi \tag{11.69}$$

where parameter ϕ is

$$\phi = \frac{k_f L \left(T_{sat} - T_w \right)}{\mu_f h_{fg} \left(\dfrac{\mu_f^2}{\rho_f^2 g} \right)^{1/3}} \tag{11.70}$$

Substituting Eq. (11.69) in Eq. (11.68),

$$\frac{\bar{h}_L}{k_f} \left[\frac{\mu_f^2}{\rho_f^2 g} \right]^{1/3} = 0.943 \phi^{-1/4} \tag{11.71}$$

The following equations are recommended for different flow regimes at the bottom of the plate.

Wave-free laminar:

$$\frac{\bar{h}_L}{k_f} \left[\frac{\mu_f^2}{\rho_f^2 g} \right]^{1/3} = 0.943 \phi^{-1/4}, \quad \phi \leq 15.8 \tag{11.72}$$

Laminar-wavy (Kutateladze, 1963):

$$\frac{\bar{h}_L}{k_f} \left[\frac{\mu_f^2}{\rho_f^2 g} \right]^{1/3} = \frac{1}{\phi} (0.68 \phi + 0.89)^{0.82}, \; 15.8 \leq \phi \leq 2530 \tag{11.73}$$

Turbulent (Labuntsov, 1957):

$$\frac{\bar{h}_L}{k_f} \left[\frac{\mu_f^2}{\rho_f^2 g} \right]^{1/3} = \frac{1}{\phi} \left[(0.024 \phi - 53) Pr_f^{1/2} + 89 \right]^{4/3}, \quad \phi \geq 2530, \, Pr_f \geq 1 \tag{11.74}$$

As an improvement to the Nusselt theory, Rohsenow (1956) considered thermal advection effects and non-linear temperature profile across the film and showed that the latent heat of vaporization, h_{fg}, should be changed to

$$h'_{fg} = h_{fg} + 0.68 \, c_{pf} \left(T_{sat} - T_w \right) \tag{11.75}$$

It may be noted that the presence of non-condensable gases in vapor degrades the heat transfer performance, as they increase the thermal resistance at the liquid-vapor interface. In view of this, the general practice in the design of a condenser is to vent the non-condensable gas.

11.7 Condensation on horizontal tubes

In surface condensers, condensation occurs on a bank of horizontal tubes, as shown in Fig. 11.8.

Nusselt derived the following mean heat transfer coefficient relationship for laminar film condensation on the outside surface of a single horizontal tube:

$$h_{D1} = 0.725 \left[\frac{\rho_f (\rho_f - \rho_g) g \, h'_{fg} k_f^3}{D \mu_f (T_{sat} - T_w)} \right]^{1/4}$$

(11.76)

For a vertical column of N horizontal tubes, Jakob (1936) proposed the following mean heat transfer coefficient correlation:

$$\frac{\bar{h}_{DN}}{h_{D1}} = N^{-1/4}$$

(11.77)

where $\bar{h}_{DN}$ is the mean heat transfer coefficient for N tubes.

The mean heat transfer coefficient for the Nth tube in the column, h_{DN}, is given by

$$\frac{h_{DN}}{h_{D1}} = N^{3/4} - (N-1)^{3/4}$$

(11.78)

where h_{D1} is the mean heat transfer coefficient for the first tube.

Kern (1958) recommended the following correlations for $N \geq 10$:

$$\frac{\bar{h}_{DN}}{h_{D1}} = N^{-1/6}$$

(11.79)

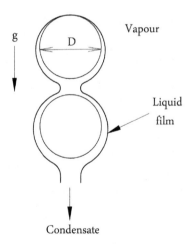

FIGURE 11.8

Film condensation on a bank of horizontal tubes.

$$\frac{h_{DN}}{h_{D1}} = N^{5/6} - (N-1)^{5/6} \qquad (11.80)$$

The above equations clearly show that the Nth tube will have a much lower heat transfer coefficient than the first, which is intuitively apparent.

Note that the Nusselt theory neglects the shear at the liquid-vapor interface. The interfacial shear becomes significant for condensation with flowing vapor and hence is an important factor that influences the heat transfer coefficient for condensation in tubes. Readers can refer to advanced texts (Collier and Thome, 1994; Stephan, 1992) for condensation in vertical and horizontal tubes.

11.8 Two-phase pressure drop

Pressure drop is an important consideration in the design of heat transfer equipment and more so for two-phase flows. Two-phase drop involves the pressure drop due to wall and interfacial shear stresses, the acceleration pressure drop for the case of evaporating flows, and the pressure drop due to gravitational head. The homogenous model and the separated flow model are usually used to estimate the two-phase pressure drop. The separated flow model, which is more complex, makes better prediction than the homogenous model but is beyond the scope of the present study. The interested reader can refer to advanced texts, such as Collier and Thome (1994) and Stephan (1992).

The homogenous model is based on the following assumptions:

(i) the vapor and liquid velocities are equal $\left(u_g = u_f = \overline{u}\right)$
(ii) the thermodynamic equilibrium exists between the phases, and
(iii) an appropriately defined single-phase friction factor can be used for two-phase flow.

The total pressure gradient is the sum of the frictional, acceleration, and gravitational head pressure gradients,

$$\left(\frac{dp}{dz}\right) = \left(\frac{dp}{dz}\right)_f + \left(\frac{dp}{dz}\right)_a + \left(\frac{dp}{dz}\right)_g \qquad (11.81)$$

The frictional pressure gradient is given by

$$-\left(\frac{dp}{dz}\right)_f = \frac{2 f_{TP} G^2 \overline{v}}{D} = \frac{2 f_{TP} G^2 \left(v_f + x v_{fg}\right)}{D} \qquad (11.82)$$

where G, v, and f_{TP} are the total mass flux (the product of density and velocity), specific volume, and the two-phase Fanning friction factor, respectively. Note that the Fanning friction factor is expressed as $\tau_w / \left(\rho u^2 / 2\right)$, where τ_w is the wall shear stress. The Darcy friction factor is four times the Fanning friction factor.

The acceleration pressure gradient is given by

$$-\left(\frac{dp}{dz}\right)_a = \frac{1}{A}\frac{d}{dz}\left(\dot{m}\bar{u}\right) = \frac{1}{A}\frac{d}{dz}\left(AG\bar{u}\right) = G^2\frac{d\bar{v}}{dz} \tag{11.83}$$

$$-\left(\frac{dp}{dz}\right)_a = G^2\frac{d}{dz}\left(v_f + xv_{fg}\right) \tag{11.84}$$

Assuming $v_{fg} \approx v_g$ and neglecting the compressibility of the liquid phase,

$$-\left(\frac{dp}{dz}\right)_a = G^2\left[x\frac{dv_g}{dz} + v_{fg}\frac{dx}{dz}\right]$$

$$= G^2\left[x\left(\frac{dv_g}{dp}\right)\left(\frac{dp}{dz}\right) + v_{fg}\frac{dx}{dz}\right] \tag{11.85}$$

The gravitational head pressure gradient is given by

$$-\left(\frac{dp}{dz}\right)_g = \bar{\rho}g\sin(\theta) = \frac{g\sin(\theta)}{\bar{v}} = \frac{g\sin(\theta)}{\left(v_f + xv_{fg}\right)} \tag{11.86}$$

where θ is the angle made by the inclined pipe with the horizontal. Substituting Eqs. (11.82), (11.85) and (11.86) in Eq. (11.81),

$$-\left(\frac{dp}{dz}\right) = \frac{\dfrac{2f_{TP}G^2}{D}\left[v_f + xv_{fg}\right] + G^2v_{fg}\dfrac{dx}{dz} + \dfrac{g\sin(\theta)}{\left(v_f + xv_{fg}\right)}}{1 + G^2x\left(\dfrac{dv_g}{dp}\right)} \tag{11.87}$$

The Fanning friction factor $\left(f_{TP}\right)$ can be evaluated using a mean two-phase viscosity, $\bar{\mu}$.

McAdams et al. (1942) suggested the following relationship for the mean two-phase viscosity.

$$\frac{1}{\bar{\mu}} = \frac{x}{\mu_g} + \frac{1-x}{\mu_f} \tag{11.88}$$

Cicchitti et al. (1960) suggested the relationship

$$\bar{\mu} = x\mu_g + (1-x)\mu_f \tag{11.89}$$

For turbulent flow, the two-phase Fanning friction factor can be evaluated using the Blasius equation

$$f_{TP} = \frac{0.079}{Re_{TP}^{1/4}} = \frac{0.079}{\left(\dfrac{GD}{\bar{\mu}}\right)^{1/4}} \tag{11.90}$$

Eq. (11.90) can be plugged into Eq. (11.87) to obtain the pressure gradient (dp/dz), and one needs to know the quality x and its axial gradient (dx/dz).

11.8.1 Total pressure drop

To obtain the total pressure drop, Eq. 11.87 needs to be integrated with respect to length. For simplicity, the following assumptions can be made.

1. The properties v_f and v_g remain constant over the length considered. The friction factor f_{TP} can be an average value, which remains constant over the length considered.
2. The term $G^2 x \left(dv_g / dp \right)$ in the denominator of Eq. (11.87) is much less than 1. The assumption of negligible compressibility of the gaseous phase is valid for many cases, especially at high pressures.

For the case of evaporation with the constant wall heat flux condition, which results in $dx/dz = $ constant, the integration of Eq. (11.87) over a length z with the inlet and outlet qualities being zero and x_e, respectively, leads to

$$\Delta p = \frac{2 f_{TP,avg} G^2 v_f z}{D} \left[1 + \frac{x_e}{2} \left(\frac{v_{fg}}{v_f} \right) \right] + G^2 v_{fg} x_e$$

$$+ \frac{g \sin(\theta) z}{v_{fg} x_e} \ln \left[1 + x_e \left(\frac{v_{fg}}{v_f} \right) \right] \qquad (11.91)$$

Note that Eq. (11.91) is based on the inlet condition being the saturated liquid (i.e., inlet quality $x = 0$). If the inlet is subcooled (i.e., the inlet temperature is less than saturation temperature), then the subcooled length can be determined and single-phase (liquid phase) pressure drop equation can be used over the subcooled length, neglecting the effect of any subcooled boiling on the pressure drop. There are models even for the accurate determination of pressure drop in the subcooled boiling condition (Ramesh and Gedupudi, 2019). Saturated water and steam properties are given in Tables 11.3 and 11.4, respectively. It is important to note that liquid properties mainly depend on temperature, and the effect of pressure on liquid properties is negligible and can be ignored. There are property tables even for the subcooled (compressed) liquid water, but they are not absolutely necessary. One can determine the liquid properties for the specified (or considered) temperature, making use of the saturated water properties (Table 11.3). For example, for water at 1 bar and 300 K, the saturated water properties corresponding to 300 K can be taken as the water properties.

Gravity head pressure drop is zero for horizontal flow. At higher qualities, the acceleration pressure drop will be high. But then the frictional pressure drop also increases proportionately, some of which comes from the interfacial shear that the homogenous model will not be able to predict (as it assumes equal velocities for both phases). Hence, there is a possibility of the homogenous model underestimating the frictional pressure drop. On the other hand, there is a possibility of the homogenous

Table 11.3 Properties of saturated water (Meyer et al., 1993; Harvey et al., 2000; IAPWS, 1994).

T (K)	p (bar)	ρ_f (kg/m³)	h_{fg} (kJ/kg)	c_{pf} (kJ/kg K)	$\mu_f \times 10^6$ (Ns/m²)	$k_f \times 10^3$ (W/mK)	Pr_f	$\sigma \times 10^3$ (N/m)	$\beta_f \times 10^6$ (K⁻¹)
273.16	0.006177	999.8	2502	4.22	1790.6	561	13.47	75.65	-68
275	0.006985	999.9	2497	4.214	1681.8	564.5	12.55	75.39	-35.5
280	0.009918	999.9	2485	4.201	1433.9	574	10.63	74.68	43.6
285	0.01389	999.5	2473	4.193	1239.4	583.5	8.91	73.95	112
290	0.0192	998.8	2461	4.187	1083.7	592.7	7.66	73.21	172
295	0.02621	997.8	2449	4.183	957.9	601.7	6.66	72.46	226
300	0.03537	996.5	2438	4.181	853.8	610.3	5.85	71.69	275
305	0.04719	995	2426	4.18	766.9	618.4	5.18	70.90	319
310	0.06231	993.3	2414	4.179	693.5	626	4.63	70.11	361
320	0.1055	989.3	2390	4.181	577.0	639.6	3.77	68.47	436
340	0.2719	979.5	2342	4.189	422.0	660.5	2.68	65.04	565
360	0.6219	967.4	2291	4.202	326.1	673.7	2.03	61.41	679
373.15	1.014	958.3	2257	4.216	281.7	679.1	1.75	58.91	751
400	2.458	937.5	2183	4.256	218.6	683.6	1.36	53.58	895
420	4.373	919.9	2123	4.299	186.7	682.5	1.18	49.41	1008
440	7.337	900.5	2059	4.357	162.8	678	1.05	45.10	1132
460	11.71	879.5	1989	4.433	144.3	670.2	0.955	40.66	1273
480	17.9	856.5	1912	4.533	129.7	659	0.892	36.11	1440
500	26.39	831.3	1825	4.664	117.7	643.9	0.853	31.47	1645
520	37.69	803.6	1730	4.838	107.6	624.6	0.833	26.78	1909
540	52.37	772.8	1622	5.077	98.8	600.1	0.835	22.08	2266
560	71.06	738	1499	5.423	90.8	570.1	0.864	17.40	2783
580	94.48	697.6	1353	5.969	83.4	534.6	0.931	12.80	3607
600	123.4	649.4	1176	6.953	75.7	495.3	1.06	8.38	5141
620	159	586.9	941	9.354	67.3	454.1	1.39	4.27	9092
640	202.7	481.5	560	25.94	55.3	414.9	3.46	0.81	39710
647	220.4	357.3	47	3905	46.9	1323	138	0.00	7735000

Table 11.4 Properties of saturated steam (Meyer et al., 1993; Harvey et al., 2000).

T (K)	p (bar)	ρ_g (kg/m³)	c_{pg} (kJ/kg K)	$\mu_g \times 10^6$ (Ns/m²)	$k_g \times 10^3$ (W/m K)	Pr_g	$\beta_g \times 10^6$ (K⁻¹)
273.16	0.006177	0.004855	1.884	9.216	17.07	1.02	3681
275	0.006985	0.005507	1.886	9.26	17.17	1.02	3657
280	0.009918	0.007681	1.891	9.382	17.44	1.02	3596
285	0.01389	0.01057	1.897	9.509	17.73	1.02	3538
290	0.0192	0.01436	1.902	9.641	18.03	1.02	3481
295	0.02621	0.01928	1.908	9.778	18.35	1.02	3428
300	0.03537	0.02559	1.914	9.92	18.67	1.02	3376
305	0.04719	0.0336	1.92	10.06	19.01	1.02	3328
310	0.06231	0.04366	1.927	10.21	19.37	1.02	3281
320	0.1055	0.07166	1.942	10.52	20.12	1.02	3195
340	0.2719	0.1744	1.979	11.16	21.78	1.01	3052
360	0.6219	0.3786	2.033	11.82	23.69	1.01	2948
373.15	1.014	0.5982	2.08	12.27	25.1	1.02	2902
400	2.458	1.369	2.218	13.19	28.35	1.03	2874
420	4.373	2.352	2.367	13.88	31.13	1.06	2914
440	7.337	3.833	2.56	14.57	34.23	1.09	3014
460	11.71	5.983	2.801	15.26	37.66	1.13	3181
480	17.9	9.014	3.098	15.95	41.45	1.19	3428
500	26.39	13.2	3.463	16.65	45.67	1.26	3778
520	37.69	18.9	3.926	17.38	50.44	1.35	4274
540	52.37	26.63	4.54	18.15	56.1	1.47	4994
560	71.06	37.15	5.41	19.01	63.34	1.62	6091
580	94.48	51.74	6.76	20.02	73.72	1.84	7904
600	123.4	72.84	9.181	21.35	91.05	2.15	1135
620	159	106.3	14.94	23.37	126.7	2.76	2000
640	202.7	177.1	52.59	27.94	250	5.88	7995
647	220.4	286.5	53340	39.72	1573	135	9274000

model overestimating the acceleration pressure drop, as it neglects the slip between the phases and assumes equal acceleration for both liquid and vapor phases. It may be noted that the homogenous model gives accurate results for dispersed flow – bubbly flow with small bubbles dispersed in liquid and spray flow or drop flow with small droplets dispersed in vapor, as the vapor velocity is approximately equal to the liquid velocity. The homogenous model also gives accurate results for higher pressures where vapor properties approach liquid properties. For accurate estimates of pressure drop over a wide range of operating conditions and vapor qualities, the correlations based on the separated flow model are recommended (Collier and Thome, 1994; Stephan, 1992; Jayaramu et al., 2019).

In summary, the preceding sections gave a step by step approach to quickly estimate heat transfer and pressure drop in boiling and condensation based on the empirical correlations reported by many researchers in the past. These should be good enough for commonly encountered engineering applications, though the reader is encouraged to take the cognizance of the assumptions under which these are valid. Should situations arise where these assumptions do not hold, more advanced correlations need to be used. If these too do not work for a problem, numerical modeling or in-house experimentation may be required to determine heat transfer coefficient and pressure drop.

Example 11.3: *Water flows upward in a vertical uniformly heated tube of length 0.9 m. The mass flux and the inlet subcooling are 500 kg/m²s and 0 K, respectively. The inner diameter of the tube is 25 mm. The system pressure is 10 bar. Determine the critical heat flux (given $F_1 = 0.478$, $F_2 = 0.662$, $F_3 = 0.4$, $F_4 = 0.0166$ for $p = 10$ bar) and the corresponding local quality at the exit.*

Solution:

Using Eq. (11.44),

$$q_{crit} = \frac{A' + \left(\dfrac{DG(\Delta h_{sub})_i}{4} \right)}{C' + z}$$

$$= \frac{A'}{C' + z} \left(since \left(\Delta T_{sub} \right)_i = 0 \right)$$

$$A' = 2.317 \left[\frac{DG h_{fg}}{4} \right] \frac{F_1}{1 + 0.0143 \, F_2 \, D^{1/2} G}$$

$$= 2.317 \left[\frac{0.025 \times 500 \times 2014.9 \times 10^3 \times 0.478}{4 \times \left(1 + 0.0143 \times 0.662 \times 0.025^{1/2} \times 500 \right)} \right]$$

$$= 3.988 \times 10^6$$

$$n = 2 - 0.00725 \, p$$
$$= 2 - 0.00725 \times 10 \ (since \ p = 10 \, bar)$$
$$= 1.9275$$

$$C' = \frac{0.077 \, F_3 \, D \, G}{1 + 0.347 \, F_4 \left(\dfrac{G}{1356} \right)^n}$$

$$= \frac{0.077 \times 0.4 \times 0.025 \times 500}{1 + 0.347 \times 0.0166 \times \left(\dfrac{500}{1356} \right)^{1.9275}}$$

$$= 0.384$$

Therefore,

$$q_{crit} = \frac{3.988 \times 10^6}{0.384 + 0.9}$$
$$= 3.104 \times 10^6 \text{ W/m}^2$$

From energy balance, the exit quality can be determined as follows:

$$G \frac{\pi D^2}{4} \left(h_f + x_e h_{fg} - h_f \right) = q_{crit} \pi D z$$

$$x_e = \frac{4 \, q_{crit} \, z}{G D h_{fg}}$$

$$= \frac{4 \times 3.104 \times 10^6 \times 0.9}{500 \times 0.025 \times 2014.9 \times 10^3}$$

$$x_e = 0.4436$$

It is instructive to note the order of magnitude of critical heat flux for flow boiling. As a rule of thumb it goes as MW/m², for water.

Example 11.4: *For the problem in Example 11.3, with heat flux equal to the critical heat flux, determine the local heat transfer coefficients and the corresponding wall temperatures at two locations: z = 0.4 m and 0.8 m from the inlet. Use the Chen correlation.*

Solution:
Saturated water properties corresponding to 10 bar are:

$k_f = 0.677$ W/mK, $Pr_f = 0.967$, $\sigma = 0.04226$ N/m, $c_{pf} = 4.403 \times 10^3$ J/kg K,
$\rho_g = 5.16$ kg/m³, $\rho_f = 887$ kg/m³, $h_{fg} = 2014.9 \times 10^3$ J/kg, $\mu_f = 148.5 \times 10^{-6}$ Ns/m²,
$\mu_g = 15.02 \times 10^{-6}$ Ns/m²

Using Eq. (11.34),

$$\frac{1}{X_{tt}} = \left(\frac{x}{1-x} \right)^{0.9} \times \left(\frac{\rho_f}{\rho_g} \right)^{0.5} \times \left(\frac{\mu_g}{\mu_f} \right)^{0.1}$$

$$q = q_{crit} = 3.104 \times 10^6 \text{ W/m}^2$$

At $z = 0.4$ m,

$$x = \frac{4 \, q \, z}{G D h_{fg}} = \frac{4 \times 3.104 \times 10^6 \times 0.4}{500 \times 0.025 \times 2014.9 \times 10^3} = 0.197$$

$$\frac{1}{X_{tt}} = \left(\frac{0.197}{1 - 0.197} \right)^{0.9} \times \left(\frac{887}{5.16} \right)^{0.5} \times \left(\frac{15.02 \times 10^{-6}}{148.5 \times 10^{-6}} \right)^{0.1}$$

$$= 2.944$$

$$Re_f = \frac{G D (1-x)}{\mu_f} = \frac{500 \times 0.025 \times (1 - 0.197)}{148.5 \times 10^{-6}} = 67612$$

Using Eq. (11.35), and since $\dfrac{1}{X_{tt}} > 0.1$,

$$F = 2.35\left(\frac{1}{X_{tt}} + 0.213\right)^{0.736}$$
$$= 2.35(2.944 + 0.213)^{0.736}$$
$$= 5.477$$

Using Eq. (11.32),

$$h_c = 0.023 Re_f^{0.8}\, Pr_f^{0.4}\left(\frac{k_f}{D}\right)F$$
$$= 0.023 \times (67612)^{0.8} \times (0.967)^{0.4} \times \left(\frac{0.677}{0.025}\right) \times 5.477$$
$$= 24555\,\text{W/m}^2\,\text{K}$$
$$= 24.55\,\text{kW/m}^2\,\text{K}$$

Using Eq. (11.33),

$$Re_{TP} = Re_f \times F^{1.25}$$
$$= 67612 \times 5.477^{1.25}$$
$$= 5.665 \times 10^5$$

Using Eq. (11.41),

$$S = \frac{1}{1 + 2.53 \times 10^{-6} \times Re_{TP}^{1.17}}$$
$$= 0.0683$$

Using Eq. (11.40),

$$h_{NcB} = 0.00122\left[\frac{k_f^{0.79} c_{pf}^{0.45} \rho_f^{0.49}}{\sigma^{0.5} \mu_f^{0.29} h_{fg}^{0.24} \rho_g^{0.24}}\right]\left(T_w - T_{sat}\right)^{0.24}\left(p_{sat}\left(T_w\right) - p_{sat}\left(T_{sat}\right)\right)^{0.75}(S)$$
$$= 0.00122\left[\frac{0.734 \times 43.62 \times 27.83}{0.2055 \times 0.0776 \times 32.58 \times 1.4826}\right]\left(T_w - T_{sat}\right)^{0.24}\left(p_{sat}\left(T_w\right) - p_{sat}\left(T_{sat}\right)\right)^{0.75}$$
$$\times 0.0683$$
$$= 0.0963 \times \left(T_w - T_{sat}\right)^{0.24}\left(p_{sat}\left(T_w\right) - p_{sat}\left(T_{sat}\right)\right)^{0.75}$$

Using Eq. (11.30),

$$h_{TP} = h_c + h_{NcB}$$
$$q = h_{TP}\left(T_w - T_{sat}\right)$$
$$3.104 \times 10^6 = \left[24555 + 0.0963 \times \left(T_w - T_{sat}\right)^{0.24}\left(p_{sat}\left(T_w\right) - p_{sat}\left(T_{sat}\right)\right)^{0.75}\right]\left(T_w - T_{sat}\right)$$

Since $q = q_{crit}$, $\left(T_w - T_{sat}\right)$ must be high.

We now face a special difficulty here. The heat flux on the left-hand side of the equation is known. The right-hand side contains T_w, which we are seeking. Unfortunately we cannot solve for T_w, as the expression on the right-hand side has the saturation pressure $p_{sat}\left(T_w\right)$, which is as yet unknown in view of the fact that T_w is unknown. Hence we have no option but to go for an iterative solution with an assumed value of T_w and iterating it to convergence.

$$\text{Let } \left(T_w - T_{sat}\right) = 50\ °\text{C}$$

$$T_w = T_{sat}(10 \text{ bar}) + 50 \text{ °C}$$
$$= 180 + 50 \text{ °C}$$
$$= 230 \text{ °C}$$
$$p_{sat}(T_w) = 27.98 \text{ bar} = 27.98 \times 10^5 \text{ N/m}^2$$
$$p_{sat}(T_w) - p_{sat}(T_{sat}) = (27.98 - 10) \times 10^5 = 17.98 \times 10^5 \text{ N/m}^2$$

$$q = h_{TP}(T_w - T_{sat})$$
$$= \left[24555 + 0.0963 \times (T_w - T_{sat})^{0.24} \left(p_{sat}(T_w) - p_{sat}(T_{sat}) \right)^{0.75} \right] (T_w - T_{sat})$$
$$= \left[24555 + 0.0963 \times 50^{0.24} \times (17.98 \times 10^5)^{0.75} \right] \times 50$$
$$= 1.832 \times 10^6 < 3.104 \times 10^6 \text{ W/m}^2$$

Let $(T_w - T_{sat}) = 65 \text{ °C}$

$$T_w = T_{sat}(10 \text{ bar}) + 65 \text{ °C}$$
$$= 180 + 65 \text{ °C}$$
$$= 245 \text{ °C}$$
$$p_{sat}(T_w) = p_{sat}(245 \text{ °C}) = 36.62 \times 10^5 \text{ N/m}^2$$
$$p_{sat}(T_w) - p_{sat}(T_{sat}) = (36.62 - 10) \times 10^5 = 26.62 \times 10^5 \text{ N/m}^2$$

$$q = h_{TP}(T_w - T_{sat})$$
$$q = \left[24555 + 0.0963 \times 65^{0.24} \times (26.62 \times 10^5)^{0.75} \right] \times 65$$
$$= 2.719 \times 10^6 < 3.104 \times 10^6 \text{ W/m}^2$$

Let $(T_w - T_{sat}) = 70 \text{ °C}$

$$T_w = 180 + 70 \text{ °C} = 250 \text{ °C}$$
$$p_{sat}(T_w) = 39.776 \times 10^5 \text{ N/m}^2$$
$$p_{sat}(T_w) - p_{sat}(T_{sat}) = 29.776 \times 10^5 \text{ N/m}^2$$
$$q = h_{TP}(T_w - T_{sat})$$
$$= 3.058 \times 10^6 \approx 3.104 \times 10^6 \text{ W/m}^2$$

Therefore, at $z = 0.4$ m,

$$h_{TP} = \frac{3.058 \times 10^6}{70} = 43685 \text{ W/m}^2\text{K}$$
$$T_w = 250 \text{ °C}$$

At $z = 0.8$ m,

$$x = \frac{4qz}{GDh_{fg}} = \frac{4 \times 3.104 \times 10^6 \times 0.8}{500 \times 0.025 \times 2014.9 \times 10^3} = 0.394$$

$$\frac{1}{X_{tt}} = \left(\frac{0.394}{1 - 0.394} \right)^{0.9} \times \left(\frac{887}{5.16} \right)^{0.5} \times \left(\frac{15.02 \times 10^{-6}}{148.5 \times 10^{-6}} \right)^{0.1} = 7.07$$

$$Re_f = \frac{GD(1-x)}{\mu_f} = \frac{500 \times 0.025 \times (1 - 0.394)}{148.5 \times 10^{-6}} = 51041$$

$$F = 2.35\left(\frac{1}{X_{tt}} + 0.213\right)^{0.736}$$

$$= 2.35 \times (7.07 + 0.213)^{0.736}$$

$$= 10.13$$

$$h_c = 0.023 \ Re_f^{0.8} Pr_f^{0.4} \frac{k_f}{D} F$$

$$= 36349 \ \text{W/m}^2\text{K}$$

$$Re_{TP} = Re_f \times F^{1.25}$$

$$= 9.244 \times 10^5$$

$$S = \frac{1}{1 + 2.53 \times 10^{-6} Re_{TP}^{1.17}} = 0.0398$$

Using $S = 0.0398$ in the expression for h_{NcB},

$$h_{NcB} = 0.0561 \times \left(T_w - T_{sat}\right)^{0.24} \left(p_{sat}\left(T_w\right) - p_{sat}\left(T_{sat}\right)\right)^{0.75}$$

$$h_{TP} = h_c + h_{NcB}$$

$$q = h_{TP}\left(T_w - T_{sat}\right) = \left[36349 + 0.0561 \times \left(T_w - T_{sat}\right)^{0.24} \left(p_{sat}\left(T_w\right) - p_{sat}\left(T_{sat}\right)\right)^{0.75}\right]\left(T_w - T_{sat}\right)$$

Let $\left(T_w - T_{sat}\right) = 65 \ °C$

$$T_w = T_{sat} + 65 \ °C = 180 + 65 \ °C = 245 \ °C$$

$$p_{sat}\left(T_w\right) - p_{sat}\left(T_{sat}\right) = 26.62 \times 10^5 \ \text{N/m}^2$$

Substituting the above values in the expression for q,

$$q = 3.017 \times 10^6 \approx 3.104 \times 10^6 \ \text{W/m}^2$$

Therefore, at z = 0.8 m,

$$h_{TP} = \frac{3.017 \times 10^6}{65} = 46415 \ \text{W/m}^2\text{K}$$

$$T_w = 245 \ °C$$

Note that this sample problem does not ask for the calculation of the heat transfer coefficient at z = 0.9 m (tube exit). It is intentional, as the heat flux is equal to the critical heat flux, which indicates dryout at the exit. The Bowring CHF correlation presented in the chapter (or for that matter any other CHF correlation) estimates the CHF that corresponds to the onset of dryout (or DNB) at the tube exit. The Chen correlation should not be used to calculate the two-phase heat transfer coefficient for dryout and post-dryout (spray flow or drop flow) heat transfer, though the correlation gives some values of the heat transfer coefficient. The heat transfer coefficient calculated at the tube exit using the Chen correlation can at best be taken as the value prior to the onset of dryout at the tube exit.

Example 11.5: *For the problem in Example 11.3, with heat flux equal to the critical heat flux, evaluate the pressure drop across the tube using the homogeneous model.*

Solution:
Fluid properties corresponding to 10 bar are:

$$\mu_f = 148.5 \times 10^{-6} \text{ Ns/m}^2, \mu_g = 15.02 \times 10^{-6} \text{ Ns/m}^2$$

$$\text{Specific volumes } v_f = 1.127 \times 10^{-3} \text{ m}^3/\text{kg}, v_g = 0.1937 \text{ m}^3/\text{kg}$$

$$v_{fg} = v_g - v_f = 0.1926 \text{ m}^3/\text{kg}$$

The tube exit quality, $x_e = 0.4436$
Fanning friction factor at the tube inlet:

$$Re_f = \frac{GD}{\mu_f} = \frac{500 \times 0.025}{148.5 \times 10^{-6}} = 84175$$

Using the Blasius equation,

$$f_f = \frac{0.079}{Re_f^{1/4}} = 0.00464$$

Fanning friction factor at the tube exit:
Two-phase viscosity can be calculated from the relation

$$\frac{1}{\mu} = \frac{x}{\mu_g} + \frac{1-x}{\mu_f}$$

$$= \frac{0.4436}{15.02 \times 10^{-6}} + \frac{(1-0.4436)}{148.5 \times 10^{-6}}$$

$$\bar{\mu} = 30.4 \times 10^{-6} \text{ Ns/m}^2$$

$$Re_{TP} = \frac{GD}{\bar{\mu}} = \frac{500 \times 0.025}{30.4 \times 10^{-6}} = 4.16 \times 10^5$$

Using the Blasius equation,

$$f_{TP} = \frac{0.079}{Re_{TP}^{1/4}} = 0.0031$$

The average value of f_{TP} over the tube length is given by

$$f_{TP,avg} = \frac{f_f + f_{TP}}{2} = 0.00387$$

The total pressure drop can be determined using Eq. (11.91). The frictional pressure drop from the homogeneous model is given by

$$\Delta p_f = \frac{2 f_{TP,avg} G^2 v_f z}{D} \left[1 + \frac{v_{fg}}{v_f} \left(\frac{x_e}{2} \right) \right]$$

$$= \frac{2 \times 0.00387 \times 500^2 \times 1.127 \times 10^{-3} \times 0.9}{0.025} \left[1 + \left(\frac{0.1926 \times 0.4436}{2 \times 1.127 \times 10^{-3}} \right) \right]$$

$$= 3054 \text{ N/m}^2$$

The acceleration pressure drop from the homogeneous model is given by

$$\Delta p_a = G^2 \, v_{fg} \, x_e$$
$$= 500^2 \times 0.1926 \times 0.4436$$
$$= 21359 \, \text{N/m}^2$$

The gravitational head pressure drop from the homogeneous model is given by

$$\Delta p_g = \frac{g \, \sin(\theta) \, z}{v_{fg} \, x_e} \ln\left[1 + \frac{v_{fg}}{v_f} x_e\right]$$
$$= \frac{9.81 \times 1 \times 0.9}{0.1926 \times 0.4436} \ln\left[1 + \left(\frac{0.1926}{1.127 \times 10^{-3}}\right) \times 0.4436\right]$$
$$= 449 \, \text{N/m}^2$$

The total pressure drop

$$\Delta p_{total} = \Delta p_f + \Delta p_a + \Delta p_g$$
$$= 3.054 + 21.359 + 0.449 \, \text{kN/m}^2$$
$$= 24.862 \, \text{kN/m}^2$$

Example 11.6: *Saturated water at 0.123 bar with a mass flux of 400 kg/m²s enters a vertical tube and flows upward. There is saturated steam at 1.013 bar condensing on the tube outer surface. The length and diameter of the tube are 0.6 m and 0.025 m, respectively. Assume laminar film condensation. Determine*

(a) The average heat transfer coefficient on the condensation side.
(b) The average heat transfer coefficient on the boiling side, which can be assumed to be approximately equal to the mean of the local heat transfer coefficients at the inlet and outlet of the tube. You may use the Chen correlation.
(c) The average tube wall temperature.
(d) The rate of condensation (total), kg/s.
(e) The exit quality on the boiling side.

Solution:
    The average heat transfer coefficient relationship for laminar film condensation on a vertical plate, given by Eq. (11.63), can also be used for condensation on a vertical tube.

$$\bar{h}_f = 0.943 \left[\frac{\rho_f (\rho_f - \rho_g) g \, h_{fg} \, k_f^3}{\mu_f \, z (T_{sat} - T_w)}\right]^{1/4}$$

    Though all liquid properties in the above expression correspond to the film temperature $(T_w + T_{sat})/2$, and h_{fg} corresponds to T_{sat}, let all the properties be evaluated at T_{sat} since T_w is unknown.

$$\bar{h}_f = 0.943 \left[\frac{958 \times (958 - 0.598) \times 9.81 \times 2257 \times 10^3 \times 0.681^3}{283.1 \times 10^{-6} \times 0.6 \times (100 - T_w)} \right]^{1/4}$$

$$= 0.943 \times 13939.67 \times \frac{1}{(100 - T_w)^{1/4}}$$

The two-phase heat transfer coefficient for saturated boiling is obtained using Eq. (11.30),

$$h_{TP} = h_c + h_{NcB}$$

Using Eq. (11.34) and the saturated fluid properties at 0.123 bar,

$$\frac{1}{X_{tt}} = \left(\frac{\rho_f}{\rho_g}\right)^{0.5} \left(\frac{\mu_g}{\mu_f}\right)^{0.1} \left(\frac{x}{1-x}\right)^{0.9}$$

$$= \left(\frac{988}{0.083}\right)^{0.5} \times \left(\frac{10.1 \times 10^{-6}}{547.8 \times 10^{-6}}\right)^{0.1} \times \left(\frac{x}{1-x}\right)^{0.9}$$

$$= 73.17 \left(\frac{x}{1-x}\right)^{0.9}$$

From Eq. (11.35),

$$F = 1 \quad \text{for} \quad \left(\frac{1}{X_{tt}}\right) \leq 0.1$$

$$= 2.35 \left(\frac{1}{X_{tt}} + 0.213\right)^{0.736} \quad \text{for} \quad \frac{1}{X_{tt}} > 0.1$$

From Eq. (11.32),

$$h_c = 0.023 \left[\frac{G(1-x)D}{\mu_f}\right]^{0.8} \left(\frac{\mu c_p}{k}\right)_f^{0.4} \left(\frac{k_f}{D}\right)(F)$$

$$= 0.023 \left(\frac{400 \times 0.025}{547.8 \times 10^{-6}}\right)^{0.8} \times (1-x)^{0.8} \times (3.56)^{0.4} \times \left(\frac{0.643}{0.025}\right)(F)$$

$$= 2521.53 \times (1-x)^{0.8} \times F$$

From Eq. (11.40),

$$h_{NcB} = 0.00122 \left[\frac{k_f^{0.79} c_{pf}^{0.45} \rho_f^{0.49}}{\sigma^{0.5} \mu_f^{0.29} h_{fg}^{0.24} \rho_g^{0.24}}\right] (T_w - T_{sat})^{0.24} (p_{sat}(T_w) - p_{sat}(T_{sat}))^{0.75}(S)$$

$$= 0.00122 \left[\frac{0.7055 \times 42.16 \times 29.34}{0.2606 \times 0.1133 \times 33.92 \times 0.55}\right] \times \Delta T_{sup}^{0.24} \times \Delta p_{sup}^{0.75} \times S$$

From Eq. (11.33),

$$Re_{TP} = Re_f \times F^{1.25}$$

$$= \frac{G(1-x)D}{\mu_f} \times F^{1.25}$$

$$= 18254.8 \times (1-x) \times F^{1.25}$$

From Eq. (11.41),

$$S = \frac{1}{1 + 2.53 \times 10^{-6} Re_{TP}^{1.17}}$$

$$h_{NcB} = 1.953 \times \left(T_w - T_{sat}\right)^{0.24} \times \left(p_{sat}\left(T_w\right) - p_{sat}\left(T_{sat}\right)\right)^{0.75} \times S$$

$$\bar{h}_{TP} \approx \frac{h_{TP,in} + h_{TP,out}}{2}$$

$$Q_{condens} = \bar{h}_f A\left(T_{sat} - T_w\right) = \bar{h}_f A\left(100 - T_w\right)$$

$$= 13145.11 \times \frac{1}{\left(100 - T_w\right)^{1/4}} \times \left(100 - T_w\right) \times A$$

$$Q_{boil} = \bar{h}_{TP}\ A\left(T_w - T_{sat}\right) = \bar{h}_{TP}\ A\left(T_w - 50\right)$$

[Note that the saturation temperature on the boiling side is 50 °C and that on the condensation side is 100 °C.]

Please note that in this problem too, T_w is unknown, and consequently, "tedious" iterations to solve the problem are unavoidable. Discerning readers will note that problems on boiling and condensation are eminently solvable with a computer code or tools like MATLAB or SCILAB.

Initial guess:

$$\text{Let}\quad T_w = \frac{100 + 50}{2} = 75\,°C$$

$$\begin{aligned}
Q_{condens} &= 146967\ A\ \text{J/s}\\
&= 146967 \times \pi D z\\
&= 146967 \times \pi \times 0.025 \times 0.6\\
&= 6925.45\,\text{W}
\end{aligned}$$

At inlet, $x = 0$ (no two-phase) and hence $h_{NcB} = 0$. Also, at inlet, $F = 1$ as $\dfrac{1}{X_{tt}} = 0$ for $x = 0$. Therefore,

$$\begin{aligned}
h_{TP,in} &= h_c\\
&= 2521.53\,\text{W/m}^2\text{K}
\end{aligned}$$

$$G\frac{\pi D^2}{4} h_{fg}\, x_{out} = Q_{condens}$$

$$\begin{aligned}
x_{out} &= \frac{6925.45}{400 \times \dfrac{\pi}{4} \times 0.025^2 \times 2382.7 \times 10^3}\\
&= 0.0148
\end{aligned}$$

$$\frac{1}{X_{tt}} = 1.6726$$

$$F = 3.748$$

$$h_c = 9338.63\,\text{W/m}^2\text{K}$$

$$Re_{TP} = 93788.79$$

$$S = 0.375$$

$$\begin{aligned}
h_{NcB} &= 1.953 \times (75 - 50)^{0.24} \times \left(38.59 \times 10^3 - 12.33 \times 10^3\right)^{0.75} \times 0.375\\
&= 3271.2\,\text{W/m}^2\text{K}
\end{aligned}$$

$$h_{TP,out} = h_{c,out} + h_{NcB,out}$$
$$= 9338.63 + 3271.2$$
$$= 12609.83 \, \text{W/m}^2\text{K}$$

$$\bar{h}_{TP} = \frac{2521.53 + 12609.83}{2}$$
$$= 7565.68 \, \text{W/m}^2\text{K}$$

$$Q_{boil} = \bar{h}_{TP} \, \pi \, D \, z \left(T_w - T_{sat}\right)$$
$$= 7565.68 \times \pi \times 0.025 \times 0.6 \times (75 - 50)$$
$$= 8912.84 \, \text{W}$$
$$Q_{condens} = 6925.45 \, \text{W}$$

Since there is a large difference between Q_{boil} and $Q_{condens}$, the assumed T_w is incorrect.

Let $T_w = 70 \,°\text{C}$. Repeating the above calculations for $T_w = 70 \,°\text{C}$,

$$Q_{boil} = 7021.49 \, \text{W}$$
$$Q_{condens} = 7940.24 \, \text{W}$$

Now $Q_{condens} > Q_{boil}$, but the difference is smaller.
Let $T_w = 71 \,°\text{C}$. Repeating the above calculations for $T_w = 71 \,°\text{C}$,

$$Q_{boil} = 7483.4 \, \text{W}$$
$$Q_{condens} = 7740.8 \, \text{W}$$

Q_{boil} and $Q_{condens}$ are quite close now.
Let $T_w = 71.5 \,°\text{C}$. Repeating the above calculations for $T_w = 71.5 \,°\text{C}$,

$$Q_{boil} = 7565.67 \, \text{W}$$
$$Q_{condens} = 7640.57 \, \text{W}$$

Now the difference between Q_{boil} and $Q_{condens}$ is very small, and it is approximately 1%.

The corresponding average heat transfer coefficients on the boiling side and the condensation side are 7467.6 W/m² K and 5689 W/m² K, respectively. Therefore,

(a) $\bar{h}_f = 5689 \, \text{W/m}^2\text{K}$

(b) $\bar{h}_{TP} = 7467 \, \text{W/m}^2\text{K}$

(c) $T_w = 71.5 \,°\text{C}$

(d) $\dot{m}_{condens} = \dfrac{Q}{h_{fg}} = \dfrac{7640.57}{2257 \times 10^3} = 3.385 \times 10^{-3} \, \text{kg/s}$

(e) $x_{out} = \dfrac{Q}{G \dfrac{\pi D^2}{4} h_{fg}} = \dfrac{7640.57}{400 \times \dfrac{\pi}{4} \times 0.025^2 \times 2382.7 \times 10^3} = 0.0163$

Note

The calculation can be repeated taking liquid properties (other than h_{fg}) in the expression for laminar film heat transfer coefficient corresponding to $(T_w + T_{sat})/2$. The calculation process needs to be repeated until the obtained T_w matches the value used for evaluating the liquid properties. The final T_w is expected to be slightly different from the one obtained above (71.5 °C) and so are the values in (a), (b), (d), and (e).

As the quality continuously changes (and hence the heat transfer coefficient) along the flow direction, accurate design of heat exchangers involving flow boiling or flow condensation needs numerical methods. This is because the local heat transfer coefficient is a complex function of quality and moreover the local heat transfer coefficient also depends on the wall superheat (for boiling) or wall subcooling (for condensation), which is unknown. Hence, for accuracy, it needs to be solved numerically (and by iteration) taking small control volumes and using the energy balance and the heat convected.

Problems

11.1 Refer to Fig. 11.4. Derive the expressions for the minimum and maximum sizes of the cavity for nucleation from a heated surface, with the liquid pool at the saturation temperature.

11.2 A hot copper sphere with a diameter of 1.5 cm and a temperature of 115 °C is dipped into a vessel containing water at atmospheric pressure and temperature maintained at 100 °C. Boiling commences in the nucleate boiling regime. Estimate the time required for the sphere to reach 105 °C. The density and specific heat of copper are 8933 kg/m³ and 475 J/kgK, respectively. Note the following assumptions:

   (i) Sphere temperature is uniform (internal conduction resistance can be neglected).

   (ii) Use the Rohsenow correlation (over the entire range: 5 to 15 K superheat). C_{sf} value in the Rohsenow correlation is 0.01. Fluid properties correspond approximately to T_{sat}.

11.3 Plot the variation of critical heat flux for pool boiling of water on a flat surface as a function of system pressure, considering the variation of saturated fluid properties with pressure.

11.4 Consider the copper sphere in Problem 11.2. During a quenching operation, the sphere initially at 300 °C is suddenly dipped into a water bath containing saturated water at atmospheric pressure. Since the temperature difference is large, boiling begins in the film boiling regime. Determine (a) the initial heat transfer coefficient, (b) the initial rate of change of sphere temperature, and (c) the initial rate of heat transfer. Neglect the effect of radiation.

11.5 Water boils on a flat polished stainless steel surface ($C_{sf} = 0.013$) at atmospheric pressure. (a) Estimate the heat flux from the surface to water if the surface temperature is 400 °C. If it is in the film boiling regime, the emissivities of the wall and liquid surfaces may be assumed to be equal to 1. (b) For the same heat flux, determine the wall temperature if it were in the nucleate boiling region.

11.6 A furnace wall riser, 19 m long, 76 mm outer diameter, and 6 mm thick receives saturated water at 86 bar and 1.5 m/s velocity. The average wall heat flux based on the outer diameter is 95 kW/m². Determine

(a) the local heat transfer coefficient and the inner wall temperature at the riser exit,

(b) the average heat transfer coefficient, and

(c) the pressure drop across the riser.

11.7 Consider Example 11.3. Assume an inlet subcooling of 40 K and a single-phase heat transfer coefficient of 5 kW/m²K. Determine

(a) the lower limiting value of the critical heat flux,

(b) the upper limiting value of the critical heat flux,

(c) the critical heat flux,

(d) the exit quality when the critical heat flux occurs, and

(e) the percentage change in the critical heat flux if the mass flux is doubled (keeping the other parameters constant).

11.8 In a cogeneration plant, steam is used for maintaining a desired drying temperature. Saturated steam at 100 bar and 0.48 kg/s enters a 5 m long vertical tube of 15 cm diameter and flows downward. Steam transfers heat (by convection and radiation) to the surroundings maintained at 50 °C. The overall heat transfer coefficient between the steam and the surroundings is 104 W/m² K. Determine the exit quality, and comment on the contributions of the frictional, acceleration (or deceleration), and gravitational head pressure drops to the total pressure drop, using the homogeneous model.

11.9 Saturated steam at atmospheric pressure condenses on a vertical flat plate of 1.25 m long, which is maintained at 60 °C by the flow of cool water on the other side of the plate. Determine the flow regime (wave-free laminar, wavy laminar, or turbulent) at the bottom of the plate, and the rate of steam condensation per unit width of the plate.

11.10 A condenser design specifies a square (inline) array of 225 horizontal tubes, each of 16 mm outer diameter and 1 m long, to condense saturated steam at 0.123 bar on the outer surface of the tubes, with a tube wall temperature of 45 °C maintained by the flow of cooling water through the tubes. Determine

(a) the heat transfer coefficient for the top-most tube,

(b) the heat transfer coefficient for the bottom-most tube,

(c) the average heat transfer coefficient for the array of tubes, and

(d) the total condensation rate.

References

Berenson, P.I., 1960. Transition Boiling Heat Transfer, fourth ed., Natl. Heat Transfer Conf., AIChE Preprint 1 8, Buffalo, NY.

Bowring, R.W., 1972. A simple but accurate round tube uniform heat flux, dryout correlation over the pressure range 0.7-1 7 MN/m2 (100-2500 psia). AEEW-R 789.

Bromley, L.A., 1950. Heat transfer in stable film boiling. Chem. Engng. Prog. 46, 221–227.

Chen, J.C., 1963. A correlation for boiling heat transfer t o saturated fluids in convective flow. ASME preprint 63-HT-34 presented at 6th National Heat Transfer Conference, Boston, 1, 1–14 August.

Cicchitti, A., Lombardi, C., Silvestri, M., Soldaini, G., Zavattarelli, R., 1960. Two-phase cooling experiments-pressure drop, heat transfer and burnout measurements. Energia Nucleare 7 (6), 407–425.

Collier, J.G., Thome, J.R., 1994. Boiling and Condensation, 3rd edition Oxford University Press, Oxford.

Fishenden, M., Saunders, O., 1950. An Introduction to Heat Transfer. Oxford University Press, Oxford.

Forster, H.K., Zuber, N., 1955. Dynamics of vapour bubbles and boiling heat transfer. AIChE J. 1 (4), 531–535.

Gedupudi, S., Zu, Y.Q., Karayiannis, T.G., Kenning, D.B.R., Yan, Y.Y., 2011. Confined bubble growth during flow boiling in a mini/micro-channel of rectangular cross-section Part I: Experiments and 1-D modelling. Int. J. Therm. Sci. 50 (3), 250–266.

Harvey, A.H., Peskin, A.P., Klien, S.A., 2000. NIST/ASME Steam Properties. National Institute of Standards and Technology, Gaithersburg, MD.

Hsu, Y.Y., 1962. On the size of range of active nucleation cavities on a heating surface. J. Heat Transfer 84, 207.

IAPWS (International Association for the Properties of Water and Steam), 1994. Release on surface tension of ordinary water substance. Technical report. Available from: http://www.iapws.org/.

Jain, S., Jayaramu, P., Gedupudi, S., 2019. Modeling of pressure drop and heat transfer for flow boiling in a mini/micro-channel of rectangular cross-section. Int. J. Heat Mass Transf 140, 1029–1054.

Jakob, M., 1936. Heat transfer in evaporation and condensation-II. Mech. Engng. 58, 729.

Jayaramu, P., Gedupudi, S., Das, S.K., 2019. Influence of heating surface characteristics on flow boiling in a copper microchannel: Experimental investigation and assessment of correlations. Int. J. Heat Mass Transf 128, 290–318.

Kandlikar, S., Garimella, S., Li, D., Colin, S., King, M., 2006. Heat Transfer and Fluid Flow in Minichannels and Microchannels, Elsevier, Amsterdam, pp. 175–226.

Karayiannis, T.G., Mahmoud, M.M., 2017. Flow boiling in microchannels: Fundamentals and applications. Appl. Therm. Eng. 115, 1372–1397.

Kern, D.Q., 1958. Mathematical development of tube loading in horizontal condensers. AIChE J. 1 (4), 157–160.

Kutateladze, S.S., 1963. Fundamentals of Heat Transfer. Academic Press, New York.

Labuntsov, D.A., 1957. Heat transfer in film condensation of pure steam on vertical surfaces and horizontal tubes. Teploenergetika 4, 72.

McAdams, W.H., Woods, W.K., Bryan, R.L., 1942. Vaporization inside horizontal tubes-II-Benzene-oil mixtures. Trans. ASME 64, 193.

Meyer, C.A., McClintock, R.B., Sivestri, G.J., Spencer, R.C., 1993. ASME Steam Tables, sixth ed. American Society of Mechanical Engineers, New York.

Nusselt, W., 1916. Die Oberflachenkondensation des Wasserdampfes. Zeitschr. Ver. deutsch Ing. 60 (54 1), 569.

Ramesh, B., Gedupudi, S., 2019. On the prediction of pressure drop in subcooled flow boiling of water. Appl. Therm. Eng. 155, 386–396.

Rohsenow, W.M., 1952. A method of correlating heat transfer data for surface boiling of liquids. Trans. ASME 74, 969–975.

Rohsenow, W.M., 1956. Heat transfer and temperature distribution in laminar film condensation. Trans. ASME 79, 645–648.

Stephan, K., 1992. Heat Transfer in Condensation and Boiling. Springer-Verlag, New York.

Tong, L.S., Tang, Y.S., 1997. Boiling Heat Transfer and Two-Phase Flow, 2nd edition Taylor and Francis, Oxfordshire.

Zuber, N., 1958. On the stability of boiling heat transfer. Trans. ASME 80, 711.

Introduction to convective mass transfer

12

12.1 Introduction

Many heat transfer problems are accompanied by mass transfer. For example, water cooling in cooling towers used in thermal power plants involves both heat transfer and mass transfer. A typical example of mass transfer is the immediate spread of a room freshener in an air-conditioned room. Convective mass transfer closely resembles convective heat transfer. The analogy between mass transfer and heat transfer is especially true for low concentrations of the species in the fluid and low mass transfer rates of the species. In this chapter, the laws that govern diffusion and convective mass transfer are discussed. This is followed by a presentation of key equations for the determination of mass transfer coefficients for a gas flow over a volatile liquid or solid surface, based on the convective heat and mass analogy. Simultaneous heat and mass transfer for the case of air blowing over a wet surface is discussed.

12.2 Fick's law of diffusion

According to Fick's law of diffusion, the molar flux of a chemical species A in a stationary binary mixture of species A and B in the x direction is proportional to the concentration gradient in the same direction. Mathematically,

$$J_{A,x} = -D_{AB} \frac{dC_A}{dx} \tag{12.1}$$

where $J_{A,x}$ is the molar flux of the species A in the x direction, D_{AB} is the mass diffusivity or the diffusion coefficient of the species A diffusing through species B; $\frac{dC_A}{dx}$ is the concentration gradient in the x direction, with C_A being the molar concentration of species A in the binary mixture. The units of $J_{A,x}$, D_{AB}, and C_A are kmol/m^2 s, m^2/s, and kmol/m^3, respectively.

Fick's law of diffusion is similar to Fourier's law of heat conduction, which is expressed as

$$q = -k \frac{dT}{dx} \tag{12.2}$$

where k is the thermal conductivity.

Heat Transfer Engineering. http://dx.doi.org/10.1016/B978-0-12-818503-2.00012-5

In terms of mole fraction of species, y_A, the Fick's rate equation is

$$J_{A,x} = -CD_{AB}\frac{dy_A}{dx} \tag{12.3}$$

where C is the total molar concentration equal to the sum of C_A and C_B for a binary mixture. The Fick's rate equation for mass flux, $j_{A,x}$, is

$$j_{A,x} = -D_{AB}\frac{d\rho_A}{dx} = -\rho D_{AB}\frac{dw_A}{dx} \tag{12.4}$$

where ρ, ρ_A and w_A are the total mass density (kg/m³), the mass density (mass concentration) of species A, and the mass fraction of species A, respectively. The unit of $j_{A,x}$ is kg/m²s. For a binary mixture of A and B, ρ is the sum of ρ_A and ρ_B.

For a one-dimensional case, the mass diffusion rate (kg/s) of species A through a nonreacting plane wall (medium B) of thickness L is given by

$$\dot{m}_{A,plane} = D_{AB}A\frac{(\rho_{A,1} - \rho_{A,2})}{L} = \rho D_{AB}A\frac{(w_{A,1} - w_{A,2})}{L} \tag{12.5}$$

where A is the area of the plane wall normal to the direction of mass diffusion, and $\rho_{A,1}$, $\rho_{A,2}$ are the mass concentrations of species A on either end of the wall, as shown in Fig. 12.1.

Similarly, the expressions for steady one-dimensional mass diffusion rates through nonreacting cylindrical and spherical walls are given by

$$\dot{m}_{A,cylinder} = 2\pi L D_{AB}\frac{(\rho_{A,1} - \rho_{A,2})}{\ln(r_2/r_1)} \tag{12.6}$$

$$\dot{m}_{A,sphere} = 4\pi r_1 r_2 D_{AB}\frac{(\rho_{A,1} - \rho_{A,2})}{(r_2 - r_1)} \tag{12.7}$$

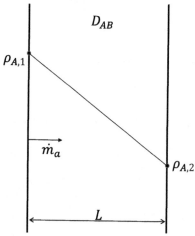

FIGURE 12.1 Mass concentration gradient of species A across a plane wall and diffusion of the species.

Note that in the case of a moving medium, $J_{A,x}$ is the diffusive molar flux of species A in the x direction relative to molar average velocity in the x direction, and $j_{A,x}$ is the diffusive mass flux of species A in the x direction relative to mass average velocity in the x direction. Table 12.1 shows some values of the mass diffusion coefficient or diffusivity $D_{A,B}$ for binary mixtures at atmospheric pressure. The kinetic theory of gases shows that, for dilute gases that follow the ideal gas law, the diffusion coefficients approximately vary with pressure and temperature as

$$D_{AB} \sim \frac{T^{3/2}}{P} \tag{12.8}$$

Table 12.1 Binary diffusion coefficients at 1 atm, Barrer (1941), Geankoplis (1972), Mills (1995), Perry (1963), Reid et al. (1977), Thomas (1991), and Black (1980).

	Component A	Component B	T (K)	D_{AB} (m²/s)
Dilute gas mixtures				
	Ammonia, NH_3	Air	298	2.6×10^{-5}
	Benzene	Air	298	0.88×10^{-5}
	Carbon dioxide	Air	298	1.6×10^{-5}
	Chlorine	Air	298	1.2×10^{-5}
	Ethyl alcohol	Air	298	1.2×10^{-5}
	Ethyl ether	Air	298	0.93×10^{-5}
	Helium, He	Air	298	7.2×10^{-5}
	Hydrogen, H_2	Air	298	7.2×10^{-5}
	Iodine, I_2	Air	298	0.83×10^{-5}
	Methanol	Air	298	1.6×10^{-5}
	Naphthalene	Air	300	0.62×10^{-5}
	Oxygen, O_2	Air	298	2.1×10^{-5}
	Argon, Ar	Nitrogen, N_2	293	1.9×10^{-5}
	Carbon dioxide, CO_2	Nitrogen, N_2	293	1.6×10^{-5}
	Carbon dioxide, CO_2	Water vapor	298	1.6×10^{-5}
	Oxygen, O_2	Ammonia, NH_3	293	2.5×10^{-5}
	Oxygen, O_2	Benzene	296	0.39×10^{-5}
	Oxygen, O_2	Water vapor	298	2.5×10^{-5}
	Water vapor	Argon, Ar	298	2.4×10^{-5}
	Water vapor	Helium, He	298	9.2×10^{-5}
	Water vapor	Nitrogen, N_2	298	2.5×10^{-5}
	Water vapor	Air	273	2.09×10^{-5}
	Water vapor	Air	278	2.17×10^{-5}
	Water vapor	Air	283	2.25×10^{-5}
	Water vapor	Air	288	2.33×10^{-5}
	Water vapor	Air	293	2.42×10^{-5}
	Water vapor	Air	298	2.5×10^{-5}
	Water vapor	Air	303	2.59×10^{-5}

(Continued)

Table 12.1 Binary diffusion coefficients at 1 atm (*cont.*)

	Component A	Component B	T (K)	D_{AB} (m²/s)
	Water vapor	Air	308	2.68×10^{-5}
	Water vapor	Air	313	2.77×10^{-5}
	Water vapor	Air	323	2.96×10^{-5}
	Water vapor	Air	373	3.99×10^{-5}
Dilute liquid solutions				
	Benzene	Water	293	1×10^{-9}
	Carbon dioxide	Water	298	2×10^{-9}
	Ethanol	Water	298	1.2×10^{-9}
	Glucose	Water	298	0.69×10^{-9}
	Hydrogen	Water	298	6.3×10^{-9}
	Methane	Water	293	1.5×10^{-9}
	Nitrogen	Water	298	2.6×10^{-9}
	Oxygen	Water	298	2.4×10^{-9}
	Water	Ethanol	298	1.2×10^{-9}
	Water	Ethylene glycol	298	0.18×10^{-9}
	Water	Methanol	298	1.8×10^{-9}
Dilute solid solutions				
	Carbon dioxide	Natural rubber	298	1.1×10^{-10}
	Nitrogen	Natural rubber	298	1.5×10^{-10}
	Oxygen	Natural rubber	298	2.1×10^{-10}
	Helium	Pyrex	293	4.5×10^{-15}
	Helium	Silicon dioxide	298	4.0×10^{-14}
	Hydrogen	Iron	298	2.6×10^{-13}
	Cadmium	Copper	293	2.7×10^{-19}
	Antimony	Silver	293	3.5×10^{-25}
	Bismuth	Lead	293	1.1×10^{-20}
	Mercury	Lead	293	2.5×10^{-19}

12.3 **The convective mass transfer coefficient**

The convective mass transfer coefficient, h_m, is analogous to the convective heat transfer coefficient, h. The mass transfer by convection will occur as long as the species mass concentration in a fluid ($\rho_{A,\infty}$) is different from the species concentration at a surface ($\rho_{A,s}$) over which the fluid flows. The species mass flux (j_A) is written as

$$j_A = h_m (\rho_{A,s} - \rho_{A,\infty}) \tag{12.9}$$

where j_A is in kg/m² s.

The total mass transfer rate ($\dot{m}_A$) on the surface of area As is

$$\dot{m}_A = \overline{h_m} A_s (\rho_{A,s} - \rho_{A,\infty}) \tag{12.10}$$

where $\dot{m}_A$ is in kg/s and $\overline{h_m}$ is the average mass transfer coefficient given by

$$\overline{h_m} = \frac{1}{A_s} \int_{A_s} h_m \, dA_s \tag{12.11}$$

The unit of h_m or $\overline{h_m}$ is m/s.
The species molar flux, J_A, on the surface is given by

$$J_A = h_m (C_{A,s} - C_{A,\infty}) \tag{12.12}$$

where J_A is in kmol/m^{2}s. The units of species mass concentration, ρ_A, and species molar concentration, C_A, are kg/m^3 and kmol/m^3, respectively. The relationship between ρ_A and C_A is $\rho_A/C_A = M_A$, where M_A is the molecular weight of species A in kg/kmol.

The gas species would remain in equilibrium with the solid or liquid phase. Hence, the species concentration at the surface is determined for the saturated condition corresponding to the surface temperature T_s. For a gas, $\rho_{A,s}$ can be obtained using the ideal gas equation as

$$\rho_{A,s} = \frac{P_{sat}(T_s)}{R_A T_s} = \frac{M_A P_{sat}(T_s)}{R T_s} \tag{12.13}$$

where R_A and R are the characteristic gas constant and the universal gas constant, respectively. The challenge in convective mass transfer is to get h_m for the engineering problem of interest.

12.4 The velocity, thermal, and concentration boundary layers

Just as a velocity boundary layer develops if the surface velocity and the free stream velocity are different and a thermal boundary layer develops if the surface temperature and the free stream temperature are different, a concentration boundary layer develops if the surface concentration and the free stream concentration of the species are different. Fig. 12.2 A–C show the development of the velocity, thermal, and concentration boundary layers, respectively, for flow over a flat plate. The species molar concentration at the surface, $C_{A,s}$, is greater than the free stream concentration, $C_{A,\infty}$. The concentration boundary layer thickness δ_c is normally defined as the value of y, which satisfies

$$\frac{(C_{A,s} - C_A)}{(C_{A,s} - C_{A,\infty})} = 0.99 \tag{12.14}$$

For a stationary surface, the mass transfer at the surface will be by only diffusion and is given by

$$J_{A,s} = -D_{AB} \frac{\partial C_A}{\partial y} \big|_{y=0} \tag{12.15}$$

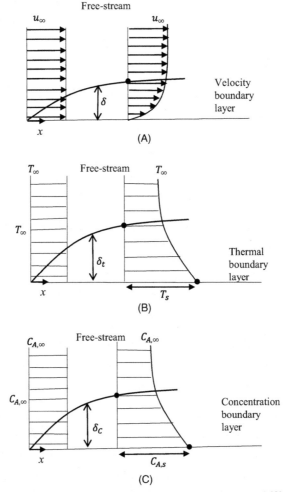

FIGURE 12.2 (A) Velocity boundary layer, **(B)** thermal boundary layer, and **(C)** concentration boundary layer.

Using Eqs. (12.12) and (12.15), we get

$$h_m = \frac{-D_{AB} \frac{\partial C_A}{\partial y}|_{y=0}}{(C_{A,s} - C_{A,\infty})} \tag{12.16}$$

On mass basis, the equations are

$$j_{A,s} = -D_{AB} \frac{\partial \rho_A}{\partial y}|_{y=0} \tag{12.17}$$

and

$$h_m = \frac{-D_{AB}\dfrac{\partial \rho_A}{\partial y}\Big|_{y=0}}{(\rho_{A,s} - \rho_{A,\infty})} \tag{12.18}$$

Note that the convective mass transfer is due to the combined effect of mass diffusion and advection (bulk fluid motion). If there is no flow, then the mass transfer is due to diffusion alone.

12.5 Analogy between momentum, heat transfer, and mass transfer

From the previous section, it is clear that there exists a similarity between the phenomena governing the growth of the velocity, thermal, and concentration boundary layers. The boundary layer equations for momentum, heat, and mass transport processes for a two-dimensional laminar steady, incompressible flow over a flat plate for a constant property fluid are

Momentum transfer

$$u\frac{\partial u}{\partial x} + v\frac{\partial u}{\partial y} = v\frac{\partial^2 u}{\partial y^2} \tag{12.19}$$

Heat transfer

$$u\frac{\partial T}{\partial x} + v\frac{\partial T}{\partial y} = \alpha\frac{\partial^2 T}{\partial y^2} \tag{12.20}$$

Mass transfer

$$u\frac{\partial C_A}{\partial x} + v\frac{\partial C_A}{\partial y} = D_{AB}\frac{\partial^2 C_A}{\partial y^2} \tag{12.21}$$

It can be seen that Eqs. (12.19)-(12.21) are similar. These equations result in two important nondimensional numbers in addition to the Prandtl number. The Prandtl number, as already discussed, is the ratio of the momentum diffusivity to the thermal diffusivity.

$$Pr = \frac{v}{\alpha} \tag{12.22}$$

The ratio of the momentum diffusivity to the mass diffusivity (diffusion coefficient) is called the Schmidt number,

$$Sc = \frac{v}{D_{AB}} \tag{12.23}$$

The ratio of the thermal diffusivity to the mass diffusivity is called the Lewis number,

$$Le = \frac{\alpha}{D_{AB}} \tag{12.24}$$

From Eqs. (12.22), (12.23), and (12.24) it can be seen that,

$$Pr = \frac{Sc}{Le} \tag{12.25}$$

The momentum, thermal, and convection boundary layers will be identical when

$$Pr = Sc = Le = 1$$

Just as the Nusselt number in convection heat transfer indicates the nondimensional temperature gradient at the surface, we define an analogous nondimensional number called the Sherwood number for convective mass transfer as

$$Sh = \frac{h_m x}{D_{AB}} \tag{12.26}$$

The Sherwood number indicates the nondimensional concentration gradient at the surface.

For convective heat transfer, the Stanton number was defined earlier as

$$St = \frac{Nu}{Re.Pr} = \frac{\dfrac{hx}{k}}{\dfrac{\rho u_\infty x}{\mu} \cdot \dfrac{\mu c_p}{k}} = \frac{h}{\rho u_\infty c_p} \tag{12.27}$$

Similarly, for convective mass transfer, the mass transfer Stanton number can now be defined as

$$St_m = \frac{Sh}{Re.Sc} = \frac{\dfrac{h_m x}{D_{AB}}}{\dfrac{\rho u_\infty x}{\mu} \cdot \dfrac{\mu}{\rho D_{AB}}} = \frac{h_m}{u_\infty} \tag{12.28}$$

In forced convective heat transfer, the Nusselt number depends on the Reynolds number and the Prandtl number. Similarly, in forced convective mass transfer, the Sherwood number depends on the Reynolds number and the Schmidt number.

Mathematically

$$Nu = f_1(Re, Pr)$$

$$Sh = f_2(Re, Sc)$$

In free convective heat transfer, the Nusselt number depends on the Grashof number and the Prandtl number. Similarly, in free convective mass transfer, the Sherwood number depends on the Grashof number and the Schmidt number.

$$Nu = f_3(Gr, Pr)$$

$$Sh = f_4(Gr, Sc)$$

Note that the Grashof number, in this case, is obtained from

$$Gr = \frac{g(\Delta\rho/\rho)L^3}{v^2} \qquad (12.29)$$

Eq. (12.29) can be used for both temperature-driven and or concentration-driven free convection flows. If no concentration gradients exist in fluids, the density difference, $\Delta\rho$, results from the temperature difference alone. In fluids with concentration gradients, the density difference results from the combined effect of the temperature difference and the concentration difference.

12.5.1 The Reynolds analogy

When the three boundary layers are identical, that is, $v = \alpha = D_{AB}$ and hence $Pr = Sc = Le = 1$, the relationship between the skin friction coefficient, C_f; heat transfer Stanton number, St; and the mass transfer Stanton number, St_m, can be expressed as

$$\frac{C_f}{2} = St = St_m \qquad (12.30)$$

This relationship is referred to as the Reynolds analogy and is used to determine skin friction coefficient, heat transfer coefficient, and mass transfer coefficient, if one of them is known. This holds the key to solving many heat transfer problems as a mass transfer study would suffice in order to obtain the heat transfer. A heat transfer study, in general, is more cumbersome than a mass transfer study, and the above analogy is handy for several engineering problems.

12.5.2 The Chilton-Colburn analogy

When $Pr \neq Sc \neq 1$, the analogy that needs to be used is the Chilton-Colburn analogy, which is of the form

$$\frac{C_f}{2} = St\,Pr^{2/3} = St_m Sc^{2/3} \qquad (12.31)$$

The above equation is valid for $0.6 < Pr < 60$ and $0.6 < Sc < 3000$.
From Eq. (12.31)

$$\frac{St}{St_m} = \left(\frac{Sc}{Pr}\right)^{2/3} = Le^{2/3}$$
$$\frac{h}{h_m} = \rho c_p Le^{2/3} \qquad (12.32)$$

12.6 Convective mass transfer relations

The relations for the mass transfer coefficient can be defined using either the Chilton-Colburn analogy or the suitable Nusselt number correlations, with the Nusselt number replaced by the Sherwood number and the Prandtl number by the Schmidt number.

12.6.1 Flow over a flat plate

For laminar flow on a flat plate, the local Nusselt number Nu_x is given by

$$Nu_x = 0.332 Re_x^{1/2} Pr^{1/3} \tag{12.33}$$

The ratio of the hydrodynamic boundary layer thickness (δ) to the thermal boundary layer thickness (δ_t) is given by

$$\frac{\delta}{\delta_t} = Pr^{1/3} \tag{12.34}$$

Similarly, for mass transfer, the local Sherwood number, Sh_x, is given by

$$Sh_x = 0.332 Re_x^{1/2} Sc^{1/3} \tag{12.35}$$

The ratio of the hydrodynamic boundary layer thickness (δ) to the concentration boundary layer thickness (δ_c) is given by

$$\frac{\delta}{\delta_c} = Sc^{1/3} \tag{12.36}$$

The expressions for the mean Nusselt number and the mean Sherwood number are

$$\overline{Nu_L} = 0.664 Re_L^{1/2} Pr^{1/3} \tag{12.37}$$

$$\overline{Sh_L} = 0.664 Re_L^{1/2} Sc^{1/3} \tag{12.38}$$

Eqs. (12.37) and (12.38) are valid for $Pr \geq 0.6$ and $Sc \geq 0.6$, respectively.

If the turbulent boundary layer exists ($Re_{x,critical} = 5 \times 10^5$) after the initial laminar boundary layer, the correlations for the mean Nusselt number and Sherwood number are

$$\overline{Nu_L} = (0.037 Re_L^{4/5} - 870) Pr^{1/3} \tag{12.39}$$

$$\overline{Sh_L} = (0.037 Re_L^{4/5} - 870) Sc^{1/3} \tag{12.40}$$

From the Chilton-Colburn analogy, the appropriate relations for mass transfer are as follows.

Laminar flow

$$St \, Pr^{2/3} = St_m \, Sc^{2/3} = \frac{C_f}{2} = 0.332 \, Re_x^{-1/2} \tag{12.41}$$

Turbulent flow

$$St \, Pr^{2/3} = St_m \, Sc^{2/3} = \frac{C_f}{2} = 0.0296 \, Re_x^{-1/5} \tag{12.42}$$

12.6.2 Internal flow

For fully developed laminar flow in tubes, the relations for heat transfer are

$$Nu_D = 3.66 \text{ for constant wall temperature} \tag{12.43}$$

$$Nu_D = 4.36 \text{ for constant wall heat flux} \tag{12.44}$$

The corresponding relations for mass transfer, as shown in Fig. 12.3, are

$$Sh_D = \frac{h_m D}{D_{AB}} = 3.66 \text{ for constant wall mass concentration} \tag{12.45}$$

$$Sh_D = \frac{h_m D}{D_{AB}} = 4.36 \text{ for constant wall mass flux} \tag{12.46}$$

For turbulent flow in tubes, the Dittus-Boelter correlation is used.

$$Nu_D = 0.023 Re_D^{4/5} Pr^{1/3} \tag{12.47}$$

$$Sh_D = 0.023 Re_D^{4/5} Sc^{1/3} \tag{12.48}$$

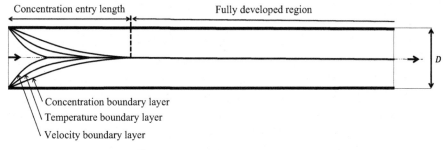

Concentration entry length   Fully developed region

D

Concentration boundary layer
Temperature boundary layer
Velocity boundary layer

Constant wall mass concentration; $\rho_{A,s}$ = constant; Sh_D = 3.66

Constant wall mass flux; $j_{A,s} = h_m(\rho_{A,s} - \rho_{A,b})$ = constant; Sh_D = 4.36

FIGURE 12.3 Mass transfer correlations for laminar fully developed velocity, thermal, and concentration profiles.

It may be noted that Eq. (12.47) is used for both constant wall temperature and constant wall heat flux conditions. Eq. (12.48) is used for both constant wall mass concentration and constant wall mass flux conditions.

The Chilton-Colburn analogy for tube flow is expressed as

$$St_m Sc^{2/3} = St Pr^{2/3} = \frac{f}{8} \tag{12.49}$$

where f is the Darcy friction factor.

12.7 A note on the convective heat and mass analogy

The analogy between the convective heat transfer and the convective mass transfer is applicable for the cases in which the flow rate of species undergoing mass transfer is low compared to the total flow rate of the gas or liquid mixture so that the flow velocity is not affected significantly by the species mass transfer between the surface and the fluid. For example, in evaporative cooling or the evaporation at the free surface of a water pool, the mole fraction of water vapor in the air-water vapor mixture at the water surface is very small, and so the use of the convective heat and mass transfer analogy will not result in significant errors. However, for the cases such as boilers, condensers, and fuel droplet evaporation in combustors, wherein the vapor fraction approaches unity, the convective heat and mass transfer analogy cannot be used.

12.8 Simultaneous heat and mass transfer

The simultaneous transfer of heat and mass occurs in many engineering applications, such as wet bulb thermometers, cooling towers, humidifiers, and dehumidifiers. Water evaporates when air blows over the water surface. The energy required for evaporation comes from the latent heat of vaporization of the water. Under steady state conditions, the latent heat transferred by the water is equal to the heat transferred to the water from the air blowing over the surface, and in the process, the air gets cooled. Using the energy balance at the water surface, as shown in Fig. 12.4, we have

$$h(T_\infty - T_s) = h_m (\rho_{A,s} - \rho_{A,\infty}) h_{fg} \tag{12.50}$$

$$(T_\infty - T_s) = \frac{(h_m / h)}{R_A} \left[\frac{P_{A,s}}{T_s} - \frac{P_{A,\infty}}{T_\infty} \right] h_{fg} \tag{12.51}$$

In Eq. (12.51), $(T_\infty - T_s)$ indicates the cooling effect. Using Eq. (12.32), Eq. (12.51) can be written as,

$$(T_\infty - T_s) = \frac{h_{fg}}{R_A \rho c_p Le^{2/3}} \left[\frac{P_{A,s}}{T_s} - \frac{P_{A,\infty}}{T_\infty} \right] \tag{12.52}$$

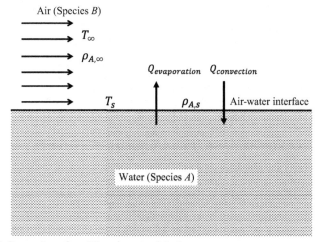

FIGURE 12.4 Energy transfer at the air-water interface.

In the above equation, the air properties, ρ, c_p, and Le, correspond to the film temperature, $T_f = \dfrac{T_s + T_\infty}{2}$.

If $(T_\infty - T_s)$ is small, T_∞ and T_s can be approximated to T_f, in which case the Eq. (12.52) becomes

$$(T_\infty - T_s) \simeq \frac{h_{fg}}{R_A \rho c_p Le^{2/3} T_f}[P_{A,s} - P_{A,\infty}] \tag{12.53}$$

If the mass concentration of species B (air) is very large compared to the mass concentration of species A (water vapor),

$$\rho = \rho_A + \rho_B \approx \rho_B = \frac{P}{R_B T_f}$$

where $P = P_A + P_B$

Eq. (12.53) can now be written as

$$(T_\infty - T_s) \approx \frac{(R_A / R_B) h_{fg}}{c_p Le^{2/3}}\left[\frac{P_{A,s}}{P} - \frac{P_{A,\infty}}{P}\right] \tag{12.54}$$

Let T_1 and T_2 be the dry bulb and wet bulb temperatures, respectively, and the corresponding partial pressures of water vapor be P_{v1} and P_{v2}, respectively.

Specific humidity, ω, is the mass of water vapor per kg of dry air.

$$w = \frac{m_v}{m_a} = \frac{R_a}{R_v}\frac{P_v}{P_a} \tag{12.55}$$

$$P_a = P - P_v \approx P \text{ as } P_v \ll P_a$$

Eq. (12.55) can now be written as

$$\omega = \frac{R_a P_v}{R_v P}$$
$$= \frac{287}{461}\frac{P_v}{P} \qquad\qquad (12.56)$$
$$= 0.622\frac{P_v}{P_{atm}} (\because P = P_{atm})$$

Using Eq. (12.56) with $R_a = R_B$ and $R_v = R_A$, Eq. (12.54) becomes

$$(T_1 - T_2) = \frac{(\omega_2 - \omega_1)h_{fg}}{c_p (Le)^{2/3}} \qquad\qquad (12.57)$$

For elaborate illustrations and more examples on convective mass transfer, readers can refer to Lienhard IV and Lienhard V (2020), Bergman et al. (2011), Cengel (2003), and Bejan (1993).

Example 12.1: *Pressurized hydrogen gas is stored at 298 K in a 1.2 m long cylindrical container made of iron. The inner and outer radii of the cylinder are 0.18 m and 0.19 m, respectively. The molar concentration of hydrogen in iron at the inner surface is 0.1 kmol/m³ and that at the outer surface is negligible. Calculate the mass flow rate of hydrogen by diffusion through the cylindrical wall of the container, assuming the mass diffusion to be steady and one-dimensional.*
Solution:
The binary diffusion coefficient for hydrogen in iron at the specified temperature is 2.6×10^{-13} m²/s (Table 12.1).
    The molar flow rate of the hydrogen through the cylindrical wall by diffusion is

$$\dot{N} = 2\pi D_{AB} L \frac{C_{A,1} - C_{A,2}}{ln(r_2 / r_1)}$$
$$= 2\pi \times 2.6 \times 10^{-13} \times 1.2 \times \frac{0.1 - 0}{ln(0.19/0.18)}$$
$$= 3.625 \times 10^{-12} \text{ kmol/s}$$

Example 12.2: *An outdoor swimming pool contains water at 30 °C. The length and breadth of the pool are 24 m and 4 m, respectively. The ambient temperature and the relative humidity are 35 °C and 30%, respectively. Wind with a speed of 0.7 m/s blows in the direction of the length of the pool. The atmospheric pressure is 101 kPa. Determine the following.*
(a) The mole fraction and mass fraction of water vapor at the water surface.
(b) The molar concentration and mass concentration of the water vapor at the water surface.
(c) The average mass transfer coefficient.
(d) The rate of evaporation of water from the pool.

Solution:

(a) The air will be saturated at the water surface. Hence the partial pressure of water vapor in the air at the water surface will be the saturated pressure of water at 30 °C. The mole fraction of water vapor,

$$y_{vapor,s} = \frac{P_{sat}(30\,°C)}{P_{total}}$$
$$= \frac{4.24\,kPa}{101\,kPa}$$
$$= 0.042$$

The molecular weight of dry air is 28.97 kg/kmol.
The mass fraction of water vapor,

$$w_{vapor,s} = \frac{0.042 \times 18}{(0.042 \times 18) + (1 - 0.042) \times 28.97}$$
$$= 0.0265$$

(b) The molar concentration of water vapor at the water surface is

$$C_{vapor,s} = \frac{P_{vapor}}{RT}$$
$$= \frac{4.24}{8.314 \times 303}$$
$$= 1.683 \times 10^{-3}\,kmol/m^3$$

The mass concentration of water vapor at the water surface is

$$\rho_{vapor,s} = M_{H_2O}C_{vapor,s}$$
$$= 18 \times 1.683 \times 10^{-3}\,kg/m^3$$
$$= 0.03\,kg/m^3$$

(c) Film temperature,

$$T_f = \frac{T_s + T_\infty}{2}$$
$$= \frac{30 + 35}{2}$$
$$= 32.5\,°C$$

The properties of air at 1 atm and 32.5 °C are
$\rho = 1.14\,kg/m^3$, $c_p = 1.007\,kJ/kgK$, $v = 1.63 \times 10^{-5}\,m^2/s$, $\mu = 1.86 \times 10^{-5}\,Ns/m^2$, $Pr = 0.706$
From Table 12.1, $D_{AB} = 2.63 \times 10^{-5}\,m^2/s$

$$Re_L = \frac{u_\infty L}{v} = \frac{0.7 \times 24}{1.63 \times 10^{-5}} = 10.3 \times 10^5$$

Since $Re_L > 5 \times 10^5$, the flow is turbulent.

$$Sc = \frac{v}{D_{AB}} = \frac{1.63 \times 10^{-5}}{2.63 \times 10^{-5}} = 0.619$$

$$\overline{Sh} = (0.037\,Re^{4/5} - 870)Sc^{1/3} = 1295$$

$$\overline{h_{mL}} = \overline{Sh}\frac{D_{AB}}{L} = \frac{1295 \times 2.63 \times 10^{-5}}{24} = 1.4 \times 10^{-3}\,\text{m/s}$$

(d) RH = 30%

$$\frac{P_{vapor,\infty}}{P_{sat}(35\,^\circ\text{C})} = 0.3$$

$$\begin{aligned}
P_{vapor,\infty} &= 0.3 \times P_{sat}(35\,^\circ\text{C}) \\
&= 0.3 \times 0.0575 \\
&= 0.01725\,\text{bar} \\
&= 1.725\,\text{kPa}
\end{aligned}$$

$$\begin{aligned}
\rho_{vapor,\infty} &= \frac{P_{vapor,\infty}}{R_{vapor}T_\infty} \\
&= \frac{1.725}{0.461 \times 308}\left(\because R_{vapor} = \frac{R}{M_{vapor}} = \frac{8.314}{18} = 0.461\,\text{kJ/kg K}\right) \\
&= 0.0121\,\text{kg/m}^3
\end{aligned}$$

The rate of evaporation is

$$\begin{aligned}
\dot{m}_{vapor} &= \overline{h_{mL}}\,A(\rho_{vapor,s} - \rho_{vapor,\infty}) \\
&= 1.4 \times 10^{-3} \times 24 \times 4 \times (0.03 - 0.0121) \\
&= 2.406 \times 10^{-3}\,\text{kg/s} \\
&= 2.406 \times 10^{-3} \times 24 \times 3600\,\text{kg/day} \\
&= 207.86\,\text{kg/day}
\end{aligned}$$

Example 12.3: *Consider Example 12.2. If the wind blows in the direction of the breadth of the pool, determine the average mass transfer coefficient and the rate of evaporation of water from the pool.*
Solution:

$$Re_L = \frac{u_\infty L}{v} = \frac{0.7 \times 4}{1.63 \times 10^{-5}} = 1.71 \times 10^5$$

Since $Re_L < 5 \times 10^5$, the flow is laminar.

$$\begin{aligned}
\overline{Sh_L} &= 0.664\,Re_L^{1/2}\,Sc^{1/3} \\
&= 0.664 \times (1.71 \times 10^5)^{1/2} \times 0.619^{1/3} \\
&= 234
\end{aligned}$$

$$\overline{h_{mL}} = \overline{Sh_L} \frac{D_{AB}}{L}$$
$$= 234 \times \frac{2.63 \times 10^{-5}}{4}$$
$$= 1.538 \times 10^{-3} \text{ m/s}$$

The rate of evaporation is

$$\dot{m}_{vapor} = \overline{h_{mL}} A(\rho_{vapor,s} - \rho_{vapor,\infty})$$
$$= 1.538 \times 10^{-3} \times 24 \times 4 \times (0.03 - 0.0121)$$
$$= 2.64 \times 10^{-3} \text{ kg/s}$$
$$= 228.3 \text{ kg/day}$$

Example 12.4: *Ethanol flows through a long tube of diameter 2 mm. Later, in order to remove the ethanol from the inner surface of the tube, water at 1 atm and 298 K is forced to flow through the tube with a mean velocity of 0.4 m/s. Assuming fully developed flow, evaluate the mass transfer coefficient inside the tube.*
Solution:
From Table 12.1, the value of D_{AB} for ethanol diffusing through water at 1 atm and 298 K is 1.2×10^{-9} m²/s.

$$Re_D = \frac{uD}{v} = \frac{0.4 \times 0.002}{0.903 \times 10^{-6}} = 886$$

Since $Re_D < 2300$, the flow is laminar. The Sherwood number, in this case, is 3.66. Therefore, the convective mass transfer coefficient can be calculated as

$$h_m = \frac{Sh D_{AB}}{D} = \frac{3.66 \times 1.2 \times 10^{-9}}{0.002} = 2.19 \times 10^{-6} \text{ m/s}$$

Example 12.5: *The dry bulb temperature and wet bulb temperature of an air stream are 30 °C and 20 °C, respectively. Estimate the specific humidity of the air.*
Solution:

$$T_f = \frac{30 + 20}{2} = 25 \text{ °C}$$

The properties of air at 1 atm and 25 °C are

$\rho = 1.16 \text{ kg/m}^3$, $c_p = 1.007 \text{ kJ/kg K}$, $v = 1.58 \times 10^{-5} \text{ m}^2/\text{s}$, and $\alpha = 22.5 \times 10^{-6} \text{ m}^2/\text{s}$.

From Table 12.1, $D_{AB} = 2.5 \times 10^{-5} \text{ m}^2/\text{s}$,

$$Sc = \frac{v}{D_{AB}} = \frac{1.58 \times 10^{-5}}{2.5 \times 10^{-5}} = 0.63$$

$$Le = \frac{\alpha}{D_{AB}} = \frac{22.5 \times 10^{-6}}{2.5 \times 10^{-5}} = 0.9$$

Specific humidity corresponding to the wet-bulb temperature is

$$\omega_2 = 0.622 \frac{P_{vapor}(20\,°C)}{P_{atm} - P_{vapor}(20\,°C)} = 0.622 \times \frac{2.33}{101.3 - 2.33} = 0.0146 \text{ kg/kg dry air}$$

h_{fg} corresponding to 20 °C is 2451 kJ/kg.

$$\frac{\omega_2 - \omega_1}{T_1 - T_2} = \frac{c_p}{h_{fg}}(Le)^{2/3}$$

$$\omega_1 = \omega_2 - (T_1 - T_2)\frac{c_p}{h_{fg}}(Le)^{2/3}$$
$$= 0.0146 - (30-20)\frac{1.007}{2451}(0.9)^{2/3}$$
$$= 0.0146 - 0.00382$$
$$= 0.0107 \text{ kg/kg dry air}$$

Problems

12.1 Air at 1 atm and 298 K flows with a velocity of 1 m/s inside a tube of 20 mm diameter and 2 m length. There is a deposit of iodine on the inner surface of the tube. Determine the mass transfer coefficient of iodine from the surface into the air.

12.2 A 1.5 m long and 1.5 m wide tub contains water at 20 °C. Air at 1 atm and 30 °C with a relative humidity of 40% blows past the water surface with a velocity of 1.6 m/s. Calculate the rate of evaporation of water.

12.3 The dry bulb and wet bulb temperatures recorded by a thermometer are 40 °C and 30 °C, respectively.
   a. Determine the relative humidity of the air.
   b. If the air is completely dry and has the same dry bulb temperature of 40 °C, determine the wet-bulb temperature.

12.4 A bottled drink needs to be chilled by wrapping it in a wet cloth and blowing air over it. If the ambient air is at 1 atm and 35 °C with a relative humidity of 45%, what will be the steady state temperature attained by the drink?

12.5 A cooling tower in a power plant cools 5000 kg/h of water from 40 °C to 32 °C. If the air enters the tower at 28 °C with a relative humidity of 50% and leaves at 36 °C with a relative humidity of 97%, determine the water evaporation rate or the mass flow rate of makeup water. Assume the ambient pressure to be 1 atm.

References

Barrer, R.M., 1941. Diffusion in and Through Solids. Macmillan, New York.

Bejan, A., 1993. Heat Transfer, John Wiley & Sons, New York.

Bergman, T.L., Lavine, A.S., Incropera F.P., Dewitt, D.P., 2011. Fundamentals of Heat and Mass Transfer, seventh ed. John Wiley & Sons, NJ.

Black, Van L., 1980. Elements of Material Science and Engineering. Addison-Wesley, Reading, MA.

Cengel, Y. A., 2003. Heat Transfer: A Practical Approach, second ed. McGraw-Hill, New York.

Geankoplis, C.J., 1972. Mass Transport Phenomena. Holt, Rinehart, and Winston, New York.

Lienhard IV, J.H., Lienhard V, J.H., 2020. A Heat Transfer Textbook. Courier Dover Publications, Mineola, NY.

Mills, A.F., 1995. Basic Heat and Mass Transfer. Richard D. Irwin, Burr Ridge, IL.

Perry, J.H. (Ed.), 1963. Chemical Engineer's Handbook. fourth ed. McGraw-Hill, New York.

Reid, R.D., Prausnitz, J.M., Sherwood, T.K., 1977. The Properties of Gases and Liquids, third ed. McGraw-Hill, New York.

Thomas, L.C., 1991. Mass Transfer Supplement-Heat Transfer. Prentice Hall, Englewood Cliffs, NJ.

Index

Note: Page numbers followed by "f" indicate figures and "t" indicate tables

Printed in the United States
By Bookmasters